D1577437

New Food Product Development

From Concept to Marketplace

THIRD EDITION

New Food Product Development

From Concept to Marketplace

THIRD EDITION

Gordon W. Fuller

CRC Press
Taylor & Francis Group
Boca Raton   London   New York

CRC Press is an imprint of the
Taylor & Francis Group, an **informa** business

CRC Press
Taylor & Francis Group
6000 Broken Sound Parkway NW, Suite 300
Boca Raton, FL 33487-2742

© 2011 by Taylor and Francis Group, LLC
CRC Press is an imprint of Taylor & Francis Group, an Informa business

No claim to original U.S. Government works

International Standard Book Number: 978-1-4398-1864-0 (Hardback)

Visit the Taylor & Francis Web site at
http://www.taylorandfrancis.com

and the CRC Press Web site at
http://www.crcpress.com

This edition is dedicated to my wife,

Joan, for her patience and encouragement.

Contents

Preface

After I left Imasco Foods Ltd., Montreal, Quebec, Canada, I taught courses on new food product development and agricultural economics at McGill University, Montreal, Quebec, Canada, as an outside lecturer. I also taught a course on communication at Concordia University, Montreal, Quebec, Canada, as a guest lecturer. This book has its origins in notes developed for these courses.

The book took form after I began my consulting firm and the lecture notes became embellished with experiences at Imasco Foods and its several companies as well as experiences gained as a consultant working with companies whose products failed somewhere in the process. I was called in to correct errors but first had to find where errors might have been made. In the course of these adventures, I met many of my ex-students at trade shows and food fairs who encouraged me to put everything together in a book—thus, this book.

In this edition, contents have been reorganized and much new material added, especially on marketing and electronic communication and their combined effect on market research. Where possible, I tried to use peer-reviewed marketing journals, but seldom do companies announce their activities in these publications; thus, resorting to business, marketing, and trade newspapers was necessary for references. Any material used had to make a substantial intellectual or technical contribution to an understanding of new food product development or illustrate a novel and innovative approach to the new product development process. The material had to describe the "real-world" environment of product development, and hence, more resort to business newspapers was necessary.

I studiously avoided "worked fictional examples" of new product development, as some reviewers suggested I include—this is a style of presentation developed by many business schools. True examples are more blatant in illustrating the elements contributing to the success or failure of a particular product situation. Real life is ever so much more educational. I have kept the confidences of my clients in the experiences I relate, but, as stated in an earlier edition, if my clients do see themselves, they should be ashamed. The anecdotes, mostly errors in the development process, illustrate particular misadventures in new product work.

The age of some of my references has been criticized, but where nothing new has been added to that provided by the older literature, I see no reason to use later works simply because they bear a later date. Besides, the older literature is often written more clearly so that principles are easily grasped. For those who may disagree, I suggest they read some older marketing literature and the later literature. There is a further defense of my literature choices in Chapter 12.

Acknowledgments

I am deeply indebted to my wife, Joan, for reading the text and making many helpful suggestions; to my son, Grahame, senior technical writer for Autodesk Canada Co., Montreal, Quebec, Canada, for preparing many of the figures, for many interesting and stimulating discussions regarding the text, and also for correcting and emending my notes on computers, communication technology, and the Internet; and to my son-in-law, Dr. David Gabriel, professor in the Department of Physical Education and Kinesiology at Brock University, St. Catharines, Ontario, Canada, for his suggestions for, and assistance with, figures. I am especially grateful to Christine Coombes, U.S. Public Relations Coordinator of Mintel International, for the data on new product introductions. Chapters 10 and 11 benefited largely from helpful discussions with Timothy Beltran—who, at the time of our discussions, was executive chef of the J. P. Morgan-Chase dining room on Wall Street, New York City, and has now formed his own catering company, Culantro Caterers, in New Jersey, and with Henry B. Heath, MBE, BPharm (London), MFC, FRPharmS., FIFST (United Kingdom), and retired president of Bush Boake Allen Corporation, Dorval, Quebec. To both of these gentlemen, my heartfelt thanks. Much is owed to James W. Baldwin for many interesting discussions on marketing and with whom I worked on the communications course at Concordia University, Montreal, Quebec, Canada, and travelled extensively in Europe looking for new product ideas for Imasco Foods; to Dr. Charles Beck, a good friend and colleague with whom I exchanged many ideas; to Dr. Sylvan Eisenberg of MicroTracers, Inc., San Francisco, California, for his thoughtful advice; and to the library staff at the University of Guelph, Guelph, Ontario, Canada, especially Judy Wanner and Michael Ridley, chief information officer and chief librarian, respectively. To all, my sincere thanks.

Author

Dr. Gordon W. Fuller has a wide variety of training and experience that he has used successfully in his consulting practice for over 30 years. He received his BA and MA in food chemistry from the University of Toronto, Toronto, Ontario, Canada, in 1954 and 1956, respectively. He also received his PhD in food technology from the University of Massachusetts, Amherst, Massachusetts, in 1962.

His work experience includes stints as a research chemist with the Food and Drug Directorate in Ottawa, Canada, and as a research food technologist working on chocolate products for the Nestlé Co. in Fulton, New York. He conducted research on tomato flavors and products at the H. J. Heinz Fellowship at the Mellon Institute for Industrial Research, Pittsburgh, Pennsylvania for two years.

Dr. Fuller served as associate professor in the Department of Poultry Science at the University of Guelph, Guelph, Ontario, Canada, where, in addition to teaching and research responsibilities into added value meat and egg products, he carried out extension work for food processors in southern Ontario. He also held a fellowship at the Food Research Association, Leatherhead, England, where he worked on water binding and reducing water losses in meat products.

Prior to forming his own consulting company, Dr. Fuller was, for eight years, vice president of technical services, Imasco Foods Ltd., Montreal, Quebec, Canada, where he was responsible for corporate research and product development programs at the company's subsidiaries in both Canada and the United States. His consulting practice has taken him to the United States, South America, Europe, and China. He has lectured on the topics of agricultural economics and food technology in North and South America, England, Germany, the Netherlands, and China. As an outside lecturer, he presented courses at McGill University, Montreal, Quebec, Canada, and was a guest lecturer at Concordia University, Montreal, Quebec, Canada, for many years.

Dr. Fuller is presently semiretired and works mainly with his old clients. He keeps himself occupied growing different varieties of hot peppers and developing new formulations of hot sauces. He now lives in Guelph, Ontario.

1

What Is New Food Product Development?

> ...The production of a new food commodity might seem to be a trivial matter unworthy of serious consideration; this is not necessarily so. The technological expertise upon which any one item depends may require the full depth of scientific understanding.
>
> **Magnus Pyke (1971)**

1.1 Introduction

To be profitable and to survive, food companies seek new products. These, if successful, give new life to a company, replacing products no longer selling well. Old bell-ringer products cannot be relied on year after year to be profitable. New product development or finding new uses for old products is essential for continued growth of a company. New products are one of a few ways a company can follow for increased profitability.

Developing new products requires talented personnel, extensive research, suitable physical facilities, and money. Such human and physical resources are expensive. Nevertheless there is no promise that any new products developed through these resources will be successful enough to justify their expense. The obviousness of the need for new product is apparent to any novice food technologist. For the novice food technologist entering new product research and development, what is difficult to appreciate is that there is a finite amount of money available within any company and an infinite number of demands to use that money. Management must see that its money is allocated to those areas in the company where the money is used to best advantage.

The marketing department wants to expand markets into new geographical territories with its bell-ringer products. It reasons that there would be perhaps some language or other label changes and no developmental expenses, and the company would be moving products with a proven customer appeal and sales record and hence the risk would be small. The finance department sees opportunities in various investment instruments that would contribute to the profitability of the company without the associated risks and costs of new product development or expansion into new market areas.

The manufacturing department would put forward its argument that with newer, faster equipment, it could lower the cost of established products, reduce energy costs, and be a better corporate citizen with a smaller carbon footprint, and more profitability for the company would result.

There are other areas of critical activity in the company that will have their own promises of how they could use the money available to best advantage: each will argue their case with senior management as to why they and they alone would succeed. And the shareholders or partners in the company also have their ideas how best to use the money.

This is the world of new product development, one where there is competition for money from within and uncertainty regarding the success of any new product that is introduced into the marketplaces.

1.2 Defining and Characterizing New Food Products

The new product development process introduces many new terms, and these terms must be defined and understood for complete understanding of the process.

1.2.1 New Products

A new food product is one that has not been presented before in any marketplace anywhere. This is a rare occurrence.

A food product may be new to a company that has not sold it before but is not necessarily new to a marketplace: other companies elsewhere might have sold it or a product conceptually similar to it before. Some characteristics of a food product providing newness for a company are tabulated in Table 1.1.

TABLE 1.1

Characteristics of a New Food Product as Introduced by a Specific Food Company

Product has never before been manufactured by that company.
Product has never before been distributed by that company.
An old established product manufactured by a company is introduced into a geographically new area by that company.
An old established product manufactured by a company is introduced in either a new package or a new size or a new form.
An old established product manufactured by a company is introduced into a new market niche, i.e., positioned as one with a new function.

Defining a new product to encompass all the characteristics listed is difficult. Two simple definitions of a new product are as follows:

1. A product not previously manufactured by a company and introduced by that company into its marketplace or into a new marketplace, or
2. The presentation or rebranding by a company of an established product in a new form, a new package or under a new label into a market not previously explored by that company.

No-name or store brand products, even those as famous as President's Choice®, are hard to pigeonhole by a definition. No-name or store brand products are purchased from a food manufacturer by a retailer or by a company that then either leaves a No-name label on them (rare but not unheard of) or puts its own branded label on it and sells it to a retailer. This company may be no more than an office with a telephone. A label owner visits manufacturers and purchases product to the grade or standard he or she wants (often very high standards—cf. President's Choice or S & W Fine Foods™): he or she has not developed, manufactured, or market-researched the product. In effect, these products piggyback on the research and development work of the original food manufacturer.

Tables 1.2 and 1.3 tabulate new food products, describe some general characteristics of each type, and provide some examples.

1.2.1.1 Line Extensions

A line extension is a variant of an established line of food products, i.e., one more of the same. Line extensions represent a logical extension of similarly positioned products.

Some care must be given to what are and are not line extensions, for example:

- Adding a canned three-bean salad product to a line of canned bean products involves a change in processing and quality control technology. Developers have gone from a high pH, low acid product to an acidified product. This is no longer a line extension; a different market niche is targeted. The product now has added value and is a new menu item.
- A potato chip manufacturer extends its product line to the manufacture of corn chips or corn puffs or roasted peanuts or popcorn. These are not simple line extensions. The new products have in common only the snack food element. These are distributed through the same channels and displayed in the same section of a retail store, but purchasing philosophies, storage facilities, and manufacturing technologies have changed extensively for the manufacturer.

TABLE 1.2

General Characteristics of Classes of New Food Products

Types of New Product	General Characteristics
Line extensions	Little time or research required for development.
	No major manufacturing changes in processing lines or major equipment purchases.
	Relatively little change in marketing strategy.
	No new purchasing skills (commodity trading) or raw material sources.
	No new storage or handling techniques for either the raw ingredients or the final product. This means that regular distribution systems can be used.
Repositioned existing product	Research and development time is minimal.
	Manufacturing is comparatively unaffected.
	Marketing must develop new strategies and promotional materials to interpret and penetrate the newly created marketing niche.
	Sales tactics require reevaluation to reach and make sales within the new marketplaces.
New form or size of existing product	Highly variable impact on research and development.
	Highly variable impact on physical plant and manufacturing capabilities. Major equipment purchases may be required if manufacturing to be done in-house.
	Marketing and sales resources will require extensive reprogramming.
Reformulation of existing product	Moderate research and development required consistent with reformulation goal.
	Generally little impact on physical facilities.
	Generally little impact on marketing and sales resources unless reformulation leads to repositioning of product.
Repackaging of existing product	The novelty of the repackaging will dictate the amount and degree of research and development required.
	Slight impact on physical facilities. New packaging equipment will be required.
	Little impact on marketing, sales, and distribution resources.
Innovative products	Amount of research and development dependent on the nature of the innovation.
	Highly variable impact on manufacturing capabilities.
	Possible heavy impact on marketing and sales resources.
Creative products	Generally heavy need for extensive research and development, therefore a costly venture.
	Extensive development time may be required.
	May require entirely new plant and equipment. Degree of creativity may require development de novo of unique equipment.
	Basically will require total revision of marketing and sales forces. Creation of a new company or brand may be required.
	Risk of failure high.

TABLE 1.3

Examples of the Different Types of New Products

Type of New Product	Examples of Category
Line extensions	A new flavor for a line of wine coolers or for a line of flavored bottled waters
	New varieties of a family of canned ready-to-serve soups
	New flavors for a snack product such as potato chips
	New flavored bread-crumb coating
	A coarser or more natural peanut butter
Repositioned existing product	Oatmeal-containing products positioned as dietary factors in reducing cholesterol
	Soy-containing products repositioned as dietary factors combating cancer
	Soft drinks positioned as main meal accompaniments
New form of existing product	Margarine or butter spreadable at refrigerator temperatures
	Prepeeled fruit or sectioned grapefruit or oranges
	Fast-cooking products such as rice or oats
	Instant coffees, teas, and flavored coffees
	Dehydrated spice blends for sauces
Reformulation of existing products	Low calorie (reduced sugar, fat) products
	Hotter, spicier, zestier, crunchier (e.g., peanut butter), smoother products
	All-natural ("greener") products, organic products
	Lactose-free milk products
	High-fiber products
New packaging of existing product	Single-serving sizes of, for example, yogurt
	Branded fruits and vegetables
	Pillow packs for snack food items
	Institutional sizes for warehouse stores
	Squeeze bottles for condiment sauces
	Pull-top containers of snack dips
	Use of thin profile containers
Innovative products	Dinner kits
	Canned snack food dips (see above)
	Frozen dinners
	Simulated seafood products
Creative products	Reformed meat cuts
	Extruded products
	Surimi and kamaboko-based products and soy bean curd (tofu) and limed corn if these were discovered recently
	Short-chain fatty acid containing products

Line extensions are not to be confused with brand extensions. A brand can be likened to an umbrella that embraces products that represent reliability, confidence, quality, and "motherhood." A brand might best be described as a concept embodying values and reliability (of a company). Customers recognize a brand as providing safety, reliability, and quality at a price they are willing to pay. Many brands have this cachet: no matter what they make, if it has their brand, it's got to be good! Brands have been described as having an essence; Barnham (2009) argues that this is wrong—brands need to be thought of as being the essence, a concept bringing greater value. Barnham presents an interesting discussion of the philosophy of brands and branding. Brand extensions, particularly brand overextensions, can be a death-knell for the products under its protective umbrella. One need only think of one's favorite brand of food and picture that brand name being extended to carry a line of women's lingerie or men's underwear or hygiene products. It can have disastrous results. Yet Marks and Spencer™, a U.K. brand better known as Marks and Sparks, had the cachet of top quality whether it was food products, clothing, or furniture. President's Choice™ has also succeeded with this in some aspects. One sees another aspect of brand as essence.

Marketing programs are usually not greatly affected by line extensions, but there can be some surprises. When a confectionery company with an established line of children's confections attempts to introduce a line of adult-flavored confections, some marketing difficulties can arise, for example, a company with jujubes and jelly beans introducing flip-top dispensed mints or humbugs. Conflict has been introduced: children do not like the new flavors; adults do not accept a child's brand of candy not knowing that it is flavored for them. Different promotions, advertisements, and store placements for the adult products are necessary. Conversely, Pez Candy® started out as a mint-flavored candy directed to adults as an aid in stopping smoking. Packed unimaginatively in a tin, they were introduced into North America in the 1950s in a cigarette-lighter-like dispenser. Poor sales led to the dramatic change to children's candy flavors, and the dispensers are now collectors' items. Promotion, repositioning, and repackaging brought new life to a losing product.

1.2.1.2 Repositioned Products

A company can be surprized to find through their consumer intercepts and letters that consumers use their product in ways that the company has never anticipated. A new market niche has been created that gives new life to an existing product. An example is ARM & HAMMER® Baking Soda finding a new niche as a body deodorant, a dentifrice, and a deodorant against food odors in refrigerators.

Repositioning an established product takes a company into new and unknown markets. Each repositioning must be considered individually. Rebranding and extensive marketing promotion may be necessary.

1.2.1.3 New Form of Existing Products

A paste product converted to a tablet (or vice versa) a sauce transformed to a sprinkle-on powder or to a spray-on product represents new forms of existing products, but they may be forms too radical a departure from the known product for consumers to accept. Such changes in form are costly processing changes for the manufacturer.

The modified product must be seen as a valued improvement over the traditional form to be successful. Profound departures in form require reeducation to new usage habits. A new form of a product, for example, prepeeled, precut french-fry style potatoes have not become successful in the retail, chilled food counter but have become remarkably successful in the food service market. Here, they offer convenience, i.e., added value.

Sampson (2010) described how the Nestlé Company found that changing the shape of the traditional squares found in most chocolate bars such as their Nestlé Noir™ bars caused better taste sensations. It was found that the chocolate in the form of waves melted better and was more "comfortable in the mouth," and the flavor lasted longer.

1.2.1.4 Reformulation of Existing Products

The "new, improved" product is typical of this category. Reformulation may be necessary for any number of reasons:

- Some improvement, such as better color, better flavor, more fiber, less fat, greater stability, fewer calories, is required in a product to meet competitive products or to fit with perceived trends.
- Raw materials or unique ingredients become unavailable or too costly. Substitutes must be found to remain price competitive.
- Ingredients with superior characteristics become available, or improved processes allow the manufacture of products of higher quality at a lower cost.
- Reformulation is needed to lower the costs of a product to meet the challenge of cheaper imitations from competitors. This is often the reason for reformulation.
- The safety of ingredients, additives, or even raw materials may be challenged. Their use is no longer permitted, or their level of usage may be restricted. Colorants, artificial sweeteners, and some nutraceuticals (for example, kavakava and St. John's Wort) have all either had their safety questioned, been restricted in their use, or been banned outright.
- Reformulation may also be necessary to create a new market niche for existing products, for example, one with fewer calories, more fiber, or lower sodium content.

Reformulation efforts vary from being inexpensive and accomplished in a short developmental time, for example, reducing salt in a product, to being very costly in research and development time such as the replacement of sugar or fat to make reduced calorie products.

A new, improved product presents a marketing dilemma for a company. Customers and consumers wonder that if this new version is an improvement, why was it not offered in the first instance?

1.2.1.5 New Packaging of Existing Products

This product sector reflects an interesting mix of simple to complex technology and change of form (see Figure 1.1). At its simplest, it is the packaging of bulk produce into smaller packages that are more convenient for consumers

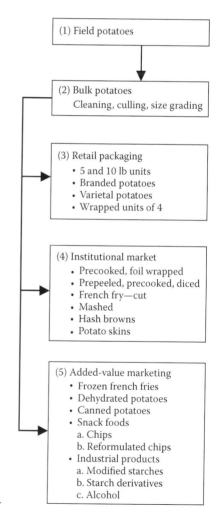

FIGURE 1.1
Adding value to the humble potato.

and can enter a new market niche, for example, yoghurt or fruit pieces packed into sizes convenient for school lunches. Manufacturing involves only purchasing, inspection, grading, cleaning with an increased emphasis on sanitation, trimming, storage, weighing, packaging, and distribution. Development costs vary from minimal to extensive.

How it becomes more complex is seen more clearly in Figure 1.1: field potatoes complete with dirt and culls were available at one time on the retail market in large 100-lb sacks. Next came the cleaning, culling, and size grading of potatoes (box 2). These were repackaged into smaller package sizes to suit smaller families and apartment dwellers with no desire or storage space for 100-lb sacks of produce (box 3). In boxes 4 and 5, greater transformations occur involving both new packaging and new forms of established products with more convenience features and serving different markets. Each step requires more sophisticated processing, quality control, sanitation, packaging and storage, and handling.

Field carrots, oversized, misshapen, and unsuitable for the produce counter, have followed a somewhat similar path but not as elaborate as that for potatoes in Figure 1.1. They, through washing, cutting, and trimming and abrasion, are now "cocktail"-sized carrots and so-called baby carrots, all attractively packed in branded packages. A product, unsaleable on the retail produce counter, has been given added value and entered into a new upscale market niche as *crudités*. The abraded pulp waste also is used and goes into feed or fertilizer.

A new merchandising concept arose: the brand labeling of packaged prepared produce and organically produced vegetables, meats, specially bred (e.g., heritage breeds), or free-range-raised animals gave traditional products a new life as new products of higher quality or convenience on which a company is proud to put its name (Gitelman, 1986). Packages of "speed scratch products," i.e., mixed, washed salad greens, prepeeled fruits and vegetables, and sliced mushrooms, fit into this niche. Previously nameless or brand-less produce and meat cuts now bear display stickers of well-known brands of food companies or naming-the-farm raised breeds. Raising heritage breeds or growing heritage crops introduces another level of complexity to this sector.

Success rests with marketing a brand, the quality of which is known and respected by customers and consumers. For suppliers to food service outlets and the outlets themselves, particularly upscale restaurants, the use of heritage breeds, organically raised animals, or local produce in their menus provides added value features for those valuing such features.

Modified atmosphere packaging (MAP) and controlled atmosphere packaging (CAP) have permitted the creation of a number of new fresh products with an extended shelf life allowing the opening up of new markets in a larger distribution area. Both technologies require extensive research for safety and shelf life stability of the products (see later). Biobased materials, another packaging development, offer a marketing ploy as environmentally sound features (Petersen et al., 1999; Shahidi et al., 1999; Rojas-Graü et al., 2009). They are new and as yet experimental, and care must be taken over the safety of these packaging films. In their review, Rojas-Graü et al. describe the incorporation of functional ingredients in edible coatings for fresh-cut fruits.

Changing from glass containers to metal ones, or vice versa, or the change-over from steel cans to aluminum cans or to plastic containers is an expensive makeover of a packaging line: it requires new machinery or the use of a co-packer. Changeover from cylindrical, conventional cans to thin profile containers (retortable pouches and semirigid trays) for thermally processed foods requires both reformulation and a new filling and packaging line. Such a change improves quality and provides added value for which customers pay more, but it does require costly plant modifications. Equipment is presently such that line speeds are slow.

Another cost associated with packaging is the cost that may be involved in the need for a returns system whereby the used containers of a product must be picked up as customers return these. What can be used for packaging restricts, in many areas, it to be recyclable, reusable, or able to be composted.

1.2.1.6 Innovative Products

Innovative products are difficult to categorize. "Innovate" is defined in both *Webster's Ninth Collegiate Dictionary* as "make changes" and *The Concise Oxford Dictionary* as "make changes in." An example of an innovative product is the Sony Walkman®, whose inventor, Akio Morita, chair of Sony, claimed the Walkman involved no new invention and no costly research and development. In his opinion, companies place too much emphasis on basic research to their own detriment. Reliance, he claimed, on basic research prevents companies from being competitive. Morita saw opportunities in various new developments and put these together with the perception of a consumer need. The Walkman was simply the putting-together expertly of established inventions into a superbly marketed product—its secret was in new packaging (i.e., putting together of ideas) and marketing. This is innovation (Geake and Coghlan, 1992).

An innovative product, then, is one resulting from making changes to an existing product or products. Are any of the foregoing product sectors innovations? Most would say "No." Generally, the more innovation (change) in a product, the riskier it is to introduce and the more costly the marketing strategies of that novelty. The Walkman was a radical, innovative, and conceptual change in communication whose customers had to be educated to its added value.

Little research or development and few production line changes are required for a frozen food processor to put a stew, some frozen vegetables, and a frozen pastry on a tray and call it a frozen dinner. Likewise, putting a can of tomato sauce, a package of dry spaghetti sauce spices, and a package of dry pasta together with a package of grated cheese to make a dinner kit requires little research and development effort. All items in these two remarkably successful innovative products are readily sourced, and their ease of combination has engendered many imitators.

New ingredients often form the basis for innovative products. Simulated crab legs, lobster chunks, shrimp, and scallops, all based on surimi technology, have allowed the development of many *faux* seafood dishes (Johnston, 1989; Mans, 1992).

The changing demographics of cities caused by immigration offer unique marketing opportunities for the putting together of ethnic dishes. Processors of chick peas, navy beans, and other lentils might satisfy ethnic tastes with such added value products as hot bean dips or hummus and hummus tahini and many variants of these. However, any venture into innovative products should only be undertaken with a brand able to carry such products. A new brand may be required.

1.2.1.7 Creative Products

A creative product is one brought into existence according to most dictionary sources: the rare, never-before-seen product. The development of surimi, a fish gel, several hundred years ago and its development into kamaboko-based products would be considered a creative product, as would tofu and limed corn meal. Today, one might consider reformed meat products as a creative development, and certainly, extrusion to produce new puffed products is creative.

The great problem for companies who have introduced successful innovative or creative products is that imitators rapidly flood the market with me-too products. Imitators telescope the time and effort (often measured in years) that was required of the developer of the creative product. Development time may be only as long as it takes the imitator to get new labels printed; marketing know-how has already been developed to educate the public by the originator. Market entry costs will be minimal.

1.2.1.8 Genetically Modified Products

Genetically modified products fit rather poorly into the above creative products category. These could be products modified from their traditional shape, taste, or color by conventional means—and there are arguments about what are conventional means—or by gene splicing technology. Gardeners are deluged with catalogues picturing red, blue, or black potatoes; yellow or green cauliflower; carrots with pink, yellow, or purple cores or with round shapes; beets with purple or white alternating interior rings or beets that are globes or elongated; tomatoes that are yellow, pink, or deep red; and red or yellow raspberries. Only their imagination limits the variations plant geneticists can develop for hobby gardeners. Seed producers and gardeners accept these initially as novelties that they can amaze their friends with, and eventually, these gradually get accepted universally. Commercial growers hint at increased phytochemical content or a vague healthy benefit from increased antioxidants.

The development of commercial crops with increased levels of beneficial phytochemicals in crops that are staples in some diets has been the target of plant geneticists; one example is a new rice variety with enhanced carotene content to combat eye disease. Other categories of genetically modified products are those modified to have a better shape—e.g., straighter bananas for easier shipping, square (or nearly so) melons for better packing, or firmer tomatoes for less damage in transit. Arguments from the anti-GMO groups are that these developments do not benefit the buying public but benefit only the food industry and are not wanted or appreciated by the public especially if there are flavor or usage changes.   .

What still appears to have a strong taboo is the eating of cloned animals. At present, the technology is extremely expensive, and many people have a strong aversion to the use of cloned meat.

There is a wealth of literature in the public domain (see Voosen, 2009) and in trade and scientific journals arguing for and against the safety, public acceptance, or ecological advisability of genetically modified crops. Many countries have banned the use of genetically modified foods or ingredients. Product developers are forewarned to research what the current regulations are.

1.2.2 Customers and Consumers

Customer and *consumer* describe two different entities: the two words should not be interchanged unless the intended meaning is clear. A customer is one who buys in a marketplace. The customer is attracted by point of sales material, promotions, or tastings in the marketplace or at food trade exhibitions. Customers make choices (i.e., purchases) according to their consumers' likes and dislikes, allergies, disposable or budgeted income, or commercial or industrial requirements. The customer is also looking for reliability (cf. brand loyalty) of delivery, quality, and price.

Dr. Kurt Lewin (reported in Gibson, 1981) described the customer as a "gatekeeper." As such, gatekeepers are

- A family member who decides what is purchased for the household
- The purchasing agent of a company, a retail chain, a central commissary for a restaurant chain, an institution, a hospital, and the like, who buy or send out tenders to suppliers with specifications required by others within their organizations
- Chefs who plan menus and decide what raw produce, ingredients, or semifinished goods are purchased for the diners in a restaurant
- Pet owners who determine what pet foods are purchased for their animals

The consumer uses (consumes) what was purchased by the customer. The consumer can also be the customer. For example, the diner in a restaurant or the eater of finger food walking down city streets is both customer and consumer.

Consumers influence customers on what is purchased or served, but as noted by Fuller (2001), "There is the conflict between the consumer's hedonistic demand of 'I want' and the customer's practical barrier of 'I need' or 'I can afford.'" This conflict is apparent by uneaten food returned to a cafeteria's waste; prisoners rioting over the quality and variety of food they are served; ingredient suppliers losing contracts when their clients' specifications are not met; or children refusing to eat their meals.

Marketing and sales personnel must distinguish clearly between the functions of customer and consumer in the marketing, promotion, and selling of new products. Sales personnel are concerned with retailing and retailers, the purveyors. Retailers are mostly interested in the customer since the customer buys. Marketing programs must attract both customer and consumer.

1.2.3 Added Value

Added value is a characteristic many new products are purported to have. The late Mae West had a memorable line in the old film *She Done Him Wrong*: "Beulah, peel me a grape!" Beulah provided added value for a consumer: peeled, ready-to-eat fruit possesses this characteristic. Added value describes a change in a product that makes that product more desirable. Meltzer (1991) rather unhelpfully defined "value added" processing ("added value" and "value added" are synonymous terms) as "…any technique that effects a physical or chemical change in a food or any activity that adds value to a product," which is another way of saying added value is added value. Whatever the value is, consumers want it. De Chernatony et al. (2000) did not do much better in trying to define added value. They interviewed 20 brand managers and as might be expected got as many different descriptions: added value meant many things to many people and served diverse roles in branding—they found the concept was "multidimensional."

Figure 1.1 pictures the concept of added value more concretely. Convenience was created by breaking 100-lb sacks of field potatoes by cleaning, culling, size grading (box 2 in Figure 1.1), and packing into more suitably sized packaging in five and ten pound units (box 3, Figure 1.1). The 100-lb sacks of potatoes were not convenient for small family units and were far in excess of the needs of the increasing number of apartment dwellers who had neither cellars nor space nor desire for such quantity. Smaller, more convenient unit packaging (foil- or glassine-packaged potatoes) introduced new market niches by targeting "live-alones" such as seniors and occasional potato users. Added value was introduced by more convenient packaging.

Adding value continued to be added with prepeeling, dicing, or slicing the potatoes. I worked closely with a potato processing operation that produced prepeeled, french fry-cut potatoes (box 4). Preparation time for institutional users such as the quick-serve eateries was minimized—waste was concentrated in a central location where it might be more profitably used. A new market niche was created providing value.

Today, potatoes and several other vegetables and fruits are offered not only by their brand or varietal name but also with suggestions for the best culinary uses of that particular variety. The customer knows what variety to buy for the purpose in mind, or the customer can buy the variety, knowhow to best prepare it, and have the assurance of quality that the brand name confers.

Adding value requires skilled labor, more sophisticated technologies, and more complex (and costly) processing equipment to safely manage the more fragile products produced. For example, peeled, precut potatoes or peeled and abraded cocktail carrots or segmented oranges are more sensitive to contamination and spoilage. The developer needs not only to prevent spoilage but, more importantly, to prevent hazards of public health significance (see Pyke's statement opening this chapter). Other needs associated with added value are new market research, the production of new marketing material, and more complex handling and distribution techniques. Adding value equals added cost that is only justified by old and new customers and consumers alike accepting that more value comes with this cost.

Meat, poultry, and fish are also sold with more descriptive names, by brand names, and with preparation instructions or recipes describing how to cook the particular cut of meat or species of fish. Added value for both the consumer and the customer has been introduced.

1.2.4 Markets and Marketplaces

Market and *marketplace* are often used synonymously. This is inaccurate. Each has a unique meaning. A market is conceptual: it depicts a need discovered in customers and consumers that marketing personnel hope to develop into a potential to sell to customers who want to buy. That is, one can say that there is a market for organically grown vegetables or for locally grown crops or for marinated (preinfused with salt and spices) pork meats or self-basting turkeys or that there is a market for low calorie foods. This means that there are customers in the many different marketplaces who preferentially purchase organically grown products or low calorie foods or premarinated pork chops. These undiscovered needs (the Walkman was a fulfillment of an undiscovered need) must be developed with skilful marketing research into wants and then satisfied with products.

Products are sold in marketplaces not in markets (despite the use in common parlance of farmers' markets). Marketplaces are real: they are not conceptual. They range from farmers' roadside stands to giant food stores, from beverage and snack bars in movie houses to coin-operated food vending machines, from mobile canteens visiting work sites to stand-alone restaurants. Even electronic food marketplaces operating from Web sites are real marketplaces; they are places where products are sold.

1.3 Marketing Characteristics of New Products

Figure 1.2 depicts three dimensionally the difficulties accompanying the marketing and selling of established products or new products in familiar and unfamiliar marketplaces and with the movement of these products into new market niches.

The y-axis is a measure of increasing marketplace complexity; the farther from the origin, the more complex the marketplace. Marketplace complexity arises with any of the following or any combination of them:

- A product is moved into marketplaces with increased activity by competitors.
- The product needs more sophisticated warehousing, distribution, and retailing treatment (e.g., a stale product return program or bottle refund program).
- New food legislation, changed local regulations, or cultural norms of a more complex marketplace (e.g., in foreign countries) need to be adhered to.
- The general economy of the country or economic developments unique to the targeted marketing area have soured.

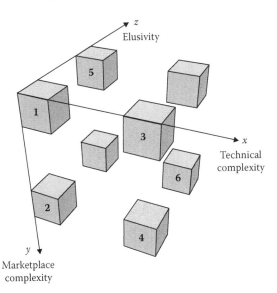

FIGURE 1.2
Product complexity, marketplace complexity, and consumer elusivity interactions characteristic of new products and their marketing.

- The marketing skills within the company or available to the company cannot meet the challenges of the new marketplaces.
- Geography of the market area is such that control over the product is diminished with possible loss of quality or safety.
- The difficulties of selling novelty or educating consumers to the value of the product present major problems.

The x-axis is an index of increasing technical complexity required for a product with innovation or creativity and added value. The farther from the origin, the more creativity, innovation, or technical complexity there is in the product. More research, development time, and costs are involved. The product is more fragile and sensitive to abuse; it requires care in handling and distribution channels to maintain its high quality throughout retailing and in the hands of the end user. Therefore, more creative marketing effort is needed to educate the customer and consumer to recognize the added value features.

The z-axis (into the page) is a measure of the recalcitrance, volatility, complexity, even fickleness or incredulity of the targeted public, a characteristic I have coined as elusivity. Is it the recalcitrance of customers that explains the growth in "healthy" and "health" food categories when there has been "a dramatic increase in the per capita consumption of high calorie desserts, salted snack foods, and high calorie confections" noted as far back as 1986 (Gitelman, 1986) and continuing to this day? Obesity has been branded as having an effect on the immune system leading to the susceptibility to infections (Falagas and Kompoti, 2006), and excess fat as measured by body-mass index is an important cause of most cancers (Larsson and Wolk, 2008; Renehan et al., 2008), yet the average caloric intake by consumers is up over what it was 20 years ago, and obesity has quadrupled (IFT e-Newsletter, October, 2005). Yet there is the apparently contradictory belief stated by many marketers that consumers are on a wellness kick—this is perversity of consumers, my elusivity.

Today, obesity is considered to be of epidemic proportions. The z-axis represents the elusivity of the recalcitrant, hard-to-find and hard-to-understand, volatile, and changeable customer who is targeted in the marketplace. Elusivity can be likened to market segmentation, the creation of a new market niche. If a product designed for the general public (a rare event—no product fits or satisfies all needs) is redesigned for the teenage market, that consumer has become more elusive and the market is being segmented. If this hypothetical product is redesigned again for teenagers of single parents, more elusivity is created. Marketing purists might cavil at this, claiming that this elusiveness is really a variant of the y-axis or marketplace complexity. It is not: it is not the volatility of the marketplace that is represented, but the volatility, fickleness, stubbornness, and disbelief of consumers and customers within these marketplaces.

Six numbered solids situated in Figure 1.2 represent typical problems faced by developers as they attempt to bring new products to market. Cube 1, at the origin, depicts the situation of an established, hypothetical product in a market regarded as home to a company.

The company moves the product into a more complex (according to any of the factors described for the *y*-axis) marketplace (cube 2). The targeted customers and consumers have not changed. There are no new development costs getting into a new marketplace. Only costs associated with marketing (labels, promotional materials, and advertising), sales (brokers), and distribution increase.

Cube 3 is the product but with added value (increased product complexity), but the product has stayed in the same (local) marketplace and targeted its regular (known) customers. Costs for the added value feature as well as for marketplace introduction of the new product (new by definition) and for promotional material to educate the old customers (and consumers) now escalate.

In the situation represented by cube 4, the new product (with its added value) is introduced in a more complex marketplace. The company has simply expanded into a new marketplace with its new product. The targeted customers remain the same but in a new playing field. Researching the new market area brings increased costs because the old marketing and promotional strategies may not be suitable. Distribution costs increase.

The company has found a new use (that is, a new market niche) for the original product: it is targeted now for an elusive consumer but in its local (familiar) marketplace (cube 5). This repositions our product for a new market. Some examples are as follows: a popular antacid is repositioned as a calcium supplement for elderly women; a hand cream proves to be an excellent insect repellent for campers; a well-known baking soda has a purpose as a refrigerator deodorant. Costs can again increase significantly to reach these new targets, which are more elusive. Such repositioning can be risky. A cosmetics company may not want their mystique to be associated with the sporting life and insect pests: a manufacturer of a habanero pepper hot sauces with a high content of phytochemicals of the capscaicinoid family may not want to enter the quasi-medicinal arena based on their product's content of this nutraceutical.

Now the company decides to move its product with its added value and reposition it for elusive customers in a market foreign to the company (cube 6). This represents the worst of all possible worlds as the following depicts; a manufacturer of Monterey Jack cheese reformulates it to contain medium chain length fatty acids (added value) for people with digestive disorders. Here, an established product has been repositioned into a healthcare market niche. Risks and development costs are high; promotion can be difficult.

Figures 1.1 and 1.2 have been compared using potatoes but one could just as easily use cocktail carrots or in-store sushi products or even prepare roasted chickens. Prepared potato products (reformulated chips, hash browns, stuffed potato skins, etc.) drive technical complexity far to the right (cube 3) and at the same time push market complexity into new market areas such as the leisure food category, food ingredient and food service markets (cube 4), and also a more elusive consumer (cube 6).

1.3.1 Product Life Cycles

Every product has a life cycle as depicted in Figure 1.3a. The horizontal axis is a measure of time. The vertical axis is an index of a product's acceptance measured as either volume of cases sold or sales dollars. Five distinct phases of the life cycle can be discerned:

1. The introductory period is heavily supported by promotions, in-store demonstrations, advertising, and slotting fees to gain introduction. Sales volume is initially low as customers and consumers are educated about the product.
2. A strong growth period ensues when first-time customers begin repeat buying and more new consumers are attracted. There is

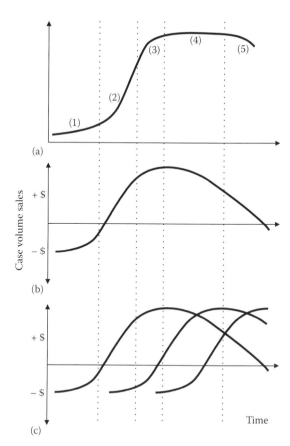

FIGURE 1.3
Characteristics of products, their life cycles, and profitability: (a) Typical product life cycle, (b) the profit picture, and (c) the contribution of new products to profitability.

positive acceleration of sales growth. Growth continues as new markets open, but continued promotion and expansion at the introductory pace are costly.

3. A decline in sales begins. Growth accelerates negatively.

4. Next comes a no-growth period. Sales are constant, a sign of a stagnating market.

5. The decline accelerates. Newly introduced competitive products adversely affect sales; customers and consumers become indifferent to the old product. Promotions cannot profitably maintain sales.

There are life cycle curves for product categories as well as specific products within a category. Instant coffee, as a product category, could be described as being still in the growth phase as manufacturers introduce flavored instant coffees, and instant coffees are used as ingredients—an activity referred to as product maintenance. Nevertheless, the leading brands of instant coffee have changed places as their manufacturers go through different stages of the cycle at any given time. The sale of flour had for years been in a no-growth phase that was only slightly ruffled by the advent of cake mixes; now, it is enjoying a modest growth as many households are returning to the art of home baking, and cook book sales have become hot items as has the popularity of TV cooking shows and cooking classes. During the 1970s, meat prices soared because of a scarcity of beef, the sale of meat substitutes and extenders grew dramatically and then plummeted drastically when meat became plentiful and prices fell. Meat extenders never reached a no-growth phase (phase 4); their life cycle is best described as a spike. Meat substitutes or retextured meats survive in the dried soup category and the soup-in-a-cup products. Life cycle curves are as varied as the products they represent.

1.3.2 Profit Picture

More revealing of the success of a product than sales is the profit brought in by those sales. The introductory phases have minimal net profit (see Figure 1.3b, phases 1 and 2). These bear the costs of past research and development, the heavy costs for promotion to get market penetration, and the retailers' demands for slotting fees to obtain shelf space. Net profits outpace expenses during the latter part of the growth phase (phase 2, Figure 1.3b). The improvement continues in phase 3 but toward the end of this phase profits begin to drop off as customer and consumer demand drops due to inroads by the competition and costs for both market expansion and for support of the product against the competition. During the no-growth phase (phase 4), the company eventually sees the product as unprofitable—it costs too much to maintain. Manufacture ceases.

To keep profits flowing and maintain viability of the company, replacement products must be ready for a launching. A sequential launch of two new products (Figure 1.3c) maintains the company's net profit picture. This tactic forces a company to have new products in various stages of development with some ready to launch; this requires an on-going research and development program. Bogaty (1974) suggests that for every one product on the national market, two should be in test marketing. For each of these, he sees four in the last stages of consumer testing and ultimately, working backward, 32 product ideas should be in screening stages. The cumulative profitability of successful new products promises a good return on investment. For these reasons, a company needs to be constantly investigating markets and marketplaces, customers, and consumers for new product ideas worthy of development.

1.4 Why Undertake New Food Product Development?

The previous section highlighted two reasons to develop new products. First, very few products last forever—they die and must be replaced, or they are rebranded into something completely new. Second, successful new products contribute enormously to a company's continuing profit picture.

Each year, new products flood the marketplaces. Figure 1.4 shows new food product introductions for the period 1964 to 2008. During this period, the data roughly approximate to a sigmoid growth curve going from approximately 1000 introductions a year to over 20,000 a year. There is an initial nascent period from 1964 until the late 1970s and into the early 1980s. This is

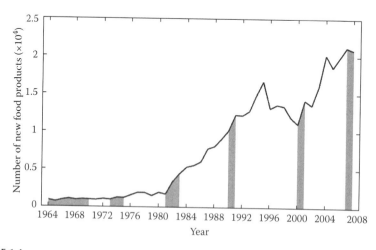

FIGURE 1.4

New food product introductions over a 45-year period. (Data courtesy of Mintel Global New Products Database [GNPD].)

followed by a rapid growth period continuing for approximately 10–15 years. Then by the mid-1990s, there is a very mixed period of decline and growth until data become unavailable.

During this 1964 to 2008 period, there are several periods of economic downturn. The criteria for estimating recessions, depressions, or simple economic downturns are varied according to which economic statistic is used: the 1960s started with an economic downturn and closed with one in 1969 with the in-between years rather difficult. There was another in the early mid-1970s; a more extensive one in the early 1980s; another in the early 1990s; a small decline in 2001; and the current economic decline we are now experiencing that began in late 2007. These economic declines are marked by gray areas in Figure 1.4. The start and finish of these periods of economic decline and recovery are difficult to mark with precision: beginnings and endings vary with the particular criteria economists chose to mark economic declines and recoveries. (Usually declines and recoveries are preceded and succeeded by periods of economic turmoil before and after the economists' signposts claim, for example, today [2009 and early 2010] economic indicators suggest a recovery, yet unemployment figures are high, retail sales disappointing, bankruptcies high, and personal debt at an all-time high.) Economic declines appear to have very little effect on new product introductions especially if one takes into consideration that any introduction may have been preceded by anywhere up to 18 months of developmental work.

Generally accepted observations of "hard times" are fewer purchases of prepared meals, that is, a return to home preparation of meals; less eating out as expense accounts are cut, and upscale restaurants start to fold; and greater cooperation among competitors occurs. Conflicting with these observations is the equally anecdotal observation that when the going gets tough, the people eat chocolate. That is, people scrimp and then indulge themselves for comfort. This was brought home to me when working in a chocolate plant where I learned their most profitable years were during the Dirty Thirties. Nevertheless, the effect of the economy on new product introductions is not clear.

The crude sigmoidal nature of the graph suggested that conversion of the new product introduction data to logarithmic notation and plotted against time might present a better understanding of influences during this 45-year period (Figure 1.5; economic downturns are marked as in Figure 1.4). There is some approach to linearity over this period that suggests no diminishing of the rate of growth of new product introductions. Harris (2002) described a decline in the latter half of the 1990s and ascribed this to several factors:

- Consolidation of many food companies diminished the numbers of product lines. This may provide a reason for the subsequent rise in introductions in the first half of the 2000s. The sale of many spin-off brands associated with the mergers and take-overs may have become the nucleus of new and smaller companies that launched new products.

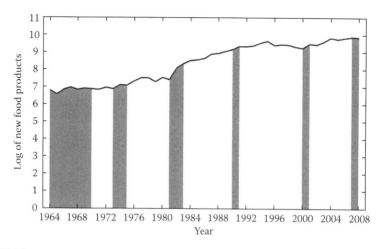

FIGURE 1.5
The rise and apparent decline of new food product introductions (logarithmic scale) over a several year period.

- Better market and consumer research have helped companies to remove potential product failures much earlier in the development process. Harris attaches great importance to efficient consumer response (ECR) technology for this early culling of losers.
- Retailers are giving more attention and space to their own private label products. This, in addition to slotting fees and promotional allowances, discourages many companies, especially small companies, from competing with new products.
- There may be product saturation in some product categories. This may be true in some categories. Or it could be argued that there is lack of customer or consumer interest in many traditional product categories. This lack of interest may reflect the growing presence of many new cultural groups who are unfamiliar with traditional product categories.

My belief was that the flattening of the introductions of new products had begun much earlier, but more extensive data prove this belief wrong. Harris' reasons do little to explain the data over this longer period of time. My interpretation now is that there has been no decline in the introduction of new products, economic downturns have little impact on introductions, and the growth of introductions is highly erratic from one year to another (see Figure 1.6). This figure is the percent increase or decrease of one year over the preceding year (see Cleveland and McGill, 1984). Using the black thread technique, one might describe a horizontal line running between zero and

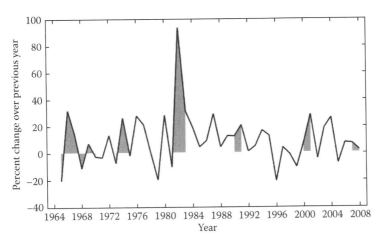

FIGURE 1.6
The percent biannual increase or decrease of new product introductions over a several year period.

a 10% increase on average year after year over this interval, with economic downturns producing inexplicable effects on introductions.

Two points (contrary to those made in the second edition) to be made here are as follows: The first is that there has been no decline in new product introductions over this longer interval. The second point is better presented as questions: Why have economic downturns (and Harris' reasons above) had such minimal effect on introductions? Are there elements in the environment (i.e., the buying public, markets, and marketplaces) of new product development that are still poorly understood? Harris' reasons for the decline that he saw may very well be valid, but more reasons to mark the decline in new product introductions in this earlier period must be sought. Besides mergers, companies at this time (late 1970s and early 1980s) were getting down to their core businesses—the buzz word of the times; that is, they were downsizing, and, often, one of the first groups to be downsized was the research and development department. Whatever the reasons are and whether or not they have some importance in understanding product development, they are not of any concern for the discussions to follow.

Customers still see several thousand new food products entering the marketplace each year. Obviously, the sheer numbers of introductions shows that companies believe that new products are important to their economic futures.

Unfortunately, few of these thousands of new product introductions will be successful and still on shelves a year from their introduction. They will fail for a variety of reasons. Estimates of the success rate of new products range widely from 1 in 6 to 1 in 20. Skarra (1998) reports that 1 in 58 new product ideas is developed into a successful new product. My own experience, gained over a 4-year period with one company's product development program, was that for each product that went into test market, 13 others had received some development at the laboratory bench level or in the pilot plant before being rejected. No count has been made of the many that had been discussed and rejected—thus occupying the time of skilled personnel—before ever reaching the laboratory bench. Clausi (1974) estimated that in one 10-year period, General Foods Corporation conceptually tested, developed, and undertook home-use testing on more than seven products to get one considered suitable for test marketing. Home-use testing is well along the developmental path and represents a significant expenditure of money. Less than half of those introduced to test markets were eventually successful. A very small number of products achieve their developers' goals of successful market placement.

Why are estimates of new food product failures so difficult to assess? There are several reasons why estimates are imprecise. At what stage of the product development process is the product determined to be a failure? Is the idea deemed a failure after taste tests of test kitchen trials? Or is the decision made after a pilot plant run, or after a mini-test market or after a regional test market is not successful? Or does management consider the product a failure after a national launch when the product fails to reach a satisfactory—satisfactory only to senior management—market share in a set time period? A "satisfactory market share" is a criterion the company's management has decided upon and may not represent an unsuccessful product for customers. Does loss of a product idea during in-house screening or home-use testing count as a failure? Is a product that is successful in a regional market but fails nationally a failure? Or is one that just plods along growing slowly and steadily but fails to reach the profit targets in the time established by management (both of which may have been unreasonable targets) a failure? At what point during development and on what basis is the decision made respecting success or failure? Therein lies part of the imprecision for calculating failure rates.

If new product development is so difficult, so costly, and so lacking an assurance for success, why go into it? As Pyke (1971) stated at the start of this chapter, much science representing time and money has gone into this endeavor. The failure rate in new product development is, indeed, horrendous. The rewards, on the other hand, can mean the continued profitability of the company.

Food companies have to afford the costs of new product development to grow and to survive: they cannot ignore new product development. "The engine which drives Enterprise is not Thrift, but Profit" (John Maynard Keynes in *A Treatise on Money*). New products are one of the major avenues open to a food company to be a profitable enterprise and to survive. Some would argue that new product development is the only path the food company can follow for survival.

1.4.1 The "Why" of "Why Undertake New Product Development?"

New product development is driven by several pressures occurring in the environment:

- All products have life cycles. As they lose profitability, they must be replaced, reinvigorated in some manner, or see consumer rejuvenation if the products or the manufacturer is to survive.
- New products provide opportunities for aggressive growth to satisfy management's long-range business goals.
- Demand for new foods by the public, for example, organic foods, functional foods, or foods designed for nutritional requirements of seniors, creates new market niches that provide opportunities for companies to enter with their products.
- Traditional marketplaces change, and new ones are created. For example, e-commerce has emerged and is maturing; the challenge to enter the new marketplace requires new products more suited to respond to the changes the new marketplaces and the new customers in them require.
- New technologies have brought new ingredients and new processes to allow food products that once were considered impossible to produce.
- Advances in the health sciences provide opportunities for new food products suited to the management of healthy lifestyles by consumers.
- Governments establish and change food legislation, establish health programs, adjust agricultural policies, or promote agricultural support programs, all of which provide opportunities for new food products.

The above pressures, singly or collectively, present opportunities that individual companies cannot ignore and must pursue according to the singular strengths of their companies.

1.4.1.1 Corporate Avenues for Growth and Profitability

Corporate business plans established by the owners or the shareholders directly provide the growth objectives and indirectly establish financial goals for the company. These objectives can be achieved in a limited number of ways:

- By expanding geographically: expansion is expensive and risky especially if the new territory has a competitor with a heavy market presence or the marketplaces in the new area are dominated by food cultural habits that the expanding company's products do not fit. Products with short shelf lives require more complex and costly handling and distribution requirements. Export markets present their own unique hazards including the need to accommodate to foreign taste preferences, foreign food and labeling regulations, cultural and religious customs, and because of the use of local agents some loss of control over their product.

- By achieving greater market penetration with a greater market share in existing markets: where there is a strong competitor presence, large expenditures for advertising and promotion as well as price reductions may be needed to battle the competition. Greater penetration is helped immensely if one's product has an added value feature giving it an edge over the established competitor's product. Thus, a better product and aggressive promotion are required.

- By developing new products: new markets can be opened up and existing markets rejuvenated with new products; these contribute to growth and profitability (see Figure 1.4c). The umbrella of brands is expanded to bring in new profits. There are, however, the costs of their research, development, and marketing to be considered.

- By acquiring a rival company or a company complementary to the buyer's long-range plans: buying the competition obviously provides an opportunity to expand a company's brands into new markets, eliminate the competition, or successfully exploit new markets with the acquired company's products. (This possibility for growth is beyond the scope of this book.)

Each avenue above comes with unique problems, but each represents a path to strengthen the company's profit position.

There are other avenues that management can use to increase profitability other than new product development. Management can also increase profitability to meet its financial objectives by reducing expenses and overhead costs. The popular expression was "getting to one's core business" or "sticking to what one was good at." Activities can be farmed out to companies who are specialists in, for example, warehousing and delivery systems, research and development, marketing and sales, accounting, legal needs, etc. Removal of activities deemed to be not core business to specialist companies permits savings and is justified as getting to one's core business. All activities that are farmed out result in some savings, but there are downside risks.

Implementation of an energy conservation program reduces energy costs, lessens the company's environmental footprint, improves company goodwill, but also costs money and research time and does not contribute to growth. Other cost-cutting measures such as reducing benefits programs and implementing wage-cuts (if union members permit such an action) and limiting or restricting travel are again only stop-gap measures at improving profitability with no impact on future growth.

Within the plant, savings can be made by improving processing efficiency, adopting a sound process and quality control program, and thus reducing production of substandard quality, losses through overfill, rework, and poor product returns. These thrift measures help the company's immediate profitability but are of limited value for long-term growth—only new product development assures future growth. Profit, not thrift, drives the enterprise to paraphrase Keynes.

When a plant operates seasonally or has a slack season, its management has an incentive to even out production throughout the year with new products. A plant operating year-round is more profitable than one idle most of the year. The slack season can be used to produce new products putting the under-utilized plant to work. This keeps trained workers employed throughout the year, reduces plant overheads, provides a more steady cash flow, and benefits the community.

1.4.1.2 Opportunities in the Marketplace for New Product Development

Warehouse stores (big box stores), big box sections in retail stores, mail catalogue shopping, and tele-shopping are a few examples of the many new marketplaces changing customers' buying habits. The traditional supermarket has become a collection of speciality food boutiques where the customer can still get meat cut to order or coffees blended, ground, and roasted to taste, yet less demanding customers still have the refrigerated counters of packaged meat cuts, packaged deli meats, and fish. There is often a pharmacy, a post office, and a small dining area, sometimes with entertainment, all to make shopping more enjoyable and convenient and to provide a more pleasant ambiance. In short, the supermarkets are becoming enclosed marketplaces where people congregate for gossip.

Small mom-and-pop stores are becoming 24-hour convenience stores, and even large supermarkets are open 24-hours a day. The abundance of restaurants, diners, take-out food outlets, mobile canteens for work sites (and for the military), and deli-counters in supermarkets makes ready-to-eat (quick serve or speed scratch) food abundantly available.

The diversity of different marketplaces brings with it a diversity of cosmopolitan customers and consumers from many different ethnic backgrounds. Retail outlets now cater to a multitude of ethnic backgrounds, and each requires special foods unique to their culture or religion. The once-a-day shopper (often, seniors or those culturally disposed to daily shopping for freshness) or the once-a-week or once-a-month shopper can be accommodated. Customers for food are also restaurant-goers, purchasing agents for the commissary of a fast-food chain, or government procurers buying for its military or its penal institutions. Because the needs of the customer and the types of marketplaces change for a variety of reasons, food manufacturers serving those buying and selling sectors need to respond quickly to changes in marketplaces and markets—for example, governments have already declared a need for foods for space travel.

The profile of customers and consumers in all food marketplaces is constantly altering and consequently resulting in changing buying habits and a need for new products for these marketplaces. Many factors cause this:

- A flood of immigrants with a variety of ethnic background bring their food habits with them. With this blend of cultures come opportunities for foods to satisfy these new customers.

- Population age. Children grow up and move away. Young couples just starting their careers move in: the economic cross-section of, and the age distribution within, a community changes. The nutritional and general food needs of the community change and opportunities for new products are presented.

- Downtown city neighborhoods, left empty because of the movement of young people with children and businesses to the suburbs, become yuppified as developers move in. These downtown areas have a rebirth as young professionals (familiarly known as "DINKs"—"dual income, no kids") with different lifestyles move in and provide opportunities for upscale eateries (that need equally upscale products) and gourmet retail food products specializing in upscale, prepared take-out (not fast) foods. Opportunities exist for rent-a-chef programs and catered dinners.

- The economic climate in any community (neighborhood, town, or city) is not stable as local companies move or are merged and disappear. The food requirements of a community change as a result of the ethnic and cultural changes the community goes through. These can offer opportunities to those companies alert to the changes.

Marketplaces in any community are in a constant state of flux as the ethnic diversity, the incomes, the education, and the lifestyles of customers and consumers in that community change. Local industries modernize, move, or fade away, and economic values change. These changes provide a framework within which new market niches for food products are developed, and they determine the format of the marketplaces in which to place products that accommodate the needs in these evolving communities.

The activity of competition influences the dynamics in any marketplace. A launch by a competitor of an improved product in any marketplace requires some retaliation from companies with similar products whose sales may suffer from this introduction. This launch requires retaliatory action. Retaliation may involve new pricing strategies and promotional gimmicks or may force the development of new products to counter the competitor's intrusion.

The fluidity of marketplaces is both a challenge and an opportunity for food manufacturers. No one product has such universal appeal that it satisfies the needs of all customers and consumers in all marketplaces. Only a battery of new products will suffice to satisfy emerging market niches. Marketplace change is a great motivator for product development.

1.4.1.3 Technological Advances Driving New Product Development

Advances in knowledge about the physical and biological world are available to any company willing to search in the many Internet-generated databases available or in academic and public libraries. This knowledge provides ideas for new products with more desirable properties and appeal to customers or to make processes more energy efficient or cleaner.

Advances in food packaging provide an example of how the impact of scientific discovery and its application have changed products: at one time, the packaging industry relied primarily on tin-coated steel, glass, and aluminum: the mainstays of packaging were the three-piece can with its lead-soldered side seam and the glass jar. The two-piece seamless can eliminated the three-piece can with leaded side seam and produced a safer can with fewer points of leakage. Other developments were tin-free steel; new plastics; coated paperboard; composites of aluminum, plastic, and paper; and even edible food cartons that gave manufacturers a variety of containers with unique properties to preserve and protect the high quality shelf life of foods with new stability concerns. Protective packaging for containers is now microwaveable, edible, biodegradable, and recyclable. Packaging made from bio-based materials with antibacterial properties introduces another convenience with its property of naturalness (Han, 2000; Petersen et al., 1999; Ravishankar et al., 2009; Shahidi et al., 1999).

Better understanding of food spoilage mechanisms has led to new preservative technologies that have given rise to minimally processed products with better stability, quality, and nutrition. Improvements in retorting technology with rotation or agitation of can contents for faster heat penetration and equipment have elevated "canning," the workhorse of food preservation, to a sophisticated process that gives thermally processed products added value through improved texture, color, flavor, and nutrition. Encapsulation technology protects and maintains the quality characteristics of ingredients through strenuous processing procedures allowing delicate flavors and textures to persist.

An awareness has grown of the role of food to health and well-being and of the importance of non-nutritive factors in foods to the prevention or amelioration of many diseases. This knowledge has led to new products based on these non-nutritive factors and to old products fortified with them— for example, probiotic yoghurts and genetically modified fruits and vegetables with enhanced phytochemical content. Such products sit in a gray area between healthy foods and health foods. Governments have become very interested in these and concerned over unsubstantiated health claims. Foods are now recognized as much for the absence of something (sodium, cholesterol, saturated and trans fats, refined sugars, or high caloric density) as for the presence of something (calcium, fiber, mono- or polyunsaturated fats, antioxidant vitamins, or nutraceuticals).

Markets, marketplaces, and the players in them are affected by the changes brought by technology:

- The public is concerned about their health and are more aware of the role of foods to their health (yet obesity and obesity-related diseases continue to rise). They are aware, vaguely and incompletely, of the role of particular food components in the prevention of disease: they want these foods.

- More discriminating shoppers are making food choices based on food nutrition labeling (when they understand this and often based on what the food doesn't contain), by using in-store computers for information about meal planning and recipes or by using home computers to comparison-shop to get best value for their money and thus widen their horizons for more shopping venues. They are using social networks to evaluate the benefits of products and services in the various marketplaces.

- Social scientists have developed better techniques to research and understand the behavior of shoppers and their emotive reactions toward products. Such technology enables marketing personnel to evaluate reactions of customers and consumers to new product concepts more accurately and to develop better communication methods that have greater impact on customers and consumers.

- Retailers are using their knowledge of customers' behavior to design stores and their traffic patterns to maximize product display, to attract customers (e.g., the use of food odors, in-store tastings, in-store delis, and entertainment) and to service their customers' needs, all to the retailers' advantage.

Food manufacturers cannot be unaware of scientific and technological advances that provide better tools for developing successful new products for the various marketplaces.

1.4.1.4 Government's Hand in Influencing New Product Development

Government's reasons for enacting food legislation are summed up by Wood (1985):

- To ensure that the food supply is safe and free from contamination within the limits of available knowledge and available at a cost affordable by the customer

- To develop in cooperation with food manufacturers, responsible consumer groups, and other interested groups standards of composition for foods and labeling standards that provide adequate information to customers in order for them to make intelligent choices respecting the food they purchase

- To maintain fair trading practices and competition among retailers and manufacturers in such a way as to benefit customers

The above requires an inspection and enforcement agency usually established as a branch of the government.

Government, whether at the federal, state, or provincial, municipal, local, or county level, strongly influences both the business and manufacturing activities of food companies. At the highest level, senior governments negotiate and establish international standards for products and trade practices among countries. These trade agreements are between a few countries (the North American Free Trade Alliance) or between many countries as in the European Union (EU). As recently as 1997, the EU adopted a Novel Foods Regulation (EU, 1997; see also Huggett and Conzelmann, 1997). Article 1 of the EU Regulation describes the regulation thusly: "1. This Regulation concerns the placing on the market within the Community of novel foods or novel food ingredients." Article 2 explains the scope of this regulation as including "foods and food ingredients which have not hitherto been used for human consumption to a significant degree within the Community." This regulation bans any food product unknown in the EU, but which could be well known and with a safe history of traditional usage elsewhere, from entry into the EU without undergoing safety checks. This is basically a nontariff trade barrier, an obstacle that new product developers need to be aware of.

Compounding the influence of these official levels of government are two more categories of governance: international trade and standards bodies and quasi-governmental agencies such as marketing boards. The first regulates international trade by establishing agreed-upon standards for commodities. Two such international regulatory bodies are as follows:

1. International Standards Organization (ISO) that has published a series of directives on quality control and management and environmental awareness (Boudouropoulos and Arvanitoyannis, 1998)
2. Codex Alimentarius Commission under the joint direction of the FAO/WHO Food Standards Programme that has issued a series of standards of identification of foods

Quasi-governmental bodies regulate agricultural production, indirectly regulate manufacturing within their countries, and directly affect imports of agricultural products from other countries. These bodies have a serious impact on new food product development. They do not have legislative powers but do have the support of national governments and hence have the effect of law. Classic examples are the various marketing boards existing in many countries that regulate the local supply, importation, and price of many food commodities and ingredients derived from them. Other examples are professional and trade associations. These establish rules of conduct for their members, members' wage scales, and regulations that participating parties must adhere to. Table 1.4 summarizes activities associated with government and other governing bodies.

A few of these require further discussion. Governments provide incentive programs to stimulate depressed regions and industries or to spur research and development for further use of underutilized commodities. These programs allow companies to undertake applied research or product development programs or to undertake retooling necessary for the new technologies.

TABLE 1.4

Various Food Business Activities over Which Governments in Different Forms and at Different Levels Exert Influence

Activity	Influence
Fiscal policy	Interest rates for development loans
	Grants-in-aid; research funding
	Taxation policy
Patents and copyrights	Copyright protection and licensing
	Research funding if guarantee of patent protection
Trade barriers	Tariffs and protectionism
	Standards of food product identity
	Availability and cost of ingredients
Environment protection	Waste disposal
	Recycling or reuse requirements for packaging materials
	Energy utilization and disposal
Marketing and trade practices	Product or advertising claims
	Billboard and advertising placements
	Zoning by-laws
	Store hours
	Container sizes
Employment practices	OSHA and worker safety
	Unemployment benefits
	Minimum wage levels
Health policy	Nutritional guidelines
	Nutritional labeling
Agricultural policy	Support programs for commodities
	Availability of commodities
Consumer protection	Product safety; safety of agricultural chemicals
	Labeling, product names, comparative advertising
	Inspection services

National health policies beget nutritional guidelines. These ultimately lead to standardized nutritional labeling legislation. Reformulation of those products suitable for export to other countries with different standards of identity and different nutritional requirement is necessary for those companies desiring to expand their export trade. For example, governments have banned or restricted content of trans fats, salt, artificial sweeteners, or colorants. When, several years ago, the U.S. government banned saccharin and cyclamates for safety reasons, manufacturers of dietetic foods and drinks reformulated with permitted noncaloric alternatives or with more intense sweeteners (fructose or high fructose corn syrup solids) that they could use less of. Some companies canned their fruit products packed in water or packed in their own fruit juices. That is, new product development was forced upon manufacturers through government regulatory changes.

A not-so-subtle program abetted by government endorsement has influenced food products in general and new product indirectly. Some governments have decreed that a certain percentage of ethanol be added to gasoline. Corn is being diverted to ethanol production for use in gasoline to spare the use of oil reserves. This has resulted in some catastrophic increases in feed prices and food products relying heavily on corn. Likewise, soybeans have found new uses in biodiesel production and in bioplastics with similar impact on soybean sources.

An excellent overview of complex of food legislation and regulations in the United States is presented in Looney et al. (2001).

2

The New Product Development Team: Company Organization and Its Influence on New Product Development

> The best of all rulers is but a shadowy presence to his subjects.
>
> **Lao Tzu,** *Tao Te Ching,* **Book 1.**

2.1 Structure of Organizations

Governments, companies, associations—even social clubs—all require an organizational structure to function in an orderly manner. The larger the organization the more complex is its structure; small companies work with little formal organization. Organization, however informal or formal, facilitates, or is supposed to, the practice of management and management techniques and establishes lines of communication, control, responsibility, and authority.

Organization is important for effective management of new product development, but an overly bureaucratic organization often hampers an essential element of development, creativity. As will become clear shortly, the practice of management and management techniques (and hence the internal structure of the organization) vary widely with who is managing whom and for what purpose. Management of the creative process is very different from management of other functions of the company's organization.

2.1.1 Types of Organizations

In the Godkin lecture, *Science and Government*, Snow (1961) describes three kinds of organizations, which he refers to as "closed politics." In the management of such closed politics, there is "no appeal [hence their 'closed' nature] to a larger assembly" such as the opinion of a formal group, an electorate, or various social forces (p. 56). These systems of closed politics are

- Committee politics
- Hierarchical politics
- Court politics

Snow's closed politics are characteristic of governance systems in all organizations from committees formed to arrange the company Christmas party to the management of the affairs of tennis clubs to the management of small and large companies; they even describe the internal workings of governments and particularly the most sensitive workings of government, that is, their cabinets. Cabinets as well as boards of directors of companies are organizations that govern without any direct appeal to any electorate. Individually, parliamentarians may be elected, but the internal committees of governments and cabinets and boards of directors are appointed: they are closed politics.

2.1.1.1 Committee Politics

In structures described as committee politics, all members of the committee have an equal voice and vote. For new product development teams, this is described as the "team" approach to organizing. In practice, this equal-voice-and-equal-vote philosophy is rarely attained: indeed, it is patently ludicrous. Those who have worked extensively on social, church, or school committees or in new product development teams will recognize this at once. There are always those who have "more equality" because of their bearing, personality, training, years of experience ("I remember when we tried to do that years ago, and it didn't work then"), rank within the group, or relationship by blood or marriage to senior management. Peters (1987) discusses position power—that is, power as the result of position, for example, the boss; interpersonal power as characteristic of the committee person showing leadership and to whom others defer; and personal power described by Peters as commitment or skills.

> I worked with a company in which meetings focussing on new products and acquisitions were held around a circular table, so that there was no head of the table. This was done in the belief, quite mistaken, that all seated around it were equals—King Arthur and the Knights of the Round Table style. When the president sat at the meetings we all knew where the power and authority was.

In small- and medium-sized companies that I have visited, committee politics is often the form of governance that prevails. I have often been told "we're all family (or equals) here"; it was seldom true. Committee politics seldom works because of its inevitable weakness; it is influenced by position power. Position power is the dominant disruptive element in this type of organization. Committee politics is found at all levels within the many subdivisions of large companies. This structure is not conducive to good managerial relationships, to team work, and certainly not to the creative spirit necessary for innovation.

2.1.1.2 Hierarchical Politics

Hierarchical politics describes the highly articulated organization typical of governance in multiplant, multicompany organizations. It is the big company structure. Power, it is assumed, is "up the line," resting in some person with ultimate, absolute power somewhere at the top of the pyramid. This assumption is seldom accurate. Snow (1961, p. 60) succinctly stated the problem regarding hierarchical organizations: "To get anything done...you have got to carry people at all sorts of levels. It is their decisions, their acquiescence or enthusiasm (above all, the absence of their passive resistance)" that decides what gets done. Attempts to get agreement or approval, especially for innovative ideas that may challenge established practices, at all levels within the hierarchical pyramid of power are difficult. As one goes up the line, other interests also vie for the attention of those higher up and the power and influence they wield.

The management of new product development in large food companies is broken up according to the company's brands and within these brands into groups led by brand or product managers. Authority, despite efforts to make the structure appear as a democratic, cooperative team approach (a committee politics), is usually well defined and linked through a brand manager by solid or dotted lines to some member of senior management. That is, it is hierarchical politics on a smaller scale.

As the paper work to justify a new venture—either a new product or a new process changeover—goes up the line of hierarchy, other interests within the company, often those holding contrary views that are sceptical of the project, review the project. Challenges are inevitable and the presenters of the new venture must convince others above them that their project is better suited to the needs of the company than their challengers' projects. (It must be remembered that the new product development team are not in personal contact with the upper echelons of power and are only in contact through proxies or a paper trail.) As Snow states it is "above all, the absence of [the doubters'] passive resistance" that decides what gets approved. Product development requires sound arguments to justify the financial support it requires. In a hierarchical system of governance, there is always competition for money for projects, each of which demands support and promises benefits to the company.

The paperwork for new product development can be daunting in large companies. Typical is the following paperwork demanded by one company: first, a description of the project with provision of examples of existing products or technology if such exist. There follows a detailed outline of the project naming those to be assigned, their overall objectives, their immediate objectives, the present status, the experimental plan, the estimated elapsed time to completion, estimated manpower requirements (by month and cost); estimated other expenses, and finally estimated capital expenditures. There follow other requirements (10 pages in all) that describe the progression (in timely reports) of the project through approval by various departments within the division initiating the request for a new product: these divisions are marketing, technical (research and development), manufacturing, legal, the divisional new product manager, and finally the division manager. The managerial levels (above) whose approvals and comments are required rise to the vice presidential level.

These forms describe the paper trail for a new venture of a major food company. It should be obvious that at many stages, it is possible for "people at all sorts of levels" who believe with equal fervor that their projects are better suited to the interests of the company to demonstrate passive or active resistance.

2.1.1.3 Court Politics

Snow's third organizational structure, court politics, is more complex but, at the same time, one that is very familiar to most people in large or small organizations; it permeates many organizations. There is power—the boss or president has it by position. There is, however, another kind of power; this one is "under the table" power, unofficial managerial authority, or an undefined ability or knack for getting things done that is exerted through some person who possesses a concentration of power or influence or contacts: this closely resembles Peters' interpersonal or personal power (1987), but it is not a return to committee politics. This person is not necessarily the boss or some person up the line of authority. This individual can be likened to the *l'eminence grise*, the unobtrusive facilitator, a reference to Père Joseph, a Capuchin monk, who was the private secretary to and very influential with France's Cardinal Richelieu—the power behind the throne. Within all organizations, there usually exists a person who subtly wields and exercises more power and influence than either title or position would warrant. They manage, somehow, to get things done, often done his or her way.

Such are the organizational closed politics to be encountered within managerial structures including new product development teams. These structures exist in companies large and small, and the best must be made of them for successful new product development.

2.2 Organizing for Product Development

Most companies structure themselves in some way that is meaningful for their operation that is based however tenuously on one of Snow's descriptions for product development. I have consulted, however, with many companies in which organization is informal to the point of being autocratic, that is, no organization was the rule:

In one company, the Research and Development manager, also the Quality Control manager—this is frequently the situation in small companies—had hastily prepared a recipe for a new product at the request of a senior manager. It was liked by the manager and other key personnel. Several hundred cases were produced for display at an international trade fair where it proved popular. Orders came for a product whose formulation had not been finalized or approved; whose raw materials and ingredients had not been characterized for purchase standards or sourced for pricing; and whose shelf stability had not been established or tested in any other fashion except by tastings of foreign buyers. Much post launch developmental work was needed when the product proved unstable in the marketplace.

This is a clear example of power by position in a hierarchical politics with hints of committee politics. Such informal systems nearly always lead to disaster, frustration, and misunderstanding.

Two questions emerge in any structuring of a product development: the organization is meant for what? and meant for whom? Answers to these two questions conflict—they present the classic Catch-22 situation as will be described.

What is the organization meant for? It is meant for the efficient management of professional human and the effective deployment of physical resources, for facilitation of communication within the group, for delegating responsibility, and for the continued successful functioning of the organization in order to produce uniform quality products that customers and consumers want to buy and use. That is, the company must be kept running in a regulated efficient manner. The leader of such a structure has the responsibility to keep direction focused, to rigidly follow established protocols, and to have authority to get "things" done. This, the physical organization, answers the "meaningful for what?" posed above: that part of the organization that pays the bills.

As has been shown in Chapter 1, making the same uniform product, day in and day out year after year, is not profitable when the needs of targeted customers and consumers change and technology changes. An old saying in the food industry is that if you are making a product the same way you were 5 years ago, you are making it incorrectly.

The answer to "For whom is organization required?" is much simpler but more difficult in practice. For product development, it creates an environment that fosters creativity for skilled professionals in many disciplines. Organization facilitates innovation and discovery. On the other hand, organizations must provide discipline, direction, and encouragement to those involved in the development process and guide firmly and wisely the project to its ultimate purpose of increased profitability for the company. The organization (its management) that is created must marry scientific and technical skills with human skills of managing people to create a harmonious and creative working milieu and to hone innovative skills to create a new product. This is Katz's three skills approach for successful management that will be discussed later (Katz, 1955). "Meaningful for whom?" has now been answered.

2.2.1 Organizing for "the What": The Physical Plant

One must separate the physical plant and its efficient operation from its management. The purpose of organization is to facilitate management's effective and profitable use of the physical plant. There is no shortage of management information replete with charts for the physical organization of research and development. With their boxes and solid and dotted lines describing lines of either authority or communication, they can be very impressive. A reader wanting these can refer to several excellent papers that are still pertinent despite their age:

- Mardon et al. (1970) discuss at some length the problems of administrating technical departments of multiplant companies, not the least of which is maintaining a balance of technical staff in productive capacities compared with those in development roles. This balance is challenged when there is a need to reduce overheads. Poor lateral communication especially in large organizations is cited as a problem.

- Head (1971) sees the role of the technical manager as one having only two resources: people and their skills; and material resources (equipment and laboratory facilities). The manager "inherits a situation, good, bad, or indifferent." Getting the best out of material resources is by far the easier task, but the people resource must be challenged with spiritual (that is, there must be a desirable objective whose achievement is vital), intellectual (the objective can be attained, the organization he is part of is efficient and his leaders worthy), and material (the best equipment and working conditions will be available) foundations.

- Aram (1973) describes those informal networks (the "undergrounds") that evolve in companies for research and development and supports their encouragement rather than making any attempt to control them with formal management techniques. For Aram, these informal networks are not to be confused with no or idle undirected research. There is a striking similarity between Aram's "undergrounds" and Snow's court politics; both involve reliance on those who get things done.

Unfortunately, organization does not always facilitate but often hampers management.

Head (1971) shows his disdain toward organizational charts with the following delightful comment, if one queries all the boxes and lines:

> You will probably be submerged in a torrent of peculiar terminology about "line control finance wise" and "inter- and intra-functional communication channels". Initially, one thing only will be clear – the appalling debasement of the English language.

One must get beyond organizational charts for effective product development. Yet many companies live by these charts and dote on the dotted and solid lines linking the various boxes. Charts allow managers to know who is in their department or for whose actions they are responsible, but charts segregate functions and permit, if not carefully watched, rivalries between "boxes" to develop with the inevitable finger pointing when something goes wrong. Head would also most surely have been a devotee of Aram's unstructured underground and Snow's court politics.

2.2.2 Organizing for Whom: The Human Side

The purpose of organization for new product development is to develop a cohesive team of diverse talents, to motivate this team, and to direct their talents toward the creation of a specified product required for the company's business plans. One is not attempting to organize rivalries.

> The future manager will become steadily more active in catalysing the participation process among his subordinates: equally, he will expect, in increasing measure, to participate with his own masters (Head, 1971).

Head's comment clearly suggests the team or committee approach but opens the possibility of conflict with positional power.

Organization is not solely for lines of command or communication but also for establishing lines of participation and facilitation. A strong element in Aram's underground research groups is the assemblage (however, unofficially and without dotted or solid lines connecting boxes) of those who can contribute, participate, and be kept focused without feeling a sense of being herded or directed in their thinking.

The difficulty for organization is threefold: there must be management to keep scientists and technologists creative and focused, management of marketing personnel, and management of the physical plant for its continued productivity of bell-ringer products and cooperation with the development team. These parties march to different drum beats. Each has within it its own pyramidal structure of organization. Organization is necessary to facilitate communication between these groups and between the multidisciplinary groups that make up the new product development team, that is, lateral communication between the pyramidal structures that develop in large corporations but also communication vertically within and without these pyramids. Communication implies participation, and this in turn implies that those who participate also contribute.

Another demand on organization is that it must gratify the personal needs for recognition of those people in the system and respect for their contribution. While keeping all feeling part of the team and not apart from it, product development managers must foster creativity, that is, thinking outside the conventional wisdom.

Hierarchical politics for product development should stop at the product development manager (or by whatever name the function might be given). The product development manager is best situated to motivate the new product development team in directions the management desires and to provide lateral communication between the members.

Large companies control their development resources through a formalized hierarchy of inter- or multidisciplinary management teams (portfolios as one company terms them). Small companies have informally structured organizations in which, as in large companies, personalities can dominate. The new product development manager (if there is one in smaller companies) must be able to ease the project through the various departments involved and smooth the way with a strong cohesive team spirit.

Managers of product development have two resources: physical plant and skilled people (Head, 1971), both must be harnessed. The physical plant is inert: it can only be used to complement the human resources who make it run efficiently. Therefore, people are the most promising resource for developing creativity and innovation. Innovative people will use physical plant facilities effectively (that is, doing it correctly) and efficiently (doing it correctly the first time) whether these be test equipment in-house, at equipment suppliers, or equipment with co-packers. The touch of the manager must be deft: too much control, too much pressure, can stifle creativity and innovation. On the other hand, too little control provides no certainty that innovative product development will ever result. Some organization is necessary; otherwise, chaos would rule in the company.

2.2.3 Organization and Management

Organization is not a synonym for management: management uses organization to manage innovation and creativity. It is here that the subtle difference between organization as most know it and Snow's governance systems of closed politics becomes apparent. Management can foster innovation and creativity; management involves people. Organization implies planned systems of predictable activities, all of which are predicated one upon the other in a controlled and controllable fashion. Ultimately, these systems are interfaced with the other systems that make companies function effectively. If the unexpected happens in this network of coordinated activities, the company is so structured that remedial procedures swing into action to control the unexpected event.

A company's organization can be likened to the human body. When the unpredictable happens to the human body, for example, an invasion by some virus, the body's immune system comes into play as a defense mechanism. That is the function of organization.

Innovation and creativity are recognized as activities that are unpredictable, uncoordinated, uncontrolled, and uncontrollable and are certainly to some degree unplanned. Managers of a company's established brand products manage those products within limits established by their brand domain: they work within a system. Those engaged in new product development work within minimal systems and may take extreme risks with brand images. Brand managers provide advice on products being developed to fit their brand's image or on how their brand's image might be expanded or complemented by new products, but product developers need not be so confined and work within minimal systems.

The above being so, does not a paradox arise whereby if organization is imposed, creativity and innovativeness are stifled? Actually less coordination, less control, less planning, and generally less bureaucracy do not lead to more creativity or innovation but to chaos and randomness (Aram, 1973). If, for example, a development group cannot be productive, innovative, creative, or inventive within time and budget constraints imposed by senior management who are responsible for the business goals of the company, there is no expectation that it will be more effective in these endeavors if no limits are provided by senior management.

2.2.4 Creativity: Thinking Differently

Interpretations and definitions of what constitutes creativity and innovation abound in the literature. For example, H. J. Thamhain (Dziezak, 1990) describes innovation as

> a process of applying technology in a new way to a specific product, service or process for the purpose of improving the item or developing something new.

Bradbury et al. (1972) put forward a more precise definition of innovation, thus consequently narrowing its meaning:

> the recognition that an opportunity or a threat exists and which is concluded when a practicable solution to the problem posed by the threat has been adopted or a practical means of grasping the opportunity has been realized.

Bradbury et al. continue with definitions for discovery: "finding or uncovering new knowledge" and, leading directly from this definition, invention is "discovery which is perceived to possess utility." Bronowski (1987) (but see Stent [1987] for comparison) used the terms *discovery, invention, and creation* in well-defined ways but added the element of "personalness." As examples of personalness, Bronowski cites Columbus who discovered the West Indies: however, these islands were already there. If not Columbus, then someone else would have bumped into them—indeed, the Norsemen and Spanish fishermen did bump into the Americas several years before Columbus. Likewise, Alexander Bell invented the telephone, but the basic elements of the telephone were there, "if not Bell, then someone else," and there are others who do claim being first. Neither event can be called a creation according to Bronowski: they were not personal enough. On the other hand, Shakespeare's *Othello* is a creation despite Shakespeare's reliance on sources written by other authors—this work is personal.

Whatever interpretation one applies, creativity, innovation, discovery, or invention requires harnessing to direct the skills into desired channels and encouragement. Companies interested in new products organize to direct their staff's activities to the companies' goals and to manage their physical resources economically and effectively and at the same time foster creativity and innovation within individuals.

There are some key phrases describing creative people: "perceiving in an unhabitual way," "have pretty good ideas that do not fit in with policy," "are eccentric," "have less respect for precedent," and "make associations of dissimilar things." These are descriptive terms that are anathema in a tightly organized system. The problem is to create free environment for creativity. Too often, one's own disciplines fetter one's mind with unwritten or even written strictures on what is proper, accepted, or the correct way to do something. A set of laws, regulations, orthodoxy, or peer pressure govern one's thinking and funnel it along acceptable (proper, orthodox, correct) lines. This direction of thought can stifle creative and innovative approaches to problems.

Children are unencumbered by this rigidity of thinking. They acquire this rigidity as they grow up, become educated, learn the accepted orthodoxies, encounter peer pressure, and begin to fear thinking and being different. Young children have a wonderful capacity to put together unrelated ideas in implausible and improbable ways as can be seen in their stories or drawings. There is no embarrassment in freely associating seemingly bizarre ideas. The undisciplined, childlike mind has not yet been confined to the path of "correct" thinking. This spirit of free association of ideas was very apparent when I worked as a YMCA instructor and youth leader:

We played a story game. I started the story, reached an impossible situation with the story characters then passed the story on to one of the children. The rules were simple: the next story teller had to start after a count of five; no magic was permitted and no violence; new characters could be introduced at any (and usually most appropriate) times. Children were eliminated when they could not either extricate the characters from whatever misadventure they were in or start the story on time. Few children were ever eliminated but they loved to eliminate, it grieves me to say, their leader. The inventiveness of these children was mind-boggling.

Creative people—writers, artists, and inventors—seem gifted and inspired compared to the ordinary people of this world. What is not fully appreciated by the rest of us is that creative people are also very hard workers who figuratively fill many a waste paper receptacle with discarded ideas, outlines, and schematics. Ideas come only after much study, thought, research, and experimentation, plus plain hard work.

2.3 Research for Creativity: What Is It?

Research means different things to different people. For instance, to the person in the street, the word conjures up images of complex laboratory set-ups with white-lab-coated scientists who are far removed from normal mundane life and activities. The favorite pose for a photograph taken by newspaper journalists is of a scientist viewed through an array of glassware gazing into a test tube of colored water looking as if the scientist knew what was in it.

Many lay persons would deny that they ever conducted research. Yet these same persons will examine brochures about cars, visit several car showrooms, hold discussions with and question numerous car salespersons, bargain with various financial institutions for the most advantageous payment terms, as well as perform an Internet search on cars and suppliers and join Web chat sites for comments from owners of particular vehicles. They would never dream of calling any of these activities research.

2.3.1 Characterizing Research

There are two broad classifications of research: basic, fundamental, or pure research; and applied research. Fundamental research is very loosely described as research for the sake of knowledge without thought of commercial exploitation (which is seldom true). Applied research conversely is research for commercial exploitation. The separation of the two categories is by no means clear cut. For example, *basic research* has also been used to describe research for which there was a possibility of exploitation. Gibbons et al. (1970) describe applied research as mission-oriented research—that is, research directed to some specific goal that did not necessarily lead to commercial exploitation.

The following classification may remove, or perhaps add to, the confusion:

- Interest-for-interest's-sake research that can best be described as research with no foreseeable application. It is the dilettante's research, just to satisfy curiosity (Gibbons et al., 1970). Muller (1980) in a very interesting paper decrying bureaucracy's stifling of innovation by pettifogging funding policies might term this *seed* research, that is, research time and effort to follow up ideas that may lead to funding or patents. There are no time constraints in this research.

- Basic (pure) research (no matter how much the academics might argue contrariwise) is always undertaken with some expectation of an application in the future—if for no other reason than to get a research grant or patent (see Muller above). Nevertheless, Muller (1980) quotes Wernher von Braun as having said, "Basic research is what I'm doing when I don't know what I am doing." If time constraints exist, and they usually do, they are measured in years rather than months, as many candidates for graduate degrees know.

- Goal-directed research is research directed to some specific mission, the application of which is quite apparent. The application need is at most a few months. This would be described by Gibbons et al. (1970) as mission-oriented research.

The readers must accept for themselves the meanings that apply within their new product development environment and how they apply these terms to what they do.

All new product developments fall in the goal-directed category. That goal is to increase profitability, to gain market share, to exploit a perceived market need, or to counter the activities of competitors in the marketplace with a new product. Small companies cannot afford any other type of research save that directed to protect and expand their profitability. Large companies may separate their research and development in two ways: very goal directed to new products; and research confined generally to basic research with some expectation of future exploitation. This latter research is usually directed to objectives to be accomplished in 3–5 years and is often contracted out to universities or other research institutes; it is often basic research cooperatively funded by several companies with common interests. Development programs are focused on goal-directed research aimed at very specific, short-term, food product objectives.

Interest-for-interest's-sake research is rarely openly pursued by food companies. It has never, to my knowledge, been engaged in for new food product development, but some proviso to this statement is required. Research with no foreseeable application is undertaken by food companies to accumulate knowledge or to gain experience in a particular topic of food science, for example, textured proteins or liquid carbon dioxide extraction of flavors. While there is no foreseeable application, nonetheless companies may have ulterior motives (not based on the science or its outcome, necessarily) in doing the work. For example, they wish to keep competitive and aware of new developments (have a leg up, so to speak). Or they develop goodwill with cooperating universities (and knowledgeable staff). They also support graduate students in some esoteric field of food science, a field of research that may, in the sponsor's opinion, bear some future value. If there is a good likelihood that those graduate students may be hired by the sponsoring companies, then is this research with no foreseeable value? Sponsoring companies have ample opportunity to evaluate their candidates and, if they are hired, they come with practical experience in the company's field of interest.

Research with no foreseeable application (interest-for-interest's-sake research) may be undertaken within a large company without management's knowledge. This is Aram's (1973) "underground research and development." Aram, in a study of a company involved in research and coincidentally in innovation, noted that informal networks developed within the organization. It was through these informal systems that innovative research occurred. Aram found

> the cross-departmental informal organization... had the connotation of an activity that was disguised, if not almost illicit. Part of its attraction and its effectiveness seemed not to be managed.

Two individuals from one particular underground group are cited, one from new product sales and the other from product engineering, whose group was responsible for several patent applications.

2.3.2 Organizing for Creative Research

An atmosphere in which strictures imposed by discipline, training, peer pressure, and peer ridicule are removed is necessary for the generation of ideas directed to the company's goals for innovation and creativity. In this atmosphere, the purpose is to glean ideas by appealing to the "childlike" in people for products meeting the needs, found through market research, of targeted customers and consumers. This appeal requires that all ideas deserve a respectful hearing regardless whether they emanate from the janitor, the technical director, the sales person restocking shelves, or the boss's wife (where many of them do come from and a source that I can speak of from painful experience).

Organization designed for new product development and for efficient plant operation seems to be a contradiction in terms. To run a food plant efficiently, one needs firm guidelines, rules, standards, regimentation, and clear lines of communication—thinking always, in today's parlance, inside the box. This inside-the-box thinking constrains the creative process. An efficient plant requires strict rules of operation to produce uniform, safe, healthy product meeting government regulation and satisfying customer's needs. The creative process integrates people and physical plant to focus on solving problems involving the future of the company. Creativity flourishes in a more relaxed, less regulated atmosphere, the opposite of that of the physical plant.

In reality, there is no contradiction here. Certainly, true innovation and creativity cannot flourish in a bureaucracy with all its pejorative connotations nor can these elements flourish without organization. Organization is required for no other reason than to keep the activities of technologists, engineers, market researchers, and production personnel under fiscal control, communicating productively with one another and provided with support resources to feed their creativity and innovative spirit.

New product development managers must create within the systems of closed politics an atmosphere that fosters creativity and harnesses the skills of creative people while encouraging them to work as a team with a common goal of meeting the company's objectives.

A brief digression into management philosophy must be introduced here. Business magazines abound with reviews of books on management philosophies by renowned authors such as Henry Mintzberg (*Managing;* Berrett Koehler Publishers). Some management techniques are applicable in all organizations but are particularly pertinent to creativity in product development. Katz (1955), in a classic paper, discussed such an approach that has come to be known as the three skill approach. The skill requirements are as follows: First, there must be technical skill, that is, a proficiency in a pursuit involving methods, processes, procedures, and techniques. Second, this must be complemented by human skills, that is, the ability to work with people and be able to create "an atmosphere of approval and security in which subordinates feel free to express themselves without fear of censure or ridicule." Third, there must be conceptual skills whereby one is able to work with abstraction and hypotheses key to creating vision and a strategic plan, that is, to see the enterprise as a whole and understand how the pieces complement the whole. (A deeper discussion of Katz's work is reviewed by Peterson and Van Fleet [2004].)

2.3.2.1 The "Unhabitual" as a Tool in Creativity

Land (1963), then President and Director of the Polaroid Corporation, on the occasion of the 50th Anniversary of the Mellon Institute, said:

> In our laboratories we have again and again deliberately taken people without scientific training, taken people from the production line, put them into research situations in association with competent research people, and just let them be apprentices. What we find is an amazing thing.... In about two years we find that these people, unless they are sick or somehow unhealthy, have become an almost Pygmalion problem; they have become creative. If there is anything unpleasant to an unprepared administrator it is to find himself surrounded by creative people, and when the creative people are not trained it is even worse. They have two unpleasant characteristics: first, they want to do something by themselves and they have some pretty good ideas that do not fit in with policy; secondly, they have the most naive, uncharming and unbecoming direct insight into what is fallacious in what you are doing, and that, of course, is a blow to policy. I do not want to romanticize these people. I am simply reporting on what we seem to find is a fact... and you have to find out what to do with these awakened people.

Here, Land (1963) latched on to the idea of putting production people into unhabitual situations where they might make associations of dissimilar things (their production background vs. a research situation) to develop such an atmosphere for creativity and innovation. Land took people out of the humdrum of fettered minds and allowed them to develop and become creative.

James (1890) describes genius as "little more than the faculty of perceiving in an unhabitual way." Another writer (Anon., 1988a) commented as follows about creative people:

> Highly creative people are eccentric in the literal sense of the word. They have less respect for precedent and more willingness to take risks than others. They are less likely to be motivated by money or career advancement than by the inner satisfaction of hatching and carrying out ideas. In conventional corporate circles, such traits can look quite eccentric indeed.

Stuller (1982) recognizes several types of creativity and provides some examples of each:

- Theoretical (Albert Einstein, Sir Isaac Newton)
- Applied (the Curies, Henry Ford, Alexander G. Bell, and Thomas A. Edison)
- Inspired (artists and composers)
- Imaginative (writers and poets)
- Prescriptive (thinkers such as Plato, Machiavelli, and Martin Luther)
- Natural (dancers, musicians, singers, and sports figures)

Regardless how he classifies creativeness, Stuller recognizes a common theme throughout: the protagonists have the ability to make associations of dissimilar things. That is, they are creative because they do not compartmentalize observations and experiences but see connectedness.

An examination of the elements of creativity (Table 2.1) seems to contradict all the precepts of good organizational structure for productivity; organization is regarded as something very much to be avoided (Stuller, 1982). Challenging assumptions, for example, especially those held dear by management, challenge management's authority and its self-assumed omniscience. A willingness to take risks can be frightening for conservative elements (the financial and processing departments) within a company. The last element, networking, is strongly reminiscent of Aram's undergrounds and Snow's court politics, that is, a social fabric of like-minded and dedicated people working "underground" toward a common goal. All are examples of thinking outside the box.

2.3.2.2 Cross-Functionality in Product Development

The development team comprises members of different disciplines, for example marketing, engineering, technology, purchasing, etc. An extensive

TABLE 2.1

Elements of Creativity in People

Elements of Creativity
Creative person challenges assumptions. They ask "why?"
Creative person recognizes similarities in patterns, events, occurrences, and concepts.
Similarly to the preceding, the creative person connects arrays of events and notes new ways to see that which is strange as familiar and the familiar as strange.
Creative people have a willingness to take risks.
They are opportunistic; they use chance to advantage.
Being wrong neither concerns them nor frightens them.
They network; they make contacts with other creative people.

Source: Stuller, J., *Sky*, 11, 37, 1982.

analysis of factors influencing cross-functional product development can be found in Sethi et al. (2001).

The cross-functional team presents an interesting influence on both innovativeness and the success of product development according to Sethi and associates. There is obviously a diversity of input that is good; too much diversity creates information overload, and decision making is hampered. The result is that the team resorts to an algorithmic approach to decision making. The nature of the cross-functional approach introduces two problems for consideration. First, there is the physical composition that has been discussed. Not yet discussed is the psychological composition that considers how the members relate to one another but also how their functional areas relate and complement goals (see Denton, 1989). A danger exists that this sociability (its physical and psychological nature) of the group could lead to group think, a very real danger with a negative influence on innovation.

The diversity of input leads inevitably to novel relationships between diverse ideas that it is hoped leads to a nonroutine approach (non-algorithmic approach). It is here that risk-taking enters into innovativeness—a nonroutine approach invites risk-taking and risk-taking is necessary for innovativeness.

An environment encouraging risk-taking requires monitoring but also facilitation by senior management to emphasize the importance of, and interest in, the project by management and to provide a reward system commensurate with risk-taking.

2.3.2.3 Fluidity as an Organizational Tool in Creativity

Seeing how the other half-lives, walking in another person's shoes, and bearing another's burden are clichéd adages used to develop understanding between people. They are all colorfully illustrative of creating cohesiveness within a group, and management has applied the concept to the new product development team. It is a direct break with the rigid organizational boxes, solid and dotted lines structure.

Many large companies create fluidity within their organizations, for example, by transferring technical people with a product as it matures from the laboratory bench through engineering and production to marketing. In this manner, all members of the team begin to talk and understand the esoteric jargon of the other team specialists or at least are understood by the other members of the team. Members of the different disciplines of the team understand their different sets of work values or interpretations of company objectives and begin to respect the others' efforts.

Communication laterally between team members and vertically within participating departments is encouraged with fluidity. This is important. Managers must be able to report clearly and comprehensively to others situated vertically and laterally within the company if for no other reason than to quell rivalries. They need communication skills to sell technology as well as the innovative skills of the group to others. The manager must make this a cohesive group. By moving members of the team with products that they developed, the professionalism of, let us say, technical people is meshed with the professionalism of other members of the team and also with the commercial and business interests of the company. A greater understanding of the contributions made by all results. It is a form of technology transfer.

The marrying of research and development personnel with marketing personnel has been described as the food industry's rewriting of the television series *The Odd Couple* (Hegenbart, 1990). (This is also the plot concept of many television police mysteries: there is the old police inspector paired with a new, university-educated sidekick.) Hegenbart discusses the turmoil that often arises between these two elements of the team. Research and development personnel are quite used to resolving problems—it's their job. What isn't so second nature to them is the resolution of relationships between differently trained people, with a different language and understandings. An early paper by Denton (1989) discusses steps to resolving conflicts between the two groups with dissimilar work habits and goals. Managers must not take sides in disputes but should define the contretemps, limit its spread amongst the team members, and get both sides talking about resolving the conflict together.

Fluidity of movement is essential within any new product development group since the skills in one group may complement the skills in another group in unexpected ways.

2.4 Constraints to Innovation

It would be an abrogation of responsibility not to mention or to not re-emphasize some constraints to creativity. They do exist in both large and small companies.

2.4.1 The Corporate Entity

The business philosophy of senior management (the head office, owners or partners, etc.) has a direct bearing on the organization and especially the creative side of organization.

2.4.1.1 Risk Capital

The reluctance of food companies and especially large multinational corporations to engage heavily in basic research is easily understood. It is expensive. Although any corporate entity would welcome a major technical breakthrough that could enhance its competitive edge, this desire and its cost must be tempered by consideration of long-range corporate financial goals, of shareholders' desire to make an annual profit, and of the need to stay within the financial constraints of an annual budget.

Senior management in large or small companies has a time horizon rarely fixed more than a year or two ahead. This horizon influences management decisions at all sublevels within the company. Most business school graduates have had it drilled into them that profit is the name of the game, and profit is viewed as short-term profit. The pressure for short-term gain forces all within the development team to look to quarterly or semiannual profits. This does two things: creative research that involves risk-taking and is expensive is given short shrift; and new product research is directed toward safe new product goals, often toward me-too products or products with only incremental improvements or advances. Short-term profit has no interest in long-term research. Dean (1974) put it more critically and succinctly concerning short-term profit "[it] is like looking for the leak in the bottom of my canoe as I drift toward the unseen waterfall."

Senior management who must look at bottom lines on a daily or at most weekly basis are ready to cut their long-term research projects when budgets appear likely to be exceeded or if predicted returns are not what had been projected. Simply put: why take risks with large expenditures of monies on innovative and creative research if the rewards cannot be assured to justify the expenses incurred? Risk capital is expensive. Only a very small proportion of research ideas ever lead to the development of some innovative product. Patents also are unlikely to be financially rewarding. Indeed, roughly 2% of all patents ever survive their full life.

2.4.1.2 Company Ego

Management's attitude to its own competitive edge or supremacy of the technology they possess can act as a constraint. Technologically successful organizations do not have a strong incentive to embark on any heavy program of new product development if their management believes that they have a proven superiority in technology over their competitors to provide products that have a seemingly never-ending global acceptance. The philosophy "if it ain't broke, don't fix it" prevails.

There is an inordinately long lapse of time between discovery or invention and an innovative application in the form of a profitable new product. One estimate puts this time interval between invention and a profitable product at roughly 11 years (Bradbury et al., 1972). Port and Carey (1997) found 15–25 years is common for what is called radical innovation. To add insult to injury, the company reaping the benefits of the new technology frequently was not the company that made the original discovery. One need only follow the changes in market leaders of instant coffees. Such observations do not encourage companies to engage in long-term research.

Comments such as "we did that 20 years ago" or "what good or use is that?" (stock phrases of one vice president of research and development I worked with) are clearly not going to promote an atmosphere in which creativity or innovation will flourish. People will hold back their ideas fearing a rebuke—peer pressure—from their associates. The NIH syndrome (not invented here) must not be allowed to prevail in the creative atmosphere where a company wants to generate new ideas. All ideas eventually should be screened, evaluated, and accepted for further development or rejected for valid, documented reasons.

Slavish attention to facts, to logic, or to reason (the refuge of technologists) stifles, at an early stage, any ideas leading to creativity and innovation. The mind-set of the dreamer is required: let the technocrat provide ways ideas can be given substance and not reasons why something can't be done. According to Sinki (1986), such technical snobbism rates high in creating a technical myopia he defines as the inability to make crucial connections between ideas and applications; the difficulty in "making the translation from abstract to concrete terms"; or "why ideas get aborted in their early stages." This is the blindfold that one's training, education, and experience can put on the free association of ideas from other disciplines to create something greater. Information is important; of course, it can assist the generation of ideas; but it can also limit the capacity for "bouncing ideas off people." Too much information can intimidate and funnel thinking into conventional channels.

Good communication between people from all disciplines within a company is essential. A lack of communication between scientists and entrepreneurs, that is, those who make the idea work, is a factor in technical myopia separating theory and practice (Sinki, 1986). Lack of correlation between unrelated disciplines ("creativeness…ability to make associations of dissimilar things"), another indication of broken communication, contributes heavily to technical myopia.

The final contributors to Sinki's depiction of technical myopia are lack of perseverance and a failure phobia, that is, not "having the guts" to take ideas on to innovative products. How much of creativity or innovation is plain, old fashioned hard work?

2.4.2 Communication

Communication problems between people, between departments in the same company, or between regional manufacturing plants within the same company are difficult to resolve at any time. In new food product development where several departments are working together, they can be particularly disruptive. Unfortunately, communication problems frequently represent conflicts of personalities or ambitions. Which came first is a moot point: the people problem, or the communication problem between the people, their departments, or their manufacturing plants.

2.4.2.1 Multiplant Communication

Communication problems are exacerbated in multiplant companies. In these companies, communication between plants (lateral communication) or even between the technical staff within these plants can be poor or even nonexistent.

I was assigned to evaluate the in-plant quality control systems of several regional plants of a large pickle manufacturing company. At one plant, a manufacturing problem affecting quality had been successfully resolved several months previously. This same problem was unresolved at a sister plant (making the same product!) not 600 miles distant where I visited a week later.

Lateral communication between the several plants of this corporate giant was nonexistent in passing or sharing information with other plants. The reason is that the plant where the problem was solved met and even surpassed their production quotas, but the other did not; ergo, the former was better run. The plant manager where the problem had been solved told me that they would get their bonuses and kudos: the other plant wouldn't. The result was duplication of research effort at the plant still trying to resolve the problem.

Large multinationals have tended to centralize their research and development resources. The reasoning is fairly easy to follow. By centralizing their research scientists and technologists, the expensive research and development equipment in one facility, together with all the pilot plant equipment and libraries to support the technical staff, there will be great economies of money, no duplication of facilities and better communications, yes? No, not always. This has frequently produced corporate ivory towers of research and development divorced from the manufacturing, technical, and developmental problems of regional plants. Centralization has often exacerbated communication problems.

Witness the following communication problems observed in large multinational, multiplant companies with centralized research departments:

1. I worked in the research and development laboratory of a regional plant of a large multinational, multibrand company. When the vice president of research of the Swiss parent company visited, we were routinely warned not to discuss with him any projects that we were engaged upon should we be asked: we were to refer everything to the laboratory director. We cleared our laboratory work benches of all working apparatuses. The fear was that our projects would be confiscated by central research and development headquartered in Europe.

2. In another company where the head office centralized research and development laboratories and their technical library rivaled that of many small universities, enquiries to this center by laboratory personnel situated at regional plants for journals, books, research reports, or information were discouraged by the local plant manager for fear that such contacts would result in "them" meddling in, or taking over, research projects in the branch operations. In short, central research and development would want to know why one wanted this information.

3. A research scientist with a multinational company told me how her company had conducted a new product development program at its corporate research headquarters in Europe on a confectionery product and test marketed the product in Europe. The product was destined for the North American market. The product failed when introduced in North America. No North American input had been asked for in the design or test marketing of this product.

4. In a similar situation, a fruit juice developed and test-marketed in Europe at the international corporate research and development headquarters of another company was packaged in material neither approved for use nor available in the North American market. It, too, was meant for manufacture by the company's subsidiaries in the North American market.

These are unconscionable, inexcusable breakdowns in communication. They depict the worst examples of the polarization of effort and lack of communication between regional plants and their large far away corporate ivory tower research centers that can occur in the management of research and development.

2.4.2.2 Technology: Its Management and Transfer

Technology transfer, that is, the communication of useful technical develop-ments within the company, is not the same as the management of techni-cal people; they are quite distinct activities requiring very different skills. To manage technology transfer within a company requires someone who can argue the advantages of the new technology or process over the older technology, who can convince a conservative, critical management of the advantages of the newer technology as an aid and not something disrup-tive or critical of old ways. This person is (often) not the laboratory manager (or supervisor or director of laboratories or project leader) but is someone who can prepare a soundly designed experiment or write a good technical report. It is not the project leader who manages technologists and scientists or has the ability to encourage and inspire people or protects them from bureaucratic intervention, and challenges them. Management of the transfer of technology, a quite different matter, should be relegated to those indi-viduals with the ability to communicate technology clearly, concisely, and convincingly. This is a vastly different skill.

Not understanding this distinction leads directly to another form of com-munication breakdown; the transfer of technology from the research and development resource to centers within the company that are able to utilize the information often fails because of the NIH syndrome (variously inter-preted as "not-invented-here" or "not-interested-here"). When people have not had an opportunity to be part of the development and have not been encouraged to see how this development might assist the objectives assigned by management, a sudden attack of NIH may, and usually does, occur. This syndrome often appears if research has been farmed out to external research and development companies and then returned to the sponsoring company.

Properly communicated to the nontechnical others in a company, technical information may spark an idea in the nontechnical community and lead to promising ideas for new products.

2.4.2.3 Personnel Issues

A failed product introduction is felt by all members of the team but espe-cially by the technical staff. Some phase of product technology, engineering or manufacturing, is often blamed although the reason may lie elsewhere. The quality characteristics demanded in the product concept may have been beyond the available technology, and if technologists and engineers are judged by their failure to attain these unattainable goals, then their contribu-tion to product development will be criticized. A product failure reflects on them and causes them to wonder about their future.

Any new product development venture is likely to fail on the basis of the history of the statistics of new product successes. This is the environment that product development teams live in. Production and engineering personnel are less vulnerable members of the team when a new product fails; their skills are needed elsewhere. Marketing and especially technical personnel stand on the front line facing the odds of failure. Brand and product managers are apart from the team and despite product failures are often unscathed by the failure and continue in time to move up the corporate ladder. Technologists must have great sympathy for a remark attributed to Churchill, a war-time prime minister of England, the gist of which is that, when England wins, the nation shouts, "God save the queen," but when England loses, they shout, "Down with the prime minister." In the event of a new product failure, corporate management must be a just and forgiving management in the face of failure in such a high risk enterprise.

Product failure must be carefully analyzed to determine what was incorrectly executed or poorly interpreted. This analysis must be conducted constructively: it is not a witch hunt but a learning experience for all to benefit by. On the other hand, errors must be rooted out and corrected. Weaknesses in the development process must be strengthened. Reassignment or retraining of staff may be necessary, and management must handle this positively to encourage their staff's development.

In the same manner, a successful product launch provides just as much to be learned with an in-depth analysis of why and how success was attained. A keen understanding and knowledge of the strengths of the total development process is invaluable for future projects. A secondary benefit derives from this analysis when management can suitably reward the achievement of the team members. This secondary benefit fosters confidence in management's support for risk taking in innovation by providing a sense of security and appreciation of one's effort.

This encouragement by management, in turn, encourages young scientists and engineers and consolidates the company's future growth requirements. The rewarding of achievement as well as the learning by analysis of both new product development successes and failures promotes the growth of new skills within the company.

Management must accept some blame if innovation has been constrained. If management is unable to defuse the conflicts that inevitably will arise between the disparate groups comprising the team and cannot oil the frictions discussed above, then the team will be unable to work creatively. Too much diversity in the team (Sethi et al., 2001) is equivalent to too many cooks spoiling the broth. A moderate level of diversity plus a high level of encouragement to take risk (management's prerogative) encourages innovation and creativity. Management must be able to provide cohesiveness for the new product development team and get them to focus on a well-defined problem. The management system and the managers may be at fault if creativity is missing.

Another human problem must be introduced here. This is more a feature of the multiplant, multinational organization. Much product development work, especially that which is involved with leading-edge technology, is multidisciplinary work. The services of specialists employed by the parent company often work at separate locations; for example, consulting academics at distant universities or research institutes will be employed on a new product development project to complement the team's efforts. Members of a team may never meet except electronically, through written reports, or by video conferencing. When this remoteness occurs, it can limit communication and dampen the team spirit of the product development group.

Companies with such development projects must make sure that all the distant members contributing to the project are assembled together frequently enough to exchange ideas personally rather than impersonally. Video-conferencing is a wonderful tool and can provide a closer association between associates than any written memos, reports, e-mails, or personal telephone or conference calls do and keep travel expenses down. But communication among far-flung team members is improved in actual face-to-face visits in the less formal moments of coffee breaks, lunches or dinners, or other social occasions that such visits provide. It is in these occasions members of the team interact by bouncing ideas off one another: it is networking.

2.5 The New Product Development Team

The new product development team requires those skills necessary to screen out ideas not suitable for the company's goals at that particular time and evaluate ideas meeting the company's objectives. The team's cross-functional resources should include the following:

- Management skills: There must be an authority that is senior enough to keep the group focused, manage disputes, provide facilities, encouragement, and resources, and is an "enabler." This so to speak is the "good cop": there is a darker side to management, and this is senior management who must decide whether a project is go-or-no-go based on this management's long-range policies.

- Engineering skills: A new product may require a new engineering technology. In small companies, the engineering skill may be entirely within the production department embodied in the mechanics and machinists who maintain the company's processing facilities; they determine processing feasibility. Outside resources are also used where in-house engineering skills or equipment are unavailable or inadequate.

- Advice and guidance from the production department: Again, the production department is an essential part of the team for the scheduling and carrying out of processing trials without production interruption; assessing processing capability and determining whether new plant or co-packers are required.

- Financial analytical skills: An ability to analyze financial data fairly and critically to monitor expenses and to make predictions of returns. This function is partly that of the manager and the "bad cop" suggested above.
- A source of legal advice: Advice respecting food legislation (labeling, naming of products, food standards, etc.), protection of intellectual property, and the company's interests in contract negotiations with research institutes, outside laboratories, and co-packers.
- Research and development facilities: Any or all of the following are required: trained food technologists, culinologists (chefs trained in food science and technology), dieticians or nutritionists, agricultural specialists.
- Marketing research capability: The ability to research and analyze customer and consumer data on which to base and evaluate new product ideas is needed. This is often an outside resource.
- An ability to source raw materials and ingredients: The team needs a competent purchasing department able to research reliable sources of necessary materials at competitive prices.
- Traffic department skills: This is essential for pricing and sourcing special warehousing and distribution needs.
- Quality control department: Safety and quality must be designed into the new product. As well, this resource provides advice on analytical methodologies for both process and quality control and assesses hazard levels that may be encountered in the processing of the new product.

Where these resources are not available in-house, outside sources can be contracted to provide the necessary skills. This adds another developmental cost.

The make-up of development teams is fluid and amorphous depending on the stage of development. Development teams in small food companies have plant managers or mechanics doubling as professional engineers of large companies; quality control managers are responsible for research and development; even presidents serve other roles perhaps as financial officers or purchasing agents and are supported by outside resources.

Communication among a small company's team members is usually good but not always effective for successful product development: it is too informal. Rigorous processing trials are seldom conducted and, often, nothing is recorded. The greatest danger in such a closely knit setting of a small company or in small product development teams of a large company is that one strong-willed individual—the company president/owner or the technologist dedicated to a pet project—will dominate (an example of position power on group dynamics; Peters, 1987). All attempts at unbiased screening are futile.

In large companies, several new product teams may be working independently on a number of different projects. Each has a team manager who acts as the recording secretary and enabler of the team. These managers report up the line of communication that often involves several levels of more-and-more-senior management. The various teams on their different projects rarely intercommunicate laterally, which means that duplication of effort can occur when one team discovers something that could benefit another team.

As work proceeds, dominant roles within the team change as major activities in the development process vary. At some phase, chef–food technologists dominate as they develop recipes and prototype products. At another stage, marketing may be conducting sensory evaluation sessions on small consumer panels, and other team members await their results. However, none of the individuals ever stops having an input into the team.

Development teams usually remain in place until the project has been released to production as a part of the company's product line. Movement of the individual team members with the project gives the members a greater appreciation of all aspects of product development. They see the whole picture and understand product development's complexities.

2.6 Phases in New Product Development

Most authors agree that new food product development can be divided into several phases (Table 2.2). There is no agreement among authors on the number, order, or names of the phases. Meyer (1984), not tabulated here, does agree with Mattson on 11 steps. Authors often then divide the stages into substages and even sub-substages. This compartmentalization of the development process is an attempt to understand the process and represents the thinking or philosophy of new product development specific to each author. Earle (1997a) describes the development process with four main stages:

1. Development of a business strategy that describes the project
2. A research and development phase of the project including the manufacturing design
3. Development of programs for marketing, production, and quality assurance
4. Organization of production and distribution for the launch and an analysis of the launch sales data

TABLE 2.2

An Overview of the Stages of New Product Development as Described by Various Authors

Holmes (1968, 1977)	*Crockett (1969)*	*Mattson (1970)*
Company objectives	Search opportunities	Idea generation
Exploration	Translate concepts into	Concept screening
Screening	products	Preliminary formulation
Business analysis	Marketing plan	Taste panels
Development	Implement marketing	Final formulation
Testing	plan	Trial placement
Commercialization		Fine-tuning
Product success		Package design
		Co-packers
		Mini-market test
		Symbiotic distribution
Oickle (1990)	*Graf and Saguy (1991)*	*Skarra (1998)*
Exploration	Screening	Assessing management
Conception	Feasibility	commitment
Modeling (prototypes)	Development	Finding the right idea
Research and development	Commercialization	Developing the business case
Marketing plan	Maintenance	Development and
Market testing		commercialization
Major introduction		

Rather inappropriately, screening is depicted as a series of sequential stages in many papers and texts describing product development. Certainly, there is some sequencing; a business plan obviously precedes all else, and closure is a successful launch with technical development somewhere in-between. However, when authors describe the phases of product development as a sequence, a one-after-the-other cascade from ideas through to a final finished product, they both misstate and misinterpret the process. A phase does not start, proceed, and then finish with the next phase then beginning. The phases are not, strictly speaking, sequential: they often overlap and are concurrent. For example, a quality control program is researched when the product concept becomes a product-in-progress, and the technologists realize some of weaknesses of the product and its processing and understand more how to make it safe. Projects might even return to the conceptual phase for a complete rethinking of their concept statements as new information arising from the development process becomes available (see, e.g., Bradbury et al., 1972).

Linnemann et al. (1998) attempted to organize the development process into an integrated system. They recognize "seven *successive* (italics added) steps":

1. Analyze "socioeconomic" developments in particular markets.
2. Translate preferences and perceptions of consumers into consumer categories.

3. Change consumer categories into "product assortments."

4. Group "product assortments in product groups in different stages of the food supply chain."

5. Identify processing technologies required for particular product groups.

6. Analyze the state of the art in required processing technologies.

7. Compare the state of the art of required processing technologies with future needs.

Their model is described in some detail; they conclude that feedback from the market is important as development is in process and state that "dedicated production systems that follow more closely market dynamics" are required (Linnemann et al., 1998). A structured, integrated approach that efficiently uses knowledge from many different areas of technology and labor is required in product development. They conclude that relationships between markets, consumer behavior, the variety of food product, and processing technologies are inadequate and require further refinement.

The obvious starting point is the establishment of company objectives followed by identification of ways and means (there are several discussed earlier) to get to these goals. Here, we assume the ways and means to attaining management's goals are through new product development. Once this is established, the development team know what is planned and why. Senior management's goals and their determination and dedication to these objectives are clear. This precedes everything else. There follows the identification of customer and consumer needs (Figure 2.1).

With this clarification of purpose, marketing personnel determine what new products would meet these company objectives and service the needs of customers and consumers. This requires market and marketplace research to gain knowledge of the needs of customers and consumers. Unfortunately, in small companies, there is a lack of understanding of these needs, and product development proceeds in a vacuum. The thick arrows in Figure 2.1 show the advance of product ideas and eventually the product itself. Thin arrows indicate flow of data and information derived from market and marketplace research.

The next phases winnow all the ideas, reducing their number to a manageable few, which are deemed to be the most capable of development successfully. There are three parallel screening criteria that are used:

1. Is the idea feasible within the time frame demanded and with the skills available? This answer should be provided by all responsible members of the development team. All need to be made aware of weaknesses in the plan and contingency plans readied.

2. Does the idea meet perceived needs of the consumer and the customer? Does it resolve the conflict of "I want" vs. the customer's "I need"? Further market research will determine how this dichotomy is to be resolved.

3. Will a financially sound business plan based on these new products stand up to critical analysis and meet objectives set by management?

The technical skills of the research and development department are brought into play as they proceed to develop benchtop prototypes that match the product statement as closely as possible. The next boxes in Figure 2.1 suggest that research and development are dominant elements. This is an incorrect interpretation. As development proceeds, new information is obtained to make more informed decisions. Such information aids in the development of standards for raw materials, ingredients, and packaging requirements and on process design and equipment requirements. All are needed for costing

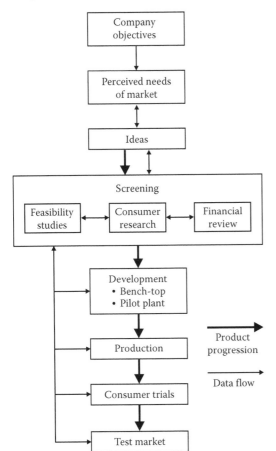

FIGURE 2.1
Phases in new food product development.

purposes. Decisions on the acceptability of prototypes through taste panel studies are constantly being fed back to the technical development group for possible reformulation.

A series of parallel events (not depicted in the figure) begins, based on the data obtained throughout the development process. An analysis of the business plan is now refined by the financial department with more complete information on ingredient, processing, and marketing costs. Sourcing of ingredients and packaging materials is carried out. Marketing people prepare draft labels and label statements, refine consumer analysis, plan marketing strategy, and develop promotional material for use in newspapers, flyers, radio, and television. The manufacturing department determines in-house production capabilities and manpower requirements.

As data produced at each phase are transformed into useful information, more positive decisions are reached. "Go" or "no-go" decisions are made. It is quite common to see many changes in the direction of product development based on new market research and competitive activity in the marketplace; this results in numerous returns to the drawing board. Development is a constantly evolving process keeping pace with those changing targets, customers, and the marketplaces.

By the time production samples have undergone successful consumer trials, management is able to decide whether to go into a test market (mini-market tests, market tests in only one or two cities) or go directly into a regional launch (Figure 2.1).

The final phase of any new product development is an evaluation of that launch and the reaction of the competition in that marketplace. Both a successful introduction and a disastrous one need to be evaluated to learn what made one profitable and the other a loss.

Development is a progression from the intangible product statement or concept to the tangible, very real product with all the attributes stated in the concept and ready to be tested in the marketplace, the final screening. A product has been created based on a concept statement and modified through sensory and consumer evaluations, the company's physical plant capabilities, safety, health, and legislated requirements, and at a cost acceptable to the targeted customer.

Figure 2.2 mimics more realistically the process shown in Figure 2.1. The upper flow in Figure 2.2 depicts efforts that are largely the responsibility of the marketing department. Food technologists formulate a tangible benchtop or prototype product based on the product statement (the middle flow in Figure 2.2). Marketing researchers then use this to stimulate consumer research for further refinements of the concept or use it in mini-test markets for the same purpose. With a prototype product in hand, sensory evaluations can begin to guide food technologists in refinements in formulation. Preliminary product safety and shelf stability testing may require further alterations both to the product and to the concept in these early phases.

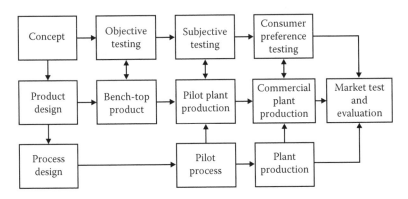

FIGURE 2.2
Idealized representation of activities flow in product development.

The bottom flow in Figure 2.2 is largely the domain of engineering and production departments. Engineering and production personnel design processes for new products incorporating changes determined by food technologists. At this stage, reliable sources for ingredients and raw materials should be found and product cost data confirmed. Potential co-packers can be evaluated if it is thought that the product cannot be made in-house, but this upsets profitability and return on investment projections. As development progresses from left to right in Figure 2.2, there is a constant interplay and exchange of ideas between the other two flows. Each shapes and molds the other streams.

All development from idea generation to test market and even to a national launch must be viewed as one long, ongoing screening process. Screening is not a stage in development—it is synonymous with development. Each stage of development brings further data that, when translated into information, provides development teams with more refined tools with which to screen. As information is gathered from marketing personnel, technologists, or suppliers, the product becomes identifiable with the needs and demands of consumers. The result is the continuing interplay of market research, technology, and financial efforts of companies to produce the right products at the right prices for customers and consumers (Blanchfield, 1988).

The purpose of the development process is to move a desirable product to market with the least amount of uncertainty respecting its probability of success in the segment of the marketplace where it is to compete. Market, financial, and technical research (and their associated analyses) are nothing more nor less than screening procedures for reducing, as much as possible, uncertainty in the development process.

3

What Are the Sources for
New Product Ideas?

> To as great a degree as sexuality, food is inseparable from the imagination.
>
> **Jean-François Revel (as cited in Robbins, 1987)**

3.1 Getting Ideas

Stimuli for new food product and ingredient ideas are found everywhere. Everybody has at some time thought up a new food product as they have fumed over some complicated recipe or been frustrated by a failed cooking exploit. There are several excellent ideas for both ingredients and food products in the scientific, technical, and trade literature; for example, Pirie (1987) was a great proponent of leaf protein, Ryan et al. (1983) extracted protein from exhausted bee bodies, Hang and Woodams (1999, 2000, 2001) made sugar and citric acid from corn cobs and husks, and Palmer (1979) produced wine from cheese whey. These are some of the more unusual ideas from more unusual sources. These products have never, to my knowledge, gone anywhere and remain just idle scientific curiosities. To be accepted by consumers and successful as new products, there was one thing missing. Who wanted them? Where was the demand? For example, with the plentiful availability of excellent wines made from grapes and other fruits, where was the need for a Château du Petit Lait du Fromage?

They were developed because "it's a shame for these by-products to go to waste." They were created to explore the utilitarian value of a waste product; this is commendable for environmental reasons where the disposal of these products is a problem. But all were product-oriented ideas and not consumer-oriented ideas; they led to producer- or technology-driven products, not to consumer-driven products. These are all technically feasible but derived in isolation from the real world of customers and consumers. Customers and consumers have to see a need for these products; the only use they see for leaves is as mulch or compost for their gardens. They are examples of what

I have described as "Little Jack Horner" research after the nursery rhyme character who "stuck in his thumb and pulled out a plumb and said, 'what a good boy am I.'"

Contrast this with a Reuters (2000) report that describes work by Brazil to create a diesel fuel from soy oil and sugar cane. This development is successful because there is an economic and political need (that is, governments need it) since the prices for both sugar cane and soy beans (major crops in Brazil) are falling, and this alternative need raises the price of these commodities for farmers. There was a political need. There is a use for the ethanol in sparing gasoline consumption. Ideas must fill a need and satisfy a want in people.

3.1.1 General Guidelines for Ideas

Thinking up ideas is not the problem. The problems are as follows:

- The ideas must satisfy the needs and desires of a company's targeted consumers and therefore attract customers, the gatekeepers to the consumers. The seed of an idea must come from those who will need and use the product.
- The ideas must also satisfy the financial and expansionist goals of the company and be within the skill level, technical capabilities, and managerial and financial resources of the company. These are often referred to as the "core competencies" of the company, that is, what the company does best. Ideas must be implementable.

Companies need not look for the "perfect idea" for new food products; there isn't one. There is not, and never has been, a perfect product, new or old, that satisfies or has satisfied all customers and consumers in all market niches. The task is to bring to fruition good ideas that will satisfy a company's customers and consumers in the marketplaces where the company is successful or in those marketplaces the company wishes to expand into, that is, specifically targeted markets for closely defined customers and consumers. The company must be able to market and sell the product (being able to manufacture the product is not necessary). The company is challenged to balance perceptions of customer and consumer needs found by its market research with the company's short- and long-term goals, its in-house intellectual skills, and its physical plant capabilities.

New product ideas will not succeed if they do not come close to satisfying the needs of the targeted customers, consumers, and the targeted market as described by all the market research.

Earle (1997b) sees two prerequisites for successful innovation: a company that is innovation oriented and a positively responsive environment. The environment both within the company and external to it must be conducive to and accepting of innovation. Earle treats the subject of innovation from a broad, conceptual, and philosophical point of view.

3.1.2 Sources of Product Ideas

Food companies have three primary sources for ideas for new products (Table 3.1), and these sources together with competent market research will uncover the perceived needs of the targeted customers and consumers, will define the marketplaces where these customers and consumers are, and will suggest product concepts that satisfy these needs.

Pressures (see Chapter 1) that drive a food company into new food product development are largely the sources of ideas outside the marketplace. New technologies or new applications of established technologies can open a wide vista of ideas for new products. Greater understanding of the principles of water activity and hurdle technology allowed the development of new product opportunities leading to semimoist foods ranging from dog food to fruit leathers to chilled foods as well as the quality improvement of others. Even the need to reduce waste by upgrading a useful by-product into

TABLE 3.1

General Sources for New Food Product Ideas

General Source	Specific Impetus Providing Inspiration
The marketplace	Market research to identify customer and consumer needs; results of customer profiling
	Retail data of buying habits of customers
	Distributors expressing their requirements for products and problems they encounter with handling and in-store customer interfaces
	Customer and consumer communication through complaint letters, 1-800 numbers, etc.
Within the company	Sales force's interaction with retail buyers, with individual customers in stores and from observations of competitive products and their placement within stores
	Government pressure or incentives to innovate
	Spontaneously generated ideas from employees
Environment outside the marketplace	National and international trade exhibitions where new machinery, food products, and ingredients are displayed
	Competitive intelligence gathering
	Competitors' new products requiring marketplace retaliation
	Food and cooking literature providing ideas on ethnic cuisine and new recipes
	Technical, trade, and scientific literature opening new horizons for development

an added value product motivates a company into doing research to find new uses for the by-product if the needs of customer and consumer alike are satisfied by doing so.

New product development is not an exercise to gussy up a waste product, nor is it a means to display a company's technical skill to the consuming world. One does not start with either a waste by-product or a new widget maker and wonder what to do with it. If these tactics do satisfy a perceived market need, this is fine, but caution is advised. Technical innovation, per se, will never sell a product except to other technocrats, nor is new food product development meant, necessarily, to be novel for the sake of novelty. New food products are meant to fill perceived needs, satisfy the desires of customers and consumers, and meet the expectations of those consumers for those products. The product must deliver what the company has promised in their promotional campaign for that product.

3.1.2.1 The Many Marketplaces

The customer, frequently with the consumer in tow, is found in many diverse marketplaces (Table 3.2). This could be a mother with her family in tow or a company president or chief financial officer visiting a trade exhibition with an engineer to see a piece of machinery that the engineer claims is needed. Logically, then, that marketplace in which the customer is located is the place to find out what satisfies the perceived needs of at least one of these entities, the multifaceted gatekeeper.

The great diversity of marketplaces is matched only with the great diversity of customers and consumers. Market researchers must first decide what market they will develop a product for and who they will target and then determine which marketplace best supports the market (markets are conceptual) where the product will be positioned. Then market researchers must determine how best to conduct research in that marketplace. How one uncovers the "perceived needs" of customers and consumers differs for ingredient, consumer product, or the food service marketplaces.

First, what is meant by perceived needs of either customers or consumers? Perceived needs are not known by the customer, the consumer, or the market researcher. Market research discovers a gap in the market, a need: customers or consumers, when presented with a mock-up or a test kitchen product or simply a presentation of the concept embodying this need, have that flash of blinding light and say, "That's what I've been looking for." Neither the customer nor the consumer may ever have realized this need themselves (cf., the Walkman). Marketing personnel did not create that need. Through their market research, they exposed the need for the customer or consumer to discover for themselves.

Market research is actually, then, a means of both discovering and screening new product ideas. Of all the ideas gleaned from all sources (Table 3.1),

TABLE 3.2

General Classification of Marketplaces Selling Food Products and Ingredients

General Classification	Examples of Types
Artisan based	Farm gate, farmers' markets, country fairs
	Specialty non-chain stores, butchers, green grocers, specialty bakeries, cheese shops, fishmongers, wine stores
	Carriage trade, specialty gourmet stores such as coffee, tea, chocolate, and confectionery stores
Grocery (food) store	Family-owned and family-run small grocery stores, large supermarkets often with artisan-based stores (butchers or green grocers) within them
	Special food stores (organic foods; health food stores; stores catering to religious observances)
	Box stores (large sizes or case lots)
	Bulk stores (products are loose in bins)
Convenience store	Paper/magazine stores, tobacconists
	Drugstores[a] (chocolate bars, snacks, and small units of canned goods)
	Department stores[a] often with a specialty food section
Food service	Full service restaurants, fast-food restaurants with sit-down and take-out service, hotels, roadside diners, happy hour bars, coffee and tea houses
	Take-out restaurants, mobile street vendors, walk-up kiosks, vending machines, cafeterias (work sites, schools), mobile canteens
	Gourmet commissaries (Chinese dim sum, Indian curries, French croissants and patés)
	Transportation meals (in-flight and on-train meals)
	Institutionalized care feeding (nursing homes, hospitals, short-term care facilities), penal institutions
	Personal chefs, rent-a-chefs, home-care feeding for elderly
	Military feeding (messes for officers and NCOs, cafeterias, field kitchen, combat rations)
Industrial arena	Food trade exhibitions with demonstrations
	Supplier-sponsored short courses demonstrating special ingredients
	Sample distribution to industrial users
Electronic and postal services	Basically catalogue-ordering services

Source: Reprinted with permission from Fuller, G.W., *Food, Consumers, and the Food Industry: Catastrophe or Opportunity?*, CRC Press, Boca Raton, FL, 2001.

[a] Local retailing legislation often limits what products these stores may carry.

only those that uncover that hidden need will be most likely to succeed and offered to the technologists for development into a tangible product. The others will be rejected.

It is within these marketplaces (Table 3.2) that food manufacturers must compete for customers and consumers with new products. Food retailing, wholesaling, and industrial sales have changed in response to the demands brought on by the changing lifestyles of customers and consumers in these marketplaces and by the very diversity of the marketplaces themselves. Each presents opportunities.

All information about the desires, buying habits, and any characterizing traits of customers and consumers helps developers refine product concepts into tangible products for a targeted segment of the buying public. As a result of data gained by market research, the needs and expectations of targeted customers and consumers are better understood.

3.1.3 Getting to Know Them: General Techniques

Census data provide the simplest information source to begin with to understand customers and consumers. Demographic and psychographic data are readily available about the vast population of customers and consumers and are essential to the understanding of the populations, but the inherent weakness of these data for obtaining targeted information must be recognized.

3.1.3.1 Census and Economic Data

Many countries provide demographic information (census data) about population size plus other population characteristics: these data are usually released at 10-year intervals. The statistical information includes the age distribution of the population, income distribution, the number of family units as well as nonfamily units, the number of children per family unit, the number of single-parent families, male to female ratios, ethnic backgrounds of different geographic regions (see, e.g., Tables 3.3 through 3.6), whether families rent or own their homes, and so on. By itself, it is not very inspiring for specific product ideas, but there are some (Foot and Stoffman, 2001) who are very aggressive at wringing information out of it for predictive purposes.

Some uses of these data are as follows: Taylor (2002) used Canadian census data to compare the changes in the top seven countries serving as sources for immigration into Canada for the years 1961 and 2001 (Table 3.3) where the data are displayed by countries from largest to smallest numbers of immigrants.

When the numbers of immigrants are examined, the extent of immigration, and its impact on the food microcosm, becomes more apparent. The Italians, in 1961, the largest group immigrating into Canada, amounted to fewer than 15,000 individuals. In 2001, the largest group immigrating into

TABLE 3.3

Top Country Sources for Immigration
into Canada, 1961 and 2001

1961	2001
Italy	China
United Kingdom	India
United States	Pakistan
Germany	Philippines
Greece	South Korea
Portugal	United States
Poland	Iran

TABLE 3.4

Top Country Sources for Immigration into the United States

2007	2000	1980	1960
Mexico	Mexico	Mexico	Italy
Philippines	Philippines	Germany	Germany
India	India	Canada	Canada
China	China	Italy	United Kingdom
El Salvador	El Salvador	United Kingdom	Poland
Vietnam	Vietnam	Cuba	Soviet Union
Korea	Korea	Philippines	Mexico
Cuba	Cuba	Poland	Ireland
Canada	Canada	Soviet Union	Austria
Dominican Republic	Germany	Korea	Hungary

Source: Courtesy of Migration Policy Institute, Washington, DC, http://
www.migrationinformation.org/datahub/charts/

Canada, the Chinese, amounted to a little over 40,000 individuals. Two obvious observations are apparent from such data:

1. The rate of immigration has obviously increased in this 40 year
   period. More people are on the move than there were 40 years ago.
2. The pattern of immigration has changed dramatically. Europe is no
   longer the origin of most immigrants; Asia has become the dominant source of immigrants.

Table 3.4 displays American immigration data, and a similar observation
about the changing profile of the immigrants emerges: Asia is the main
source of immigrants into the United States.

For new product developers, this observation gleaned from census data
should suggest possibilities for ideas for new products. Immigrants bring
a diversity of cultural traditions and ethnic food cuisine wherever they

TABLE 3.5

Top 10 States to Which Immigrants Migrate Compared
to Bottom 10 States Least Frequented by Immigrants

States Most Frequently Immigrated to		States Least Frequently Immigrated to	
1990	2007	1990	2007
California	California	Arkansas	Delaware
New York	New York	Alaska	Mississippi
Florida	Texas	Delaware	Alaska
Texas	Florida	Mississippi	Maine
New Jersey	Illinois	Vermont	Vermont
Illinois	New Jersey	West Virginia	West Virginia
Massachusetts	Arizona	Montana	Montana
Pennsylvania	Massachusetts	North Dakota	South Dakota
Michigan	Georgia	South Dakota	Wyoming
Washington	Washington	Wyoming	North Dakota

Source: Courtesy of Migration Policy Institute, Washington, DC,
http://www.migrationinformation.org/datahub

TABLE 3.6

Cultural Diversity in Five Centrally Located
Montreal Schools

District	Students from Other Cultures[a] (Mother Tongue)
St. Kevin's	95.2% (Tagalog: 53%)
Henri Beaulieu	94.3% (Arabic: 79%)
Barclay	94.8% (French and English: 57%)
Camille Laurin	94.3% (French and English: 53%)
La Voie	93.6% (Tagalog: 75%)

Source: Lampert, A., Montreal schoolchildren speak in
many tongues, *The Gazette, Montreal*, A1, A2,
October 9, 2002. With permission.

[a] Children born outside Canada or born to at least one
immigrant parent.

settle. Restaurants and food stores want to provide foods reflecting their food habits and cooking styles. Here, there are ample opportunities for new food product ideas for any food manufacturer. The impact of immigrants on the food supply is such that eventually their foods, for example, Italian and Chinese cuisine, are no longer considered exotic ethnic cuisine but have become mainstream.

Table 3.5 provides further demographic data valuable to developers: where are the immigrants settling? If ethnic cuisine is the desired development route, then market research will be directed to where the targeted customers and consumers are. It will be here in these regions of high immigrant density that test markets will be conducted and market research carried out to uncover consumer needs.

Table 3.6 takes the available data down to a finer level, albeit for Canadian data. Demographic data provided from school registration records give insight into population profiles and hence directly into new product ideas at the city and school district level. Here, the cultural diversity for five centrally located schools in Montreal is presented in Table 3.6 (Lampert, 2002). Any baker, butcher, grocer, restaurateur, delicatessen owner, or supermarket owner in these school districts who ignores the richness of the cultural diversity in their immediate vicinity by not providing food products—especially child- or teen-oriented snack food items—reflecting the community profile does so at their risk. If there are children from a diverse ethnic background, there must also be parents who shop for foods required in their cuisine. This opens the door for consumer food product manufacturers and suppliers of ingredients and raw materials for these ethnic products.

These census figures are typical of population changes noted in many countries and reflect movements of peoples bringing their ethnic cooking traditions with them. They modify the available local products to suit their tastes. Whether this is called the localization of ethnic tastes or the ethnicization of local foods is immaterial. They represent new product ideas and opportunities for the introduction of new products meeting their needs.

Per capita consumption figures are calculations based on what food has been produced and imported minus what was exported minus what, if any, was stockpiled and divided by the population. They are not consumption figures per se but are "disappearance" figures; these foods disappeared or were used somehow and somewhere in the country. Table 3.7 (and Table 3.8 in part), using U.S. data, is an example of such information. They show, for example, that milk and red meat consumption have declined since 1970, and the other food groups have increased (note the dramatic increase in soft drinks). But why? They provide a poor and indirect reflection on either customers' purchasing habits or consumers' usage habits and none at all on the habits of specific segments or particular regions of the population. They purport to pertain to the average consumer, an entity that does not exist as there is no knowledge of who or what is using up the food or how it is used.

TABLE 3.7

U.S. Consumption (Disappearance) Data of Some Major Food Groups

Year	Milk (gal/Capita)	Soft Drinks (gal/Capita)	Vegetables (lb/Capita)	Fruit (lb/Capita)	Red Meat (lb/Capita)	Fish (lb/Capita)
1970	31.3	24.8	336.5	242.3	131.9	11.7
1980	27.6	35.1	337.5	270.5	126.4	12.4
1990	25.7	46.2	387.3	272.2	112.3	15.0
2000	22.6	49.3	425.1	279.4	113.7	15.2

Source: U.S. Census Bureau, Statistical abstract of the United States: 2003 (extracted from Table No. HS-19). With permission.

TABLE 3.8

Percentage of Sample Populations Indicating
Some Degree of Vegetarianism

Year	Vegetarians (%)
1984	2.1
1985	2.6
1986	2.7
1987	3.0
1988	3.0
1990	3.7
1993	4.3
1995	5.4
1997	5.0

Sources: Vegetarian Resource Group, Baltimore, MD,
http://www.vrg.org/nutshell/faq.htm#poll,
2003; Vegetarian Society, http://www.vegsoc.
org/info/statveg.html, 2003. With permission.

This term, average consumer, is still often seen in the literature; it is so highly inaccurate that it is meaningless. An average is a statistical construct with significant meaning only if it is representative of a homogeneous group, that is, the members of the group to be averaged must have recognizably similar traits. Even a superficial reflection on the customers and consumers seen in any marketplace will reveal that they are not recognizably similar. There are children who can be subdivided into babies, toddlers (for whom purchases are made by parents, grandparents, and godparents), tweenies (a well-branded group at their age), and teenagers. Adults are subclassified by so many niches as to be almost indefinable; by age, by ethnic groupings, by geographic area, by income, by profession, and so on. Compound these with ethnic, religious, and cultural diversity and a world of market niches appears.

A facet of elusivity was discussed in Chapter 1. There is yet another facet; this quite destroys the false concept of the average mass. For example:

- As pointed out above, the mass is more variable. It has disintegrated as a recognizable entity.
- The disintegration of the nuclear family into sub-sections, each with its own interests, has already been mentioned.
- The public has become immune to sales pitches with the endless advertising it is deluged with. In short, it has tuned out.
- The mass now has the option to watch cable networks and satellite TV. They are harder to pinpoint.
- They now download their entertainment and thus hide from commercials. They are more elusive.

3.1.3.2 The Fallacy of Privacy

Two points must be kept in mind: First, one cannot speak about the average shopper or how much they spend and when they spend it for reasons discussed above. Second, new product developers are not developing a product for an average consumer nor are they or can they develop a product with universal appeal locally, nationally, or internationally—this universal product does not exist.

A product is developed for specific customers and their associated consumers, that is, a targeted, identifiable group, who represent niches in which these targeted individuals are to be found. These are entities that census and economic data cannot identify satisfactorily for developers. Therefore, market researchers need to define their targeted customers and consumers better with more data.

An astonishing amount of information can be obtained about customers and consumers using other demographic data that have been broken down for cities and towns and then used with other databases (Foot and Stoffman, 2001). For example, retailers often issue their own frequent customer cards or credit cards. The enrolment data are held in privately owned data banks or stored in various Web sites. These data banks may hold people's health, credit, educational, employment, and marital status; they contain records of telephone calls (to whom, length of call, etc.), library and video rentals, drugs, and medication; and indeed anything e-mailed to interest groups, for example, newsletters, is archived. And then, there are loyalty (reward) cards—every time a loyalty card is swiped, banks of highly personal data are accumulated and organized such as when one shops, where one shops, and how much is spent on what. When customers enlisted for these cards, they provided a lot of personal data, such as address, telephone number (plus area code), and other personal information including often ticking a box indicating their earning range. Some cards even asked how much they might be spending on certain items—to better serve you and ensure the safety of your information, of course. The "mining" of these data provides an amazingly complete picture of customers. The "gunshot" approach to getting one's message to customers is over: targeting customers is much easier by mining these data. While the sites do claim to provide privacy of any information they contain, one wonders what happens when such trackers of information go bankrupt or are bought and their data banks have new owners?

A bit of trivia: Canada developed a no-call list of telephone numbers to thwart telemarketers. Those who put their names on this list are receiving more telemarketing calls than before. The no-call list was invaded by telemarketers calling from the United States (exempt from Canadian law), certain that they now have a "live" telephone number. So much for privacy lists.

Magazine subscription lists or even magazine sales in local area stores (supplied on the sales slip and the courtesy card) provide information on individuals' interests (e.g., topical magazines about hobbies, cooking and barbecuing, outdoor sports activities, etc.) and indirectly provide an indication of financial status. These data combined with other mined data suggest the financial prosperity of areas. Membership lists for trade and professional associations provide addresses, educational levels and, again, an indirect estimate of the income levels of the area.

Consumer information as well as personal information about consumers are located in databases all over the country. When all sources are cross-referenced, surprisingly accurate pictures of customers and consumers in general, even individuals, can be obtained. An address provides information about income status when used in association with access to property tax records that identify expensive residential areas. This collation of data is called consumer profiling. The lack of privacy it represents in the computer age is a fact of life and a frightening concern for many people.

There are stories, perhaps apocryphal, of seminar leaders teaching competitive intelligence techniques who have challenged their students to find out as much about that teacher as possible by legal means: many have been embarrassed about what has been discovered.

3.1.3.3 Data Mining

Data mining (geodemographics, cluster research) and its sister term "knowledge discovery" were comparatively old hat when I first came across an article by Brodley et al. (1999) describing the process of using software applications to look for patterns in data. Essentially, the algorithms upon which the software is based search for anomalies in data; that is, they are looking for patterns in the data. Customers buy in patterns, and consumers use in patterns. By segregating these out, much can be learned of what customers might want. This leads to product ideas. Wisely used, such customer data can provide a company with trigger points (customer needs) to whet customer interest. No single bit of data is very useful, but the aggregate of data is useful. There are companies who are proficient at data mining technology, the drawing together and analysis of data.

Demographers group people into clusters with the assistance of census data (income, ethnicity, education, etc.) and overlay this with other neighborhood data (e.g., spending patterns). This allows mapping of areas of like-peoples allowing the narrowcasting of advertisements, promotions, products, or spam. An interesting use of data mining was described by Anderssen (2004) to demonstrate differences between Americans and Canadians that point to why products, advertising, and promotional campaigns cannot cross borders:

- In Canada, ethnic groups are more tightly knit, whereas in the United States, there is more assimilation of these groups.
- In the United States, the richest citizens live in high-end communities outside urban centers, but in Canada, this same group tends to live downtown in cities.

- Canadians cycle more than Americans but play far less basketball than Americans. That is, there are different sport interests.
- Americans belong to automobile clubs, but very few Canadians do. Canadians spend far less on cars than do Americans. In Canada, an SUV is considered a family car, but in the United States more SUVs are driven by younger females for protection.
- Americans eat twice as many doughnuts as Canadians per capita despite our ubiquitous Tim Hortons™ outlets.
- Poptarts™ are popular in the U.S. South, but in Canada they are popular only in rural Newfoundland.
- Americans are inclined to take more pills than Canadians.

If these differences point out the dangers of crossing borders of very similar neighbors, it also highlights the possibilities of customizing marketing efforts and product introduction efforts to specific geographic areas—that is, geosurfing.

The above demonstrations of differences point to the power of information retrieval that can be obtained when the buying patterns are superimposed upon a much smaller demographic area. Information is obtained suggesting new products; best areas for new product introductions, promotions, and advertising; and the best targets.

Such mining activities can also be used to analyze a company's competitors and their activity in the marketplace: here, it is called competitive intelligence or simply market research, never industrial espionage.

3.1.3.3.1 The Improper Interpretation of Data

Mined data can be misinterpreted, badly misinterpreted. The data must be integrated with other sources of information and carefully analyzed, but see Foot and Stoffman (2001). If one were to utilize the following demographic data isolated from other databases, the result could be disastrous:

- The over-55s are the fastest growing segment of the North American population.
- They have the most wealth of any age category.
- Females make up the greater part of that older population because statistics show women outlive men. Consequently, many over-55s are either widowed or single.
- Women over 55 years of age need extra calcium as a preventive against osteoporosis.
- Other physiological changes in ageing for this population are general frailty, poor circulation, high blood pressure, a general loss of flavor and taste sensations, often a loss of teeth, and constipation.

All the above-observed bits of data are true, but they can be put together to provide very misleading information. A product developer incorrectly saw this as an opportunity here to bring out a high calcium, low sodium, low fat (simply because the percent of calories in the diet due to fat is acknowledged to be too high), added value product (older people can pay), bland, soft food fortified with fiber and vitamins (for good measure) in an easy-open (over-55s lose their strength and dexterity), single-serve (they live alone) container. Such a product was developed based on the exact findings noted above; it failed.

Certainly, there are frail and incapacitated over-55s needing home care and in nursing homes. For them (an example of an elusive population), new products designed for their special needs will come from astute product developers in food service companies. But the over-55s are also those finishing schooling interrupted by nurturing children. They join travel groups and travel to exotic places on conducted tours, for example, with Elderhost™ or on tours replete with lectures sponsored by their old university alma mater. They initiate their own Olympics for the elderly: they consult either professionally or voluntarily with small local businesses or with projects in third world countries; they volunteer for community activities.

The aging market, as one example, must be seen clearly as highly segmented. New product ideas will come from recognizing and understanding the needs of these different niches to capitalize on their diversity. More than demographic data is required for this understanding.

3.1.3.4 The Internet: Social Networking, Blogging, Tweeting, and All That Buzz

The familiar "cookie," a tiny identity tag first created in 1994 and dropped onto online advertising, permits online profiling of Internet surfers by tracking their movements. One company doing such tracking bragged, in 2000, that it had more than 100 million profiles on file. The cookie droppers claim that by profiling surfers, they can serve surfers better by dropping advertising of products that the surfers want. Cookies also track a surfer's movements through the Internet and provide the observer with a more complete picture of who the surfer is and what that surfer's surfing habits are. Companies can follow a surfer's route and provide links at appropriate places and change advertising of products to suit the surfer's habits.

The Internet has brought many changes and has forced companies to rethink their marketing strategies. The rules for doing business have changed online. In effect, there are no rules. Via online resources, a company can sell worldwide, and this may contravene local distribution agreements when contracts are made. An interesting side dilemma occurs: When is a contract accepted? On the push of a computer button, or on the receipt of goods?

3.1.3.4.1 Twitter®, a Service to Communicate with Friends via Tweets

Tweeting is a form of blogging—micro blogging. Twitter provides real-time communication, but this also means real-time information that marketers can take instant note of. There being a limited number of characters, a tweeter must get to the message quickly. Twitter can be interest group based and like-interested people can tweet opinions, views, and trends; that is, twitter is beyond two-way broadcasting and is more like multicasting. This is grist for the mills of marketers as they can follow with the search function feedback about what customers are saying about their products and those of their competitors. They can also follow via tweets the so-called tribal leaders (that is, the taste makers) and so determine trends quickly. A breakdown by age of twitterers (twits?) in Canada is as follows:

- 13% are between 12 and 17 years of age.
- 42% are between 18 and 34 years of age.
- 28% are between 35 and 49 years of age.
- 17% are 50 years of age and older.

It is obviously a network with a higher percentage of a younger demographic.

For companies, twitter is also a mechanism for keeping noticed and as such is a powerful tool for brand awareness. If a product or brand has not been mentioned by others, people wonder why. It is also an equally powerful tool for brand sabotage, and companies must keep aware of what people are saying about them.

3.1.3.4.2 Social Networking as a Communication Tool for Large Masses

Some companies have blocked social networking sites such as Facebook in the workplace and thus prevented their employees from using them. This is not a sound business practice according to Dr. Nicole Haggerty, (Ivey Business, London, Ontario). As workers get e-competence, they become more adept at using information and communication technologies; they learn how to build virtual relationships and trust and, finally, they build the confidence needed to persist and be strategic in developing of ideas. Despite all the knowledge sources (libraries, catalogues, journals, etc.), people still turn to colleagues, those with whom they have made networking contacts and other people interfaces for assistance in problem solving. These are all assets in product development (Qureshe et al., 2009).

As an idea-gathering tool, social networking is invaluable as it presents opportunities to collate information about people's opinions on products, companies, interests, etc., without the intervention of a questionnaire or questioner or interpreter of the results of a survey. In short, it is unbiased although companies must always be aware of deliberate maliciousness.

By knowing what people think of their products both daily and virtually, a company can quell bad or malicious information before it goes viral.

3.1.3.4.3 Blogs and Blogging

A blog (web log) is a micro Web site on which a person, company, or interest group can post comments, items of interest to other like-minded individuals, images, videos, opinion pieces, etc., and many blogs have been enabled to add comments so, in a way, they are interactive. They are therefore means of communication by which bloggers have a soapbox that others can just follow or link to with comments. For companies, blogs can be a means of obtaining comments, opinions, and ideas from potential customers as they add to the blog. Blogs are also a means of gauging the public's poor opinion of a company or dissatisfaction with its products or both.

Blogs are organized and can be searched by topics. They do not have the real-time communication of tweeting. However, bloggers can subscribe to Real Simple Syndication (RSS) feeds to know when blogs that they subscribe to have been updated.

3.1.3.5 Just Looking and Being There

Useful data and information are received first hand from customers and consumers through surveys, polls, observation (customer interfaces), and social networking. Kraft Canada, Inc. is reported to have tried a different tack. It approached social anthropologists for an interesting study (Flavelle, 2010) in understanding the markets of the future. The plan paired senior executives with households that were not the traditional mother and father, the obligatory three children, dog, two-car garage in a middle-income salary bracket in a middle class neighborhood. Executives were paired with households such as

- Canadian native married to Japanese immigrant
- Four fraternity young men
- Filipino nanny raising her children and her employer's children
- Baby-boomer couple working opposite shifts
- Same-sex couple
- South Asian family
- Conventional family of four

Kraft executives found the results astonishing even to finding one household without a single ubiquitous Kraft product. Plans were initiated to experiment in Kraft's test kitchens with, for example, South Asian flavors and other foods more appropriate to the changing "family" unit.

The Proctor & Gamble Co. has opened an "eStore" as a consumer research laboratory in its efforts to get to know its customers and consumers and as an aid to retailers. It is an effort to see how shoppers respond online and in stores to digital ads, coupons, "store promotions, and other factors." The chief executive in charge of the site has described it as a giant sandbox in which the company's brands can play. Procter & Gamble state, rather cryptically, it is being run for the data it will provide about shoppers.

An interesting form of being there has been developed by a company called Cognovision, which has developed a camera to be attached to the ubiquitous TV cameras found on subway stations, elevators, or shopping malls. It watches the people watching the TV. It can note with good accuracy whether a male or female is watching, estimate the age, and estimate how many faces are watching and for how long—all important demographic data (Falk, 2010). It is claimed that it can change ads according to the demographic it "sees." This represents another piece of armament for communicating and gauging product interest.

3.1.3.5.1 *Interviews, Surveys, and Polls*

An initial word about surveys: by far the best and most interesting of several courses in statistics I have had was given by a professor who made the memorable comment about surveys that has stuck with me as a cautionary principle: "Don't believe them." By way of explanation, he described what is often heard on the radio or TV news or read in newspapers: "This survey result is correct within ± x percentage points 95% of the time." My professor would get very excited, saying, "No, no, no! That is an incomplete statement. It is wrong. They should say that this result is either right or wrong. If it is right, then it is correct…."

Then he explained. The survey questionnaire could have been poorly prepared, the survey could have been improperly carried out, or the results could have been misinterpreted. Without the assurance of knowing how a survey has been conducted, no survey results should be accepted as fact but merely as a news item filler.

A survey is a collation of many interviews. Interviews are simply one-on-one encounters between an interviewer and an interviewee conducted person to person or through a written questionnaire in which attitudes of the interviewee are sought. There are two types of interviews:

1. Structured interviews, in which the interviewer has a very specific list of questions to follow faithfully and poses these to the interviewee. The interviewer then records the answers by checking off the appropriate squares.

2. Unstructured interviews, in which the interviewer uses a prompt sheet rather than a list of questions with which to conduct the interview and record answers and opinions.

They can be conducted person to person, by telephone, by e-mail, by mail, or gleaned by social networking by consumers directly to the company. Cohen (1990a) discusses interview techniques as a tool in product development.

Whether a formal, structured questionnaire or a cue sheet of topics is used, surveys permit interviewers to get personal opinions, comments, and reflections on a wide variety of topics of interest to the client provided the right questions are asked. Unstructured interviews give greater freedom to interviewers to probe very deeply into subjects but require skilled interviewers. Herein is one of their weaknesses—interviewers in unstructured interfaces may prompt or force "don't know" or "no opinion" responses into specious answers or opinions from interviewees through overzealous prompting or body language or by the interviewee's innate desire to be helpful.

Converse and Traugott (1986) discuss at some length the errors that can arise in the analysis and interpretation of polls and surveys of the general public; these are equally valid for polls and surveys of targeted customers and consumers. There are several sources of error:

- There is a volatility of opinion amongst respondents that can vary with time and according to the predominant stimuli in the environment. These sources of error are especially important since development does take time—time during which subjective opinions can change or be changed by external events in the environment. Developers must be aware of the changing customer and consumer.

- There is an inherent difficulty in measuring subjective states of respondents. How does one measure, for example, "How likely are you to buy…?" which can be found in questionnaires? It is a conditional question prompting a conditional answer, that is, "is likely to buy if such and such happens."

- Incomplete data such as "no response" or "won't respond" compromise sample coverage.

- Measurement errors can be introduced by the interviewer or the respondent or be inherent in the questionnaire itself.

- Market research houses have their own individual procedures for weighting responses when calculating results and, more importantly, different objectives in interpreting data for their clients. These procedures can generate bias in the interpretation of results.

Interviewers, using either structured questionnaires or prompt sheets, can sway respondents in many subtle ways. Voice, clothing, body language, physical appearance, age, or sex of the interviewer can trigger this influencing effect. To avoid any bias being introduced during an interview, an interviewer requires skill, training, and experience.

Unfortunately, many interviewers used for polls are not skilled, trained, or experienced. They are, perhaps, college students earning money on school breaks. An incident observed in the shopping mall of a small city provided an excellent example of inexperienced interviewers:

> Two college-aged men were conducting a survey in the mall. I was curious as to what was being surveyed and frankly wondered why I wasn't being interviewed since males and females seemed to be the targets. I edged close enough to them to overhear one young man say to the other, "It's my turn to get the next (good looking girl) to interview."

The results of this survey from this particular portion of the interviewing team might be biased toward attractive young women. Properly trained interviewers add to the expense, but their use is worth the investment. For companies employing market research houses, the caveat is "buyer beware" regarding how the survey or poll is conducted. If the survey is improperly conducted, the cost is twofold: the money spent on the survey is wasted, and the information obtained is misleading or erroneous.

The structure of questionnaires or their counterpart, prompt sheets, requires careful design to remove ambiguity in the questions and to minimize dishonest answers (Waters, 1991). The language of the preambles of surveys, and even the order of questions, must be designed to avoid any bias: indeed, any preamble should avoid letting the respondent know what the subject of the survey is. The following hypothetical example illustrates how a change in a questionnaire could introduce a bias:

> Potential voters are being polled for their voting preferences for three political parties. Mr. Right leads the Right Wing Party. Ms. Left leads the Left Wing Party. Mr. Middle leads the Middle-of-the-Road Party. All the lead-in questions pertain to the platforms and public issue policies espoused by each of the parties. The last question, however, asks: "Would you *vote* for Ms. Left, Mr. Middle or Mr. Right?" From delving into political philosophies, the survey has jumped to leadership personalities, introducing a very different slant to the survey. The respondent may hate the party leader but not the party. The proper question should have been "Would you vote left wing, right wing or middle-of-the-road party?"

Consciously or subconsciously, respondents express opinions or answers as directed or influenced by the wording in the questionnaire. Clients hiring market research houses want unbiased and objective data; they do not want data that support preconceived concepts about customers and consumers that either they may hold or that is held by the market research house. If clients have preconceived, but erroneous, ideas of what their targeted customers and consumers want, their desire for vindication of their view may influence the wording of the questionnaire. Different research houses may have their personal reasons for weighting responses (Converse and Traugott, 1986).

Data based on interviews must be cautiously interpreted. The investigator attempts to use data from a small (compared to the entire targeted population) sample of targeted consumers to describe broader populations: this is valid only if the small sample is truly representative of the whole. The results, without the hyperbole of marketspeak introduced by the market research house or even by the client's own marketing department, should be seen as providing only a very general reflection of the targeted population. Waters (1991) describes the design of questionnaires to weed out false or dishonest answers. Hawkins (1991) discusses some problems encountered with buffers such as answering machines, caller I.D., and no-call legislation, including high refusal rates encountered in surveys using numbers picked randomly. Hawkins describes the attitudes of Canadians to surveys including attitudes to "sugging," that is, selling under the guise of conducting a survey, a practice used by some companies and one that is understandably making enemies of interviewees.

An assignment given to my students for a course on new food product development required that they keep a diary of all food purchased and eaten during a 1-week period. In class, the purchases were discussed with respect to quality, desirability, satisfaction (as student consumers), ease of preparation with limited dormitory kitchen facilities, ease of storage, packaging and its ease of disposability or recyclability, cost vs. satisfaction value, and so on. They were asked for ideas for products that they would have preferred or that would have fitted in better with their lifestyle as students; that would have made life, their "food life," more enjoyable.

The students, in effect, surveyed themselves. When confronted with a what-would-I-prefer situation, they began to look at their life style: students are cash poor, have long working hours, limited food preparation facilities, limited time for preparation, and academic pressures. They saw beyond a bare set of data a survey provides and began to fit the data into lifestyles. They came up with ideas that fit their needs. They looked more closely at themselves as both a customer and a consumer; they were no longer faceless. They had an identity; they became, on a small scale, market researchers, and they were the targeted consumer.

Psychographic data obtained through polls and surveys reflect behavior and attitudes of customers and consumers. From an analysis of the data developers can define the qualities that must be designed into products to meet the needs and expectations of targeted consumers. If, based on demographic data, people have more leisure time and a desire for a healthier life style; this suggests ideas for nutritious snacks or low calorie beverages such as fruit beers, wine coolers, or low alcohol beers to accompany leisure activities. The suggestion still requires research to determine its strength and firmness as an avenue for development.

The 1960s, 1970s, and 1980s saw a growing public concern about nutrition in a very general way. Nutrition was good for one's well-being without consumers being very sure why. During and since then nutritional knowledge and with it ancillary areas such as diet and health have burgeoned sometimes in confusing and chaotic ways with contradictory findings about this or that food or this or that diet. Today, there is more focus on specific health, diet, and nutrition issues that complement one another, such as

- Products to fit a healthy diet that are generally, but not exclusively, meant to mean a weight-loss diet in combination with a healthy lifestyle.
- Foods that provide protection against diseases such as cancer, heart attacks, high blood pressure, depression and memory loss, and a host of other maladies. These emphasize foods and products with, or enhanced with, nutraceuticals (functional foods, probiotics and prebiotics, phytochemicals), for which a body of reputable technical literature is slowly appearing that suggests these foods may help people with specific disease conditions and neurological conditions.

Today's developers must conflate these psychographic data with the observational data of fast food chains that healthy choices do not sell. Many have dropped their healthy diet lines. There is still interest in low calorie, low-salt (sodium), high-fiber, and low-cholesterol foods and an interest in natural or organic foods (by which consumers generally mean additive-free foods) but not in the food service industry. This dichotomy of information leads to a confusing but interesting array of new product ideas.

Consumers are slowly changing cooking and eating habits. Consumer magazines are enjoining their readers to fry less and broil more. This is especially true with the locavore movement with its push to see the use of local products used in home cooking. Some large companies have adapted some products to fit this desire.

Meal patterns have changed. In Victorian days and earlier, four and five meals were the order of the day, but today's work routines have reduced that to three. My observations of eating habits of most office workers, students, and manual laborers today show that five and six meals are more common than thought. The following meal periods with typical fare were common with university students and particularly young female office workers:

1. Breakfast was fruit juice, coffee or tea, cold cereal or toast, muffin, croissant, or Danish. The beverages and the baked items were frequently eaten on their way to work or school or in class or at their work station.
2. Mid-morning: coffee or milk or juice, a baked item (Danish pastry or muffin) often with fruit, or fruit alone.
3. Noon: sandwiches, yogurt, hamburgers, or hot dogs (the students' fare), cole slaw, and fried potatoes.

4. Mid-afternoon: similar to mid-morning with the possible addition of machine-dispensed foods such as soft drinks, peanuts, and chocolate bars.

5. Evening: students relaxed over beer, chips, and other snacks in the campus buttery; single office workers went home to a light deli snack (paté, small pizza, cole slaw, bread—often items picked up on the way home) before going out for the evening.

6. Students ate supper in the cafeteria or at apartment residences.

7. Late night: both office workers and students usually ate some snack before retiring.

These are "eating periods," "snacking periods," or "grazing periods." Today many people follow no traditional meal pattern regulated by the clock but eat when they are hungry. Psychographic data combined with demographic data provide the inspirations for ideas for food products.

3.1.3.5.2 *Telephone Surveys*

Telephone surveys reach a very dispersed population in a large geographic area. They are not selective for respondents, since they are randomly generated and not all people in the geographic area (area code) have telephones, but with the ubiquitous cell phone or wireless communication device, that is changing. The ability to move one's telephone number with one as people move out of defined area codes (a factor with some land lines and certainly one with cell phones) confounds an ability to keep a survey to a tight geographic area.

Questionnaires used in telephone surveys must be brief and simple; the calls are intrusive and interrupt respondents at home, at work, or at play. Questions cannot be detailed or require lengthy explanations as these prolong the interruption and increase the respondent's irritation. An advantage is the ability to key in the respondents' answers into computers, and the results of the survey are available in real time. Interviewer bias in telephone surveys is much less since facelessness of both interviewee and interviewer removes any visual-biasing elements.

Vicente et al. (2009) found that modern mobile telephone technology brought problems. They compared Spanish surveys conducted by mobile phones versus fixed lines using randomly picked numbers. Surveys by mobile phone were more difficult to conduct and differed from those using land lines. Nearly 60% of mobile numbers were not in use as compared with only 26% of fixed lines. As a consequence, it took much longer to screen and communicate with mobile phone users. The demographics of the study were different. Mobile phone users tended to be younger, better educated, and employed whereas those with fixed lines were older, poorer, and less educated.

3.1.3.5.3 *Postal and E-Mail Surveys*

Electronic mail and postal mail surveys have no geographic bounds. Selectivity of respondents can vary widely from good—market research companies keep mailing lists (both e-mail and postal addresses) of respondents identified by income, religion, ethnic background, and many other characteristics—to poor (surveys mailed "to the householder"). Respondents can be selected, but they are far from either randomly selected or representative of an elusive population. They are that market research house's private listing.

Mail surveys are the least costly technique of gathering data. They have no interviewer bias but that does not mean the survey questionnaire itself will be without bias. Questions on mail surveys (electronic or postal) must be clear and self-explanatory. Explanatory text and wording of questions themselves must be without bias, or objectives of the survey can influence responses. Difficult-to-understand questions discourage respondents from finishing a survey or may elicit false data because of the respondent's confusion. Information that companies can obtain in postal or electronic mail surveys is limited by the need for simplicity.

A further disadvantage to mail surveys (electronic or postal) is a notoriously poor mail-back response rate. Filters on e-mails may remove anonymous sources of e-mails, and an e-mail address provides no inkling of the geographic whereabouts of the respondents. Response rates of 45%–50% are considered excellent: lower rates of return are the norm. The slow and staggered return of completed questionnaires delays data analysis.

Nonresponses of targeted populations whether in personal, telephone, or mail surveys contribute significantly to sampling error (Converse and Traugott, 1986). The intrusive nature of surveys of any sort can be an annoyance and may invite negative survey results.

In Table 3.9, the advantages and disadvantages of the various surveys are compared.

3.1.3.5.4 *The Delphic Oracle (Predicting the Future)*

A novel method of surveying, especially to spot trends that might lead to new product development ideas, is the Delphi method of forecasting. It is based on questionnaires mailed to addressees whose opinions are valued (e.g., senior company executives, eminent authorities in specific fields, etc.). These selected respondents are assumed to be acknowledged experts in the topics under consideration or are leaders in the various industries for which forecasts are sought. The process is much like conducting a think-tank by mail at this initial phase of enquiry.

Opinions from the first survey are collated. A second questionnaire is formulated based on a consensus of the responses given in the first one. The same respondents are canvassed again, but this time the questionnaire is based on the expert opinion derived from the first survey. The respondents

TABLE 3.9

Advantages and Disadvantages of Various Survey Techniques

Survey Characteristic	Type of Survey			
	Personal Interface	Electronic Mail	Postal Mail	Telephone, Land Line or Wireless
Cost[a]	High	Low	Low	Low to medium
Geographic scope and specificity	Limited, usually confined to one area	Limitless but no specificity re: area[b]	Very specific to area but not to respondent	Variable[b]
Potential for bias	High	Low	Low	Low to medium
Availability of results	Generally slow	Rapid as responded to	Slow	Rapid
Selectivity[c] of respondents	Nonselective to limited selectivity	No to very limited selectivity	Medium to good selectivity	No to limited selectivity
Questionnaire complexity	Can be complex and lengthy	Simple	Simple	Short and simple due to intrusiveness

[a] Costs for any type of survey can be variable. Generally, personal interviews are more expensive techniques, especially if specific interviewees are targeted. Long-distance charges can add to telephone surveys unless done locally in several cities.

[b] Electronic mail addresses are often portable. An addressee may no longer be in a targeted market area. Similarly, many cell phone numbers are portable, and the holder of the number may be anywhere. Some land lines despite their area codes are also portable.

[c] If respondents in each survey technique are preselected from screened lists, then the survey can be highly focused. Random selection of telephone numbers from a directory provides no selectivity. Accosting shoppers in a mall also provides little selectivity beyond selecting for obvious physical traits such as age, sex, skin color, weight, etc.

may or may not revise their opinions in the face of this new information. Again, the respondents' answers are collated. The researchers then decide whether to continue with yet another questionnaire digested and distilled from the additional, rethought information gained in the second survey. Rarely are more than two mailings performed. By this second round, a semi-quantitative analysis of future trends in particular fields of interest based on expert opinion has been obtained. The Delphi technique of surveying is a very powerful tool in evaluating and forecasting trends: it is not suited to a rapid need for new product ideas. A good example of a Delphi survey is the Food Update survey described by Katzenstein (1975).

3.1.3.6 *Using Acquired Knowledge to Source Ideas*

3.1.3.6.1 *Customer Trends: Less Processed, Natural, Organic, and Local Foods*

Environmental and health concerns have led many consumers, indirectly and directly, to want less processed, more natural foods. There is a growing reluctance to accept highly processed foods—for which the consumer understands "ersatz" and "somehow not good" or with lists of ingredients reading like a list of chemicals. Lee (1989) called these people "food neo-phobes"; they are those who consider new food technologies, food additives, and ingredients as untried, unnatural, and somehow not good for one. Extreme food neophobes want to go back to foods as their mothers and grandmothers prepared. This feeling is not new: Busch (1991), when discussing the public's concern over the growing application of biotechnology to the food microcosm, put it this way:

> They desire foods that have been prepared in traditional ways, that contain few or no additives, that are "natural" and that are made neither from transgenic plant or animals nor via new fermentation techniques.

This distrust of processing and high technology presents a challenge to developers who may dismiss food neophobia as irrational, but nevertheless, customers and consumers believe it, and processors are well advised to satisfy this need. There is a dichotomy in that consumers espouse biotechnology in all its subdisciplines in matters of health and vanquishing disease.

Also, the demand for organically grown products, even added value products processed from organically grown crops, has increased. Both Jolly et al. (1989) and Campbell (1991), some 20 years ago, noted concerns about residues and processing chemicals in foods. Obviously, there is a perception that natural or organic produce has advantages, advantages that are highly disputed, except that these command higher prices, and many in the public believe that in some ill-defined way they are better. These sum up to "naturalness," a trend that can be noticed to this day.

Several large companies have taken naturalness in a new and exciting direction. Haagen-Dazs Five advertises itself as using the same ingredients used at home: milk, cream, sugar, eggs, and cocoa. Pillsbury Simply professes its cookies to contain simple, wholesome ingredients you and your family know and love; the Campbell Soup Company's soup Créations have followed a similar route, and the vice president (Canada) of marketing is quoted as saying "There should be nothing on the ingredient panel that you wouldn't recognize or even add in if you were making the soup yourself" (Leeder, 2010).

Despite what scientists preach, other people—nonscientists, customers, and consumers—do believe pesticide-free, organic, or natural foods have advantages for both health and safety. Many retailers in several countries maintain food sections catering to this need. However, there are differences of opinion about what constitutes organic, organically grown, or raised foods.

"Big Firms Get High on Organic Farming," read headlines in *The Wall Street Journal* (Nazario, 1989). Many farmers and large companies with extensive farm holdings (Sunkist Growers, Inc. and Castle & Cooke, Inc.) have joined the movement to natural or organic farming. They find there is a demand for their organically grown products. Cargill, Inc. announced its Good Nature organic pork brand in a press release (July 9, 2008) "sourced from hogs that are raised on family farms in the Midwest." Tyson Foods, Inc. is reported in the Institute of Food Technologist's (IFT) Newsletter of June 10, 2009 to have adopted a "We Care" responsible pork initiative for the establishment of ethical practices to ensure food safety and animal well-being. Many realize that organic farming can be as profitable as so-called chemical farming. As Nazario (1989) wrote, "...case studies have found that yields and profits can be just as high on an organic farm as on a non-organic farm." Avery (1998) and Avery and Avery (2002) of the Hudson Institute dispute this and other statements of the advantages of organic produce with rather partisan views.

Product development technologists owe themselves a visit to a natural food store or to wander up and down the aisles of the organic products section of a chain store to examine the products and fresh produce, all available at elevated prices. People are buying despite the price differential so deep is their distrust of nonorganically grown or nonnatural produce. New product ideas can be found. Since large farming companies have moved into organic or natural farming with their more efficient agricultural practices, prices for fresh produce are falling. The organic food business is no longer "...a counterculture business run by flaky hippies" (Nazario, 1989). McPhee (1992) reported that organic food introductions have seen an increase of 400% since 1986, and the organic beverage category has increased 1450%. The organic food industry is not without its problems. As McPhee noted, many companies are still testing the market cautiously, well aware that any movement into this new market could have repercussions with their regular product lines.

The IFT published a *Scientific Perspective* (Newsome, 1990) in which an objective assessment of organically grown foods was presented. It was pointed out that the claim that organically grown foods are healthy is scientifically baseless: Newsome highlighted problems with organically grown foods citing the *Listeria* outbreak in cabbages caused by the use of sheep manure (natural fertilizer usage). Newsome concluded that

- Organically grown foods were not superior with respect to quality, safety, or nutrition to conventionally grown foods.
- They are more expensive.
- They provide less variety of products for consumers.
- Diets based only on organically produced foods may "...present the risk of possible loss of balance and variety...."

Twenty years ago, Newsome's observations were true and still mostly are, but times and thinking have changed. It is true that organically grown foods are more expensive. It is true that their carbon footprint is not necessarily smaller than conventionally chemically grown produce, but there is a wide variety of produce as growers realize there is a market for organic produce, and third world countries see organic culture as profitable for them to exploit as developed countries demand them. Developers are not concerned about this argument; it is not their fight. Their task is to satisfy the needs and desires of customers and consumers. Developers must explore this market for new product ideas. Hauck (1992) described a ranching and meat operation in the United States that provides organic beef and lamb whose sales of $25 million in 1991 increased to $50 million in 1994. There is clearly a market for organic foods because it targets those who believe these products are superior.

A very strong consumerism movement has been spurred by a better informed, better educated, and more vociferous consumer. The result is that scientists and their technologists are viewed with somewhat more skepticism than before. Their science and technology have apparently outstripped scientist's communicative skills. This poor communication has created a consumer mistrust, not only of science but also of big business as a major supporter of research and, equally, big government, a strong supporter of research directed to practical goals (see O'Neill, 1992). Communication between the public, on the one hand, and science and technology interests, big business, and big government, on the other, is not well developed (Busch, 1991; Lee, 1989). Until communication has been improved and the vitriolic rhetoric that has developed is curbed, consumerism and all its manifestations may be a prominent consideration of new food product developers in the future.

Developers need to pay attention to many elements in canvassing for ideas for new products. The results of the next scientific finding as well as environmental concerns, social, ethical, and religious considerations will shape idea development, market research goals, and company philosophy.

3.1.3.6.2 Changing Food Habits and Lifestyles

Cooking, particularly "gourmet" cooking, has become a hobby for many consumers, and many more are becoming generally more interested in food and cooking and staying home to eat for economy reasons. Proof of their interest is the burgeoning numbers of cookbooks describing a wide gamut of ethnic cuisines, the growing popularity of cooking schools, and the popularity of television cooking shows. Consumers are doing more cooking from scratch especially with local produce and cheaper cuts of meat (reflecting the economic times): that is, consumers have gone back to using traditional recipes for home cooking especially for special festive occasions. Experimentation with new cuisines has led to the use of more exotic ingredients. This awakening to exciting foreign foods and cooking styles has provided food manufacturers with opportunities for new products and ingredients, especially Asian sauces, Indian curries, and fermented vegetables.

Families eat fewer meals together because of their individual schedules, and more meals are eaten away from home. In addition, grazing (snacking) is an established and preferred eating pattern for many people who accept several small meals a day as normal practice. Such a change in eating habits opens up opportunities for finger foods that are tasty and nutritious. For years, food pushcarts have appeared on the streets of many of the larger metropolitan areas to take advantage of the snackers and grazers. They serve the function of the fast-food restaurant without the overhead of expensive real estate. Many of these pushcarts sell ethnic foods, which seem very suited to finger foods, for example, empañadas, mini-pizzas, calzones, tacos, etc. In addition, they have available fresh fruits and prepeeled raw vegetables; nutritious snacks have become a natural consequence of this change in eating habits. Take-away and full service take-away is a major shift (Sloan, 2008). A sidebar to this is the observation of a U.K. cutlery that finds it is selling fewer knife and fork sets—sales of knives are drastically down while those for forks and spoons are steady. Does this portend more meals in front of the television, more snacking on-the-go or pushcart eating (see below)? Or is the future already here?

There is a serious side to meal patterns and their influence in daily life. Chrononutrition, an off-shoot of chronobiology, the study of biological rhythms, has emerged as a study of the "time-dependent features of nutrition" (Arendt, 1989). How does time or timing of meals and the consequences of digestion affect food selection or other biological rhythms in the body? The practical application for these studies is the concern for the safety and efficiency of shift workers, airline crews, or military personnel when on patrols; the daily rhythms of these people are disrupted. The gastrointestinal problems of shift workers can lead to work loss, fatigue, and inattentiveness. Worker safety is endangered.

Foods, particularly snack foods, could be so designed as to provide the proper nutrition based on the findings of chrononutrition to workers whose biological rhythms are disrupted by their work schedules.

3.1.3.6.3 Pursuing Health

Vegetarianism is a dietary regimen, and those practicing it represent a highly segmented market who may practice it for health, ethical, or religious reasons. They range from those rigorous in their practice to sometime vegetarians with all stripes in between. Developers are not concerned with the reasons for adopting a vegetarian diet but only for its market opportunities.

Silver (2003b) quotes the vegetarian market to be worth about $1.25 billion with the sale of meat alternatives having grown nearly 40% annually for the past 8 years. An interesting characteristic of this market is that many meat eaters often use vegetarian main dishes as a side dish on occasion. One area of special activity for the food service sector is that many fast food chains, college and university cafeterias, and sit-down restaurants now include vegetarian food items.

According to a 1992 survey by Krizmanic (1992), 7% of Americans were vegetarians at that time. More recent statistics set the number of vegetarians at approximately 3% of the population. These discrepancies reflect the variations in proportions of vegetarians in the population by area, age, ethnicity, and sex. What there is not agreement on is whether this is a growing group or one that is relatively static. Data in Table 3.8 obtained from the Web sites of the Vegetarian Resource Group (2003) and the Vegetarian Society (2002) suggest that there is modest growth in this area. The former group estimates the market potential as approximately 20%–30% of the population! Unsubstantiated reports put the number of people practicing some form of vegetarianism as somewhere between the 3% of Table 3.8 and 20%. A more detailed breakdown of the vegetarian market and vegetarianism is provided by Ginsberg and Ostrowski (2010). It is a market that cannot be treated as the fad diet for "tree huggers" that it was once considered.

A health reason for vegetarianism was certainly headlined several years ago in a respected science magazine, "Surgeon General Says Get Healthy, Eat Less Meat" (Anon., 1979). Such reports are then taken up by every science editor of every newspaper chain. This view has received support more recently from Pollan (2008), a popular food writer, with his plea to "eat food, not too much, mostly plants."

This is a market with many niches (ethical, religious, and environmental) and especially health as many proponents believe a diet with greater dependence on fruits and vegetables provides health-giving pre- and pro-biotics (phytochemicals) that will benefit them. The many attractions for developers are

1. It straddles menus two ways, satisfying main dish needs of vegetarians and nonvegetarians as a side dish.
2. Many ethnic cuisines are, by nature, vegetarian. As such with an ethnic-inspired flavor addition, vegetarian dishes fit many cultural niches.
3. There is already an established market niche for vegetarian foods, albeit a small one, so acceptance is already present.
4. It satisfies many market niches (main course, side dish, ethnic cuisine, health or healthy food) for shoppers, and, thus, many diverse consumers can be attracted to it.

A wider selection of many different varieties of fresh vegetables and fruits has become available. The locavore movement (the 100-mile diet) emphasizes local produce, and many retail chains and food manufacturers have hooked onto this movement as a selling point with products manufactured from local produce. Vegetables and fruits, whether part of the locavore movement or not, are having a small resurgence and becoming a more important and attractive part of meals, joining, but not supplanting, meat as a main menu item. Several influences are contributing to this trend, not least of which is the popular writings of journalist Pollan (2008), some quotes from whom are "eat mostly plants, especially leaves," or "the more meat there is in your diet...the greater your risk of heart disease and cancer." Disappearance figures for fresh fruit and vegetables (Table 3.7) show consumption of fruits and vegetables is growing; the growth of vegetarianism and the migration of food cultures based more heavily on vegetables, cereals, and fruits than standard North American diets as well as the locavore movement perhaps are contributing in lockstep to this growth.

Vegetarian cuisine requires new ingredients to replace the texture and taste meat provides and to replace the properties of gelatin. This has been accomplished with soy-based products, textured wheat proteins, high protein vegetables, and rice starch.

3.1.3.6.4 Obesity and Other Health Conditions as Spurs to New Products Ideas

CNN News flash viewed February 4, 2003: "London theater seats too narrow for U.S. behinds."

In the IFT Newsletter for July 8, 2009, the following headline for a news item appeared: "F as in Fat: How obesity policies are failing in America." The item described the findings of a report by the Robert Wood Johnson Foundation and Trust for America's Health that reported adult obesity rates in 23 states had increased and that obesity rates had not decreased in any states. A further finding was that the percentage of obese or overweight children in 30 states is at or above 30%.

Governments are very concerned about obesity as it represents a major expense for the medical and health care systems. MacAulay (2003) describes obesity as an issue of national importance in the United States. She mentions several programs and campaigns that the American government is implementing including some programs by individual states that might possibly tax high-fat, high-sugar foods (Fletcher et al., 2009; Karlin, 2009). These according to the IFT's Newsletter have failed or are failing. Car, airplane, and theater seats have to be redesigned to accommodate wider bottoms and bigger bellies (see above news flash). Obesity is directly a factor in many health and social problems (Lachance, 1994; Birmingham et al., 1999; Larsson and Wolk, 2008; Prospective Studies Collaboration, 2009). According to Dr. P. James of World Health Organization (WHO), obesity is on the increase, "doubling every 5 years" (Reuters, 1996), and James has described this as an epidemic.

The National Center for Health Statistics (NCHS, 1999) reported, on the basis of the 1999–2000 National Health and Nutrition Examination Survey (NHANES), that the proportion of overweight or obese (body mass index [BMI] greater than 25.0) U.S. adults is estimated at 64%, up from the previous NHANES III report (1988–1994) that had estimated the proportion to be 56%. Since the NHANES II survey (1976–1980), there has been a steady increase (age-adjusted) from 47% to 64% in increase in 20–74 year old category.

The data (NHANES, undated) for children, whether boys or girls, male or female adolescents, or white, Hispanic, or Mexican American also show an increase in obesity. The American Obesity Association (A.O.A.) has obesity data that are broken down by history, age, gender, educational level, and geographic distribution in the United States (A.O.A., 2002).

The evidence is clear. The North American populace is, on average, fat and getting fatter—an increase that results in health and social problems later in life. It is occurring despite a declared interest by the public for wanting good health, eating healthy foods, and living a healthy lifestyle. The meaning of these contradictory findings is unclear. It demonstrates the public's ignorance of nutrition, abetted by misleading or ambiguous advertising and overzealous promotion by food manufacturers, fast food service outlets ("supersize me"), or the "bunless" burger ascribed to Kentucky Fried Chicken (Parry, 2009). Gitelman (1986) commented on the same issue nearly 30 years ago. A colleague of mine—a biochemist no less—exclaimed to me when fat-reduced chips made their appearance, "Now I can eat all I want without worrying about calories!"

At the same time, skinless chicken breasts appeared and are now disappearing from menus, and extra rich creamy foods and extra strong (i.e., cream) drinks are making their reappearances. There seem to be market niches for a diversity of interpretations of health foods, healthy foods, and hearty foods. There is still a strong hedonistic market, which can be catered to side by side with healthful foods.

Sloan (2003a) suggests this confused market demonstrates four rather confusing product opportunities:

1. Foods for either control or loss of weight and foods for health, that is, disease prevention
2. Foods that are low carbohydrate and high protein
3. Foods that are minimal caloric density since the desire to lose weight is a high priority
4. Repositioning existing "healthy" foods back to what they are, low fat, low carbohydrate foods

The above have an obvious common theme running through them: healthy food choices.

Lobbying by vested interest groups enters into any recommendations for dietary change or suggestions that this or that food in one's diet is good or bad. Eilperin in the *Washington Post* (2003) reports the sugar association in the United States to be angry at the World Health Organizations (WHO) recommendations to curb sugar intake to no more than 10% of the daily calorie intake. Likewise, the American Meat Institute in a press release (2005; 2007) has challenged the findings of a published European study on a positive response found between meat consumption and colon cancer. Development teams should be alert to governmental and nongovernmental agencies' actions through their programs and perhaps to changes in taxation policies if "empty calories" should come under government scrutiny in this still confused arena.

There is a seemingly contradictory report (Orpana et al., 2009) that found that the relationship between obesity and risk of mortality in over 11,000 adults over a 12 year period was questionable. They found an increased risk of death in underweight individuals while overweight adults whose BMI (body mass index) ranged from 25 to 30 was associated with a significantly reduced mortality risk. A contradictory article had prominent display in news syndicates embracing the United States, the United Kingdom, and Canada and described 20 year studies with Rhesus monkeys that showed those on restricted diets lived longer than those allowed to eat as much as they wished.

3.1.3.6.5 Ethics, Social Responsibility, and Religion as Sources of Ideas

These issues are more obscure than the above obesity issues. They are highly polarizing issues between people that may carry over into food products.

There is a small but growing market for foods that are not classifiable by their nutrition or health-giving properties but on what may broadly be called ethical or social grounds. That is not to say that their adherents do not *believe* these to be more healthy for them nor that they necessarily are more healthy but simply the world is a better place (somehow) because of them. The marketing issue is that there are *consumers who believe this to be so,* and there is a market for them. The growing organic food market, one small part of this ethical question, proves this to be the case associated as it is with so-called responsible agriculture. For developers, this segment is full of paradoxes.

Environmental, ethical, social, and religious concerns as ideas-generating elements for food products are confusing. Some equate the concerns to the developed world's proprietary attitude to third world nations; others see a confusion of issues involving agricultural practices such as organic farming, factory farming, animal rights, and health concerns.

Arguments against the locavore movement are mounting—opponents claim it does not lessen the carbon footprint. Organic farming with its reduced yields of produce resulting from organic farming technology (not to mention the insufficiency of manure and compost to sustain it) is being condemned in face of increasing food prices, a depressed economy, and a scarcity of food. Catering to these concerns with new food products can be difficult: somebody's concerns will be ignored.

Ethics will have a greater role in shaping the purchasing habits of future consumers (Wilson, 1992). Guides are available for shoppers that provide information on the companies behind the food brands: for example, *The Ethical Shopper's Guide to Canadian Supermarket Products* authored by Helson et al. and the staff of EthicScan Canada and published by Broadview Press. These guides describe a company's environmental record, its policy on women's issues, labor relations, and consumer issues in general. There have been similar developments in the investment community where stock investment plans are devoted entirely to holding stocks of companies including food companies whose policies are ethical and "green." Kraft™ reported in one of the IFT October (2003) E-Newsletters that they plan to sell coffee certified by the Rainforest Alliance.

Social concerns are shaping food product development; old products are being redesigned to be more ethically responsible. Shade-grown coffee beans and products derived from them are being promoted as new products, and many coffee houses tout that they use shade-grown coffee beans for their coffee beverages. A very recent development is biodynamic agriculture, and several wines have appeared in markets claiming to be biodynamically grown and fermented. A vintner (interviewed on Here and Now, Canadian Broadcasting Corporation, Radio 1, April 22, 2010) described some aspects of biodynamics and the growing popularity of such products. New products and marketing and marketing policies are changing to accommodate these shifts.

It is still a confused area, and development teams are well advised to tread warily.

3.1.3.6.6 Pollution and Associated Concerns Spurring Ideas

Pollution and the environment are major concerns amongst people in the developed world and are an emerging concern amongst indigenous peoples in the lesser developed world. Since the "Earth Summit" in Rio de Janeiro, Brazil, in 1992 and subsequent summits in Tokyo and Copenhagen (2009), governments see that they must act, or be seen to act, to reduce pollution. The major concern appears to be directed to carbon dioxide emissions and warming, but there are already indications that water is emerging center stage as an environmental problem. Dumping waste into oceans, lakes, and rivers or burying it in landfill sites is no longer acceptable. Medical wastes, dumped at sea, have washed up on the eastern seaboard beaches of the United States and have been clandestinely dumped in foreign countries (Patel, 1992). Cities and nations can no longer find landfill sites for garbage disposal. Disposal of waste by separation into recyclable material, compostable material, and plain waste is costly for municipalities. Composting facilities require noise and odor prevention controls and suffer from not-in-my-back-yard reactions. Waste represents a significant cost for food processors who pay to have waste hauled away and then pay a tipping fee at a land disposal or composting sites.

To manage waste, less must be produced in the first place. New products or processes will be rejected if they produce excessive waste unless the desirability of the new product by the public is such that the public or the government or both are willing to accept the problem (cf., the automobile and carbon emission or the ever present take-away coffee container). Companies are forced to divert some of their technological skills to designing new processes and products to make these less wasteful or to upgrading the waste into useful by-products. Packaging must be designed to provide protection to the product yet be easily opened and be large enough to allow a label with its requirements of nutritional labeling, ingredient listing, and preparation instructions; the print must be large enough to be legible (and perhaps in more than one language, be compostable or recyclable). These hurdles present formidable challenges to printers and package designers. They are becoming the demands of governments at all levels and the wishes of customers.

Food containers strewn along highways, dumped into streams, or littering city streets are the most obvious features of processed or take-out foods. Yet, the container is a necessary evil. Without the package, according to Akre (1991), nearly half the available food stuff in the developing world would be lost to human consumption whereas developed economies lose no more than 2% of food available for sale. Waste and its ever-present companion, pollution, are inevitable in the food microcosm; the mechanisms to deal with these problems effectively and efficiently are often lacking. Developers need to consider their actions respecting waste production and consider alternatives or devise systems to reduce waste and pollution. Environmental activists are enough of a political force that governments listen to when the environmentalists become concerned, active, and vocal (Akre, 1991).

Pollution and other environmental concerns caused by agriculture (especially large-scale factory farming operations that produce tremendous amounts of animal waste), fish farming, food manufacturing, and especially the food packaging industry are highly visible.

Product developers are, therefore, compelled by public and governmental pressure either to use recyclable or reusable packaging or to use biodegradable material for packaging. Both alternatives present developers with a dilemma. Recycling or reusing requires a collection system, an added cost. Some experts argue that recycling is not energy efficient, requiring energy to return the recyclables, energy to pick-up the recyclables to deliver them to a recycling center, and energy to process the recyclables. The news that biodegradable packaging material is more expensive and is not readily biodegradable in most landfill sites makes manufacturers reluctant to use these (Lingle, 1990). The use of degradable packaging materials raises several problems, not the least of which is the reaction of regulatory officials to the uses of the degradable materials in contact with foods. A similar challenge may be put to the edible films that are emerging.

Headford (1996) describes efforts for a database of the "global food industry" that will include environmental issues and a study of the reporting strategies respecting environmental issues of major food manufacturers and retailers. Her studies indicate that the major environmental issues for companies are energy, packaging, recycling, and pollution. Control of these issues may not result in new product ideas but possibly will provide an element of goodwill for a company.

3.1.3.6.7 Elusive Populations

There are populations, elusive populations (see Figure 1.2), that present unusual problems for market researchers to get to know. Some examples of elusive populations are (Sudman et al., 1988)

- Racial, ethnic, or religious groups within larger populations. Those following religiously, philosophically, socially, or culturally based diets such as halal, kosher, Bahá'í faith (Solhjoo, 1994), or Zen Macrobiotics typical of Zen Buddhists could offer ideas for products that conform to teachings. These diets prescribe, loosely or rigidly, foods to be consumed and their processing (e.g., see Erhard (1971) and Glyer (1972) for extensive reviews and discussions of various cultural dietary philosophies).
- Income levels of various groups (professionals, trades people, retirees, etc.).
- Persons with specific illnesses or disabilities.

An expansion of the concept of elusive populations that can suggest product ideas for food product developers is as follows:

- Users of organic foods and vegetarians (Nathan et al., 1994)
- Those practicing ethical or socially responsible food selection (e.g., buyers of shade coffee beans or free-range chickens)
- Those with specific food allergies or metabolic dysfunctions (Blades, 1996; Brown, 2002; Cullinane, 2002; Steinbock, 2002) resulting from surgery or prophylaxis (e.g., immunosuppressed individuals)
- Those who are overweight, underweight, or anemic and require special, nutritionally designed diets
- Those who use dietary supplements, such as athletes and the elderly
- Teenagers, the elderly, and pensioners

(The articles by Cullinane, Steinbock, and Brown discuss allergy problems and preventive action by a supermarket chain, an industrial caterer, and a company in the hotel, restaurant, and leisure activity markets, respectively.)

Each elusive population above presents different levels of difficulty to survey. Obviously, listings of elusive populations are not always readily available to researchers. Hospital or medical records, where accessible, or self-help or social associations to which members of these groups might belong may have listings of those with specific disorders. Some groups, such as teenagers and ethnic groupings, can be geographically located within cities, for example, in high schools or at religious establishments.

By any measure, sampling these populations is difficult and costly and requires novel techniques. Sudman et al. (1988) discuss sampling techniques for groupings that are geographically clustered; more general techniques (network sampling) for non-geographically grouped populations; and sampling techniques by capture–recapture methods for the most elusive of all populations, mobile populations. Demographic data (census data) can assist market researchers by identifying where some of these groupings, particularly ethnic peoples, have settled or predominate in cities or states.

3.1.3.7 Using Retailer/Distributor/Manufacturer Interfaces for Ideas

3.1.3.7.1 Analysis of Purchases

Much information about customer shopping habits is available through the universal product code printed on foodstuffs and recorded at the time of purchase. This includes the following:

- What items are purchased and, through analysis, which items are not moving well. Stock keeping is simplified and opens the door to efficient consumer response (ECR) and its progenitor, supply chain management (SCM). ECR is simply keeping the shelves stocked with no bare spaces, and SCM is the link between retail outlet and either manufacturer or distributor whereby the former tells the latter that stock is low.
- Which items are purchased as a percentage of the total dollar amount and hence those most worthy of keeping well stocked.
- Which stores have the highest average purchase per receipt. This is important for large chains in assessing whether stores should be enlarged or their cross section of products displayed changed or which real estate could profitably be sold.
- What combinations of products are purchased together. Ideas for new product development (or for piggy-back couponing) come from frequent purchase combinations. If product A and product B are purchased together frequently, then combining the two in a single product incorporating both concepts or packaging them together as a single unit could be a logical product idea.

Caution is needed in interpretation of data of this nature: products purchased together are not necessarily used together. Simple movement of products in combinations requires further research.

A similar caution applies to the interpretation of data describing case movements of goods between warehouses. Developers use these case movement figures to spot trends in customer purchases. These movements are what they are: product is being moved around, nothing more, nothing less. Case movement is only an indirect measure of a trend in customer interest, much like per capita consumption figures, to be interpreted with caution.

A note is necessary here about supply chain management and its resultant efficient consumer response. The data provided are what is selling: this by itself is good competitive information. The same cautionary admonition respecting this information as initiating ideas is necessary as with case movements, interpret with caution. The obvious question arises: what else is happening in the marketplace to cause the movement?

The condition of product as it arrives and is stocked on retailers' shelves (poor or weak packaging, spoilage, scuffed and damaged labels) as well as customer complaints can be reported. This information helps packaging engineers and technologist make product and package improvements.

3.1.3.7.2 Marketplace Analysis

After studying all categories of products available in all marketplaces, developers record the gaps they find—that is, the products or categories of products not available. GAP analysis is another technique for generating ideas for product development. Simply put, marketing people select a particular product category and then examine the marketplace for empty spaces in that category. When gaps are discovered, new product ideas are suggested. Then, through more consumer research, the desire or need of consumers for that product is explored. A gap does not necessarily mean a need. There may be a very good reason for a gap; nobody wants such a product.

In a typical GAP exercise, a grid is drawn. Each row of the grid describes some product attribute such as texture, flavor, color, particle size, function, application, etc. Columns might be labeled such as used above—that is, solid, liquid, or gas; or ready-to-serve, condensed, solid as in frozen, or solid as in dehydrated for terms pertaining to soups. When the grid is filled in with data from the marketplace, ideas for new products may be revealed by any empty spaces on the grid.

In Figure 3.1, an example of a grid for gap analysis using a hot pepper condiment sauce is shown. Horizontally, the form of the condiment sauce has been expanded from a liquid to a solid with two possible forms, a concentrated paste (squeeze tube dispensed) and a ground powder. The gaseous forms are also divisible into two, dispensed as a foam or dispensed by gas

Hot sauce flavor	Solid ——————————————————→ Gas				
	Sprinkle-on	Paste	Pour-on	Dispensed as foam	Pressure-dispensed
Hot pepper					
Ginger					
Garlic					
Mustard					
Horseradish					
Blends (e.g., curries)					
Black pepper					

FIGURE 3.1
Example of a GAP grid for hot condiments in general.

as a liquid or paste, the latter pushing the product into more sophisticated packaging techniques.

Vertically, one can adapt various flavors based on a single variety of hot spices from habaneros peppers through ginger, garlic to black pepper, or with other spice and herb combinations complementing the heat principle; or one can follow up with ethnic varieties such as Indian, Thai, or Japanese cuisines. The grid can be very complex and include child-oriented or adult-oriented or gourmet sauces.

GAP analysis, a form of attribute analysis, looks at the marketplace for product vacuums. A gap may simply mean that neither customers nor consumers see an advantage or a need for such a product. On the other hand, is there an undetected need that has never been fulfilled because such a product had never been presented? The main purpose of this exercise, however, is to stimulate thinking about unfulfilled market opportunities as the grid is stretched beyond its limits both horizontally and vertically.

GAP techniques can be applied broadly to a number of business problems from financial to marketing matters. For example, the columns of a grid could represent different sales regions of the country and rows the product categories missing in a particular area that marketing needs to introduce into those markets to keep them strong. The grid allows an analysis of market needs.

3.1.3.7.3 *Alternative Formats for Selling as Ideas*

The influence of the so-called alternative formats for retailing food, the warehouse outlets, direct mail selling of food items, as well as direct sales (selling directly into the home) has not yet been fully assessed as to the changes these will bring to the food marketplace (will such alternatives even survive in the future?) or to prompting the creation of new products. However, these alternatives will influence the development of new package formats (bulk packs or loose packs allowing customers to bag what they want). Traditionally, the container has been used to attract or communicate a message to the customer; with the big box items, what must be done to get the customer's attention? How will e-tailing influence conventional retailers? Developers have an opportunity to research product concepts that fit e-tailing uniquely. With the power of social networks, finding elusive customers in elusive markets has become much easier. The task is to develop products for these markets.

3.1.3.8 **Other Environments as Sources of Ideas**

Ideas generated exclusively from within companies or from internal resources hired by companies may bring an introspective range of ideas for development. Exploration of opportunities derived from sources outside the company will balance this.

3.1.3.8.1 The Competition

One marketing axiom states that a wise company knows what its competitors are doing. To state the blatantly obvious, the competition also competes for the customer's attention and their dollars and for the tastes and preferences of the consumer. Any competitive activity by a company in the marketplace requires some retaliatory action by a rival company. This action can be the impetus for an accelerated product development program activity requiring the generation of new ideas for products to counter the competition's new introductions.

There are several ways (see, e.g., Table 4.1) in which the activities of the competition can be followed. Each one provides little pieces of information that when integrated reveal a broad picture of the competition's direction. This activity is called competitive intelligence gathering and is discussed more fully in Chapter 4.

Many food companies omit even simple intelligence-gathering activities like walking the aisles of marketplaces. When management do not know what is happening on store shelves in the marketplaces that they service, they do not know who their competition is, how their products look compared with their competitors' products, or how their products are judged by customers, consumers, and retailers. Management need to know to provide some direction to development. Are there ideas here for improvement of existing products to stand out better against the competition? Are there ideas for demonstrating a perceived edge or difference over the competition that is discernible and desired by the consumer?

Both inspections of competitive products on store shelves and cuttings in laboratories that include sensory and compositional analyses as well as functional testing provide data that allow

- A comparison of ingredients to provide an approximation of ingredient costs and therefore an estimate of the competition's cost margins (i.e., competitive intelligence gathering).
- A comparison of the quality characteristics (taste and flavor, texture and mouth feel, color, etc.) of the competitions' products and a determination of the characteristics most appealing to the competitors' customers.
- An evaluation of package and label appearances to gauge what the customer sees before a package is picked up in a supermarket.

This police work helps a company to maintain a strong awareness of the competitive products it must constantly be active against.

Information obtained by comparisons provides ideas for new products by indulging in some free association of ideas to generate new product concepts. The putting together of an attribute of one leading competitor's product with features of another market leader combined with something from a third product may lead to a new product concept. Such free thinking can be very exciting for the developer. Competitive products on the store shelves are probably the source of many new product ideas for many companies. This activity has now been graced with a name: *benchmarking*.

3.1.3.8.2 *Food Conferences, Exhibits, Trade Shows, and Research Symposia*

Attendance at domestic and international food and equipment trade fairs is an essential activity for members of the development team; these are places where ingredient and equipment companies showcase their new developments and advanced technologies. Attendees have access to vast arrays of new food products, ingredients, and the latest developments in processing equipment. They see and sample a variety of consumer products and have demonstrated for them the properties of new ingredients and food equipment from around the world. They network with international business people to discuss export market opportunities and discuss their processing problems with technical sales representatives. In other words, they get maximum exposure to a broad range of ideas, products, and contacts from many countries. Such awareness of the wide variability and availability of food products internationally plus creative thinking can stimulate ideas for new products.

Conferences where technical papers are presented and discussed reveal novel research and development activity that open up for receptive minds possibilities for new products and could also reveal the research activities of competitors. For example:

- Technical papers delivered at conferences provide access to research and development findings a year or more before these results are published in journals and several years before they appear in books. This has changed somewhat with the Internet with many technical journals having online editions of their journals. Articles are now identified through their DOI (digital object identifier).

- Authors of these papers can be quizzed about details of the research methodology and about directions that more recent, as yet unreported developments have taken. Such discussions are particularly valuable in poster sessions where more probing, directed, self-interested questions can be asked. Useful contacts (networking) with experts active in a company's interest areas and with academics can be made.

Who should go to shows and conferences? The obvious answer is those who will be the most interested in personal development and who will return the most benefit from the experience to their companies.

3.1.3.8.3 *Public Libraries*

Public libraries have sections on food and cooking with an extensive collection of cookbooks filled with recipes. Recipes based on local, national, and international cuisines provide ideas for new food products or serve as starting points for benchtop test products. More importantly, public libraries have librarians professionally trained in getting sources of specialized information for those companies without their own facilities.

Computer-assisted information retrieval systems, common in all libraries, give subscribers access to databases to fit any need for information. These services are reasonably priced and can complement the resources of small library collections.

3.1.3.8.4 Specialized Libraries

Without access to information, businesses would be helpless. Information about markets, food legislation, trade statistics, financial analyses of companies, consumerism, and consumer trends are necessary tools for strategic planning. Equally valuable for food technologists is access to technical information. However, most food manufacturers do not have a staffed on-site library (Goldman, 1983). One way around the problem of lack of library facilities is to subscribe to databases that are available from publishers online.

Williams (1985) provides a general discussion (now sadly out of date) of the basic types of databases available: bibliographic databases; full-text databases; and numeric databases that can still provide a base of understanding for new search material. As well, she describes and discusses aids to online retrieval such as user-friendly front ends; intermediary systems, which help the searcher with questions; and gateway systems, which assist the searcher to reach other databases within the system. Access to technical literature through these databases reduces the time spent in literature reviews; provides a wide access to literature beyond the budget means of most food companies; and provides scientists in development with the technical literature support they need.

Buxton (1991) describes the equipment needed to go online with databases and the terminology used (in user-friendly language) in going online. Some general databases available in the United Kingdom are described.

Hill (1991) describes International Food Information Service (IFIS), which is a database producer. Its main product is the well-known *Food Science and Technology Abstracts (FSTA)*®, which is available in printed form, online, on magnetic tape and on CD-ROM. Its database contains original abstracts, authors' summaries, or, in approximately 10% of the cases, title-only entries. Classification in *FSTA* is tabulated, and online hosts for accessing the database are described.

A different database compiled for retrospective searching was described by Mundy (1991). The Institute for Scientific Information (ISI), based in the United States, developed citation indexes. Well known to most food technologists is the *Science Citation Index (SCI)*™. This index comprises four interrelated indexes. *SCI* can be used to find recent papers in a particular topic. The subject index uses key words or phrases to provide a list of authors using these reference phrases in their titles. The corporate index provides information on what has been published by researchers at a particular company, and this leads to an index of authors from that location. *SCI* is available on CD and online and was previously available on magnetic tape. ISI also produces the popular *Current Contents*, which is available online and was available on magnetic tape.

The Information Group of the Leatherhead Food Research Association (United Kingdom) has produced *Foodline®*, which comprises scientific and technical, marketing, and legislation databases (Kernon, 1991). These can be accessed online. The scientific and technology database with the acronym *FROSTI* (Food Research on Scientific and Technical Information) dates from 1974. All entries contain a concise abstract of the article with key word cross-referencing. The marketing database (*FOMAD* for Food Market Data) is supported by two databases, *FLAIRS UK* (Food Launch Awareness in the Retail Sector) and *FLAIRS NOVEL*. The latter contains information about products described as novel with respect to attributes and that have been introduced anywhere in the world. *FOREGE* (Food Regulation Enquiries), legislation database, provides information concerning permitted food additives worldwide (Kernon).

An interesting demonstration of the value of computers for storing and sorting data for application in predictive techniques in microbiology is discussed in Gibbs and Williams (1990), Cole (1991), Walker and Jones (1992), Williams et al. (1992), and Buchanan (1993).

O'Brien (1991) provides an overview of available food and food-related databases. For readers interested in the topic or perhaps interested in subscribing to a service, some difficulties and problems associated with food databases are discussed by Pennington and Butrum (1991), especially the virtually inescapable translation problems associated with food nomenclature and food descriptions as well as taxonomy of foods as these vary from country to country. These are major problems in trade between countries. For example, fish and fish products have a rich variety of local, regional, and taxonomic names that present difficulties in translation. Klensin (1991) describes the use of multimedia (and hypermedia) techniques to answer some of these difficulties described by Pennington and Butrum (1991).

Stored technical information is useful only if it can be accessed. If that information could be organized and structured so as to be available to assist the user just as the apprentice in real life has the expert to lean on, it would be a valuable tool. Anderson et al. (1985) describe developments in intelligent computer-assisted instruction that are "programs that simulate understanding of the domain they teach" and can interact with the student according to a strategy. This is, in essence, an expert system.

The use of expert systems in the food industry is comparatively new. An oft-cited example is the Campbell Soup Company's retrieval of the expert knowledge of its hydrostatic cooker operator before his retirement and the development of this information into an expert system (Whitney, 1989). The chief value of expert systems is in improved training programs for personnel; in having an "expert" on-site in times of crises to prevent the introduction of hazards into the food; in the design of new processes; in developing better control systems for safer food; and in the preservation of the accumulated wisdom of experts. Bush (1989), McLellan (1989), and Herrod (1989) describe in detail what is involved in the development of expert systems and the immense amount of effort required in their production.

Specialized libraries, for example, business libraries, reference libraries, technical (medical, engineering, etc.) libraries, and patent libraries, provide information and statistics on a wide variety of subjects valuable to developers. Most magazines, trade journals, and technical and scientific journals can be accessed via the Internet.

The patent literature can be an interesting source of data to investigate what is patented and what patents are pending. Pending patents reveal directions in research and development that competitors are pursuing.

Technical libraries are equally profitable in providing a company with ideas for new products or new developments in the science and technologies of foods, food processing and preservation, and nutrition (see Table 3.10: this is presented to show only what is available). Specific ideas such as these, ranging from snack foods to ingredients, may not have direct interest for a developer, but they can provide guidelines for formulations or processing, experimental protocols or suggest products that can be adapted from these ideas.

The value of reviewing scientific and technical literature is threefold:

1. The subject matter of articles has value for what it describes about a food or a process. Babic et al. (1992), for example, describe stabilization systems for chilled ready-to-use vegetables (with carrots as the example); Slade and Levine (1991) and Best (1992) describe water relationships in foods, their implications for quality and safety of many foods, and their use to stabilize foods.

2. Information on the authors, and where the work was performed, provides contacts with whom the product developer may communicate in future.

3. There are acknowledgements at the end of the article listing the supporters of that particular research. This information plus the authors' addresses may identify who sponsored the work and thus reveal competitive activity in a particular field (all part of competitive intelligence gathering).

3.1.3.8.5 Trade Literature

Ingredient and equipment suppliers provide trade literature, newsletters, and bulletins describing their new products and their applications. This literature can be surprisingly fertile ground for new product ideas or information

TABLE 3.10

Examples of New Processes and Products That Can Be Found in the Technical Literature

Type of Product	Description of Product	References
Jellies	Complete process for extraction of juices from wild native fruits and their manufacture into jellies.	Mazza (1979)
Vegetable pie filling	Carrot pie filling preparation described.	Saldana et al. (1980)
Beverage	Nutritious, chocolate-flavored shake-type drink prepared from chick peas. Novel alcoholic beverages.	Fernandez de Tonella et al. (1981) Gómez et al. (2009)
Snack	Nutritious West African snack food, akara, prepared from cowpeas.	McWatters et al. (1990, 1992)
Fruit leather	Fruit sheet made from prickly pear with formulation, composition, and sensory data included.	Ewaidah and Hassan (1992)
Fruit leather	Detailed manufacture including packaging and storage of fruit leathers from mango, banana, guava, and mixed fruits.	Amoriggi (1992)
Snack and bakery item	Process for explosion puffing of bananas described.	Saca and Lozano (1992)
Ingredient	Synthesis of valuable ingredient, vanillin, is described using methodology allowing vanillin to be called natural.	Thibault et al. (1998)
Extruded snack item	Formulation of corn flour, green gram dhal, and gum snack food. Primarily a demonstration of response surface technique for formulation optimization.	Thakur and Saxena (2000)
Gruel-based foods	Processing conditions described for preparation of ethnic gruels such as Scottish porridge, atole, kishk, and trahana.	Tamime et al. (2000)
Alcoholic beverages	Production of spirits and liqueurs using melon fruits.	Gómez et al. (2009)
Fermented foods and beverages	Review of fermented foods and beverages from tropical roots and tubers.	Ray and Sivakumar (2009)

about new ingredients that will inspire ideas for product development. For example, in a very timely review published in *Dragoco Reports*© published by Dragoco, Inc., Jones (1992) surveyed food labeling directives for EEC flavoring that would be invaluable for developers who may be contemplating entering the EEC with flavored foods. This publication may no longer exist; computer searches for it have been fruitless. Back issues may be available in libraries.

Ingredient suppliers often provide sample recipes employing their products with their promotional and technical bulletins. For example, The *California Raisin Report*, Winter 1993 issue, describes kosher foods and provides recipes for baked goods suitable for Jewish traditional holidays. The

American Spice Trade Association supports a column, *Flavor Secrets*, published in *Prepared Foods*. The March, 1993 issue, entitled *"Comfort" Pies*, describes meat pies from various countries, provides a recipe for one, and advertises the availability of pilot recipes for others. The June (2009) issue describes curries from the Pacific Rim and has a recipe for Thai pork and vegetable curry.

3.1.3.8.6 Government Publications

There is a wealth of new product ideas in the literature available from various governmental departments and agencies. Governments regularly promote the use of agricultural commodities or underutilized crops. They provide recipes using the foodstuffs with manufacturing directions and occasionally market test data. Where these fit the manufacturing capabilities and resources of a food company and the demand of a company's targeted consumers, this readily available source for increasing a company's ability to generate new food product ideas should not be overlooked.

Government publications are also valuable sources of much demographic data such as population movements, age composition of the population, incomes, food and nutrient consumption per day (USDA, 1980), meal patterns, and so on. Disclosure on companies who have received research monies, grants-in-aid, or development loans provides information on activities in food research and plant construction among competitors.

Descriptions of developments in regional food research laboratories and agricultural research stations are published regularly. For example, the Food Research and Development Centre at St. Hyacinthe, Quebec, Agriculture Canada, published an information bulletin in one issue in which Gélinas (1991) described work on frozen bread doughs. The National Academy of Sciences (NAS, 1978) (United States) published a book describing the properties of underexploited food plants with promising economic value. Included in this are descriptions of the vegetable chaya, now readily available in supermarkets; winged beans; the cereals quinua (quinoa; now regularly available on store shelves) and grain amaranths, and the oilseeds of jojoba and of buffalo gourd. Specialty stores in many large cities have all of these presently available. Another example from the Food Development Division (FDD, 1990), Agriculture Canada, describes an extensive study on modified atmosphere packaging for the developer interested in using this technology for the development of new products.

3.1.3.9 Internal Sources of Ideas for Development

3.1.3.9.1 Ideas from within the Company

Ideas for products that satisfy the needs and desires of targeted customers and consumers in specific market niches can arise internally within any food company. The qualification is that these internally sourced ideas cannot be based on the personal whims of dominant individuals within the company but on consumers' needs.

First and foremost as sources of ideas for new products are the company's retail and industrial sales personnel and technical sales representatives: They interface with customers and consumers. They act as sensors in the marketplaces they serve where they, as the eyes and ears of the company, review the store shelves for the pricing, condition, and placement of competitors' products or in-store promotional displays. They also scout exhibitions and trade fairs. These people meet customers and consumers. They know what the competition is doing at several different marketplace levels. Their discussions with store management can reveal weaknesses in products, their packaging, or deliveries. Store returns may point to problems with formulations, and complaints by customers to store management may uncover faulty preparation instructions (compare Section 3.1.3.3).

The sales force is the strongest resource a company has for determining what is happening in the marketplaces they service; they provide confirmation of marketing's more sophisticated research. They provide competitive intelligence by reporting the earliest signal of competitive activities in the marketplace, which needs to be heeded by the company. By listening to retailers, customers, and industrial users, sales representatives can present their company with ideas of what is wanted in a new product, and they can provide information about how much customers are willing to pay for this innovation.

A novel approach is used by one company to give this consumer- and marketplace-awareness to its engineering, research and development, and distribution departments. Members of these departments are required at regular intervals to accompany a sales representative on customer-contact visits. Such visits served several purposes:

- Retailers and industrial customers were impressed by the concern displayed by the company.
- Staff of these sometimes insular engineering and research and development departments were exposed to the environment of the marketplaces where their products competed with others for customers' attention.
- These individuals see opportunities for improving their products and packaging or for new product ideas as they interface with customers, consumers, and retailers.

Even CEOs and COOs can benefit from such visits (Flavelle, 2010).

3.1.3.9.2 *Ideas from Customer and Consumer Contacts*

When consumers or customers take time to e-mail, write letters to, or telephone food manufacturers, those people are expressing a need to be heard. They may want to

- Vent anger about a failed product. This provides clues regarding faulty preparation instructions or the need to reformulate to improve a product.

- Express pleasure about a favorite product. This suggests ways to improve a product by enhancing those pleasurable features associated with the product.

- Enquire about a product's suitability for some diet. Is this perhaps an unexpected opportunity for product extension or product maintenance?

- Seek clarification about cooking instructions. This suggests cooking instructions are not clear.

- Offer useful information or new uses they found when using the product, thus suggesting possible line extensions for the product. These could lead to product maintenance opportunities.

Customers and consumers need to be respectfully listened to or their letters responded to in some constructive manner. Their letters provide valuable product information.

Hotline telephone numbers (1-800 numbers) are used by many companies on their labels and advertising brochures; these let customers and consumers call directly to the consumer relations department where their needs are addressed immediately. In addition, companies with staff skilled in consumer relations can elicit background information on the callers to get user profiles. Valuable psychographic information so obtained assists food companies in their marketing strategies and new product development plans. Better still are Web sites; for example, Kraft Kitchens has a site where customers who sign on are "personally" corresponded with. In addition, food companies have started blogs and tweets as means to get to know customers and consumers.

Product complaints are the tip of an iceberg: more is hidden beneath the surface than shows above. Estimates to evaluate the significance of complaints vary widely. Ross (1980) estimates that for each consumer who complains, there are eight who did not and they will not try that product again. Other estimates provided by colleagues suggest that 20 or more consumers do not complain for each consumer who does. Graham (1990) reports that for each consumer who complains, there are 50 who do not. The impact that complaints can have can go "viral" with the impact of such social networking tools as Facebook.

All complaints should, first, be acknowledged. How it is handled publicly between customer and company is based on the company's policy. Providing cents-off coupons is potentially damaging as word spreads and floods of phoney complaints may be received. Then internally, the complaints should be classified according to their nature, the product involved and its code identification, identification and location of the complainant, place of purchase, and the time of the year when the purchase was made and the defect noted.

From these records, the astute, consumer-oriented company looks for common threads. For example:

- A hidden product defect (a common recurring complaint) indicating a need to improve a food product by reformulation or new process technology. Similar complaints from across broad marketing regions suggest something serious is happening in manufacturing or in the distribution channels.

- A package weakness (poorly positioned labels, scuffed labels, overgluing and glue stains, dented cans indicative of rough can handling, leaking lids, etc.) indicates a poorly managed packaging line or rough handling in transportation and distribution or both.

- Customer complaints can lead indirectly to ideas for a new product. If products consistently fail to meet consumers' needs as evidenced by complaints, then an analysis of complaints may find ideas for new products more closely designed to solve the failure problem and meet these needs (Cooper, 1990).

Daniel (1984) describes a very elaborate computerized system for organizing consumer complaints of a hypothetical food company with a 1-800 telephone number. It is designed for a quality control function but could be adapted for an approach more oriented to consumer relations. Cooper (1990) describes the planning and implementation of a program to organize consumer complaints to provide

- Weekly summaries of all complaints
- Complaints broken down to product or brand
- Complaints broken down to a specific factory

Such information in the hands of responsible managers as well as senior management alerts all to potential weaknesses in the entire manufacturing system, can also be a stimulus to process or product improvement, and can lead to new product ideas.

The Internet has brought a new means for airing complaints: web or blog sites on which people can post complaints anonymously—and sometimes maliciously. These sites are commonly called "gripe sites." Such sites require monitoring by a company as they may contain malicious and misleading information about a company or its products that needs to be countered early. Damaging information may arise from disgruntled employees, cranks, or competitors.

Complaint files should not be handled facilely with a form letter, coupons, or products. All consumer complaints, indeed any consumer communication, should be catalogued, cross-referenced under the respondent's name, street address, city and postal code, and the nature of the communication, and responded to with a meaningfully written letter addressing the problem. These files contain information about consumers and how they use or value a company's products. They identify product or package defects that require attention. They pinpoint the source of crank calls by those seeking free coupons (caller identifications on the receptionist's line help here). In new introductions and product development, such gathered intelligence is useful.

3.1.3.9.3 In-House Product and Process Research and Development

All food companies conduct experimental trials in their research facilities or on their processing lines to test the use of new equipment, new ingredients, and new suppliers of raw material. These studies and experimental trials require documentation, cataloguing, and cross-referencing and then storage in a central repository where they can be accessed. Unfortunately, so simple a task is rarely undertaken in many companies. My experience in both small and large food companies has indicated a dismal record of maintaining files describing a company's experimental plant trials or other research projects. These were either not written up at all, or if they were written up, they were hidden away in privately held files scattered throughout the plant. This unavailability of information represents a significant loss of research and development dollars.

These records have more than historical value; they may contain clues to ideas for products that will be found by future technologists. At the very least, they are records of past product ideas and formulations that were rejected. Reasons for rejection of products or processes 10, 5, or even 2 years ago may not be valid today. Times and technology change as do customer and consumer needs. Projects impossible a short time ago could be within present processing skills through advances in process, package, and ingredient technology.

A later research team will be doomed to waste time and money "reinventing the wheel" when they could have been building on the past research. If a test is worth doing, it is also worth recording and filing the results, so that others can find the data and understand the information contained in the documents.

Information retrieval can be confounded by bureaucratic red tape all in the name of security:

In one large multinational company where I worked, all projects did require write-up and were deposited with management. Indeed, even laboratory work notebooks were confiscated at the completion of projects. I had reported on several projects mainly on the rheological properties of molten chocolate, fondants, and syrups. I needed one of my reports on previous work for reference on a current project. To my amazement, I was refused access. I did not have security clearance for access to my own research work! Corporate life can bear a Kafkaesque resemblance.

All reports of process and product experimentation that have been conducted need to be collated, organized, and catalogued with computer access within an information management system that permits ready accessibility.

3.1.3.9.4 Collective Memory

In-house reports for computerized information management retrieval systems as presented in the previous section can be developed further to create a collective memory of what happened when problems were encountered and how those problems were solved. In the history of any company, people have encountered and solved problems in manufacturing; they have overcome short supplies of raw materials by either a combination of reformulation or novel processing steps; and they have resolved failures in products or packaging. Ideas and experiences that company employees have had are important to the company as a resource. If the knowledge and experience of the cadre of old-timers and pensioners can be organized into an accessible body of catalogued information, then companies have an asset valuable for the clues to problem-solving and suggestions for new product ideas or process improvement contained within them.

The collective memory of their key personnel is a resource that no company should allow to be lost due to their retirement. Many companies debrief personnel about to retire and catalogue any pertinent information they may provide concerning their work activities that may not be covered in company policy books. Whitney (1989) describes the development of expert systems using the skills and experience of company personnel. A general discussion of expert systems is provided by McLellan (1989).

3.2 Criteria for Screening Ideas

It is obvious a company has many sources for new product ideas. Not all of these ideas will be suitable for all companies to develop nor will all sources of ideas inspire or be suitable for all product categories in all market niches.

There are dangers to following up on development of products whose merits are based on the unproven benefits of phytochemical-laced foods. The dangers of fortification of foods with phytochemicals blur the distinction between medicines (health foods) and healthy foods, especially so when leisure foods (popular foods for teens and children) are fortified.

Most idea-generation techniques are accessible to all companies, and now the ideas must be evaluated against criteria (usually financial) established by senior management, the findings of market research about the desires of the customer and consumer in the market niche the company has targeted, and the competency of the company to meet the criteria. General criteria for screening are outlined in Table 3.11. These should be applied only loosely as each product presents a unique situation that requires individual application.

TABLE 3.11

General Criteria for Screening New Food Product Development Ideas

Criterion	Comments
Marketability	Is the product profitably marketable within the company's umbrella of products or brands?
	Does the company have the marketing skills in-house to market the product?
Technical feasibility	Can a quality product be developed within reasonable cost and time constraints?
	Does the company have the technical skills in-house to develop a quality product within time and money constraints?
	Should outside development resources be employed to complement in-house development process?
Manufacturing capability	Does the plant have the manufacturing capability to make the product at a cost and quality desired by management?
	Is the purchase of equipment to manufacture the product justified or could a company with the equipment and skills be purchased?
	Is the service of a co-packer justified since increased costs and decreased profitability would result?
	Is there justification for entering into a partnership with another company?
Financial capability	New product development costs money. Is the company financially healthy enough to assume the burden? The financial department must monitor costs and project profits and keep management aware of the financial risks of the project.

3.2.1 Environment in which Criteria Are Applied

The ideas have now been gathered and the market research done: before any physical work is started, the ideas should get an initial screening. What can be rejected before too much, if any, time and effort are expended and what are worthy of further creative study and work? The facile answer to why some ideas are rejected is simple: market research found no interest displayed by any definable market segment. There are other reasons, however, for rejection: these can be summed up in one word, *capability,* but capability as a characteristic has different meanings if seen from different parties.

Each member on the development team brings knowledge, experience, and training that contribute to a collectivity of criteria for screening. These criteria are applied to product concepts and prototypes, and they are also used at all stages of development to narrow a broad range of product ideas down to a few that appear capable of success. Thus, screening of ideas is based on criteria: satisfying the needs of the targeted user as understood from the reality of market research; satisfying the goals established by management for the company's growth; and on those limitations imposed by the company's physical plant and intellectual innovativeness of the development team.

These determine whether ideas should be moved along to more advanced, and more costly, stages of development or not. If it does proceed, how?

Criteria for competent screening depend on the elements of capability, but capability can mean different things in this context. A clearer definition of some of these different meanings emerges through an examination of Table 3.11. None of these criteria present insurmountable problems, but they do require that management ultimately make the go-or-no-go decision.

Conflict inevitably arises over how the criteria are weighted in decision making and by whom the criteria are to be applied. Challenging skills and opinions of others introduces conflict. In large companies with their multiplicity of divisions and departments, managers of new product development teams face challenges as friction may build between people or department leaders. Managers must control group dynamics to apply criteria evenly, justly, and without the personal prejudices of the individual resource persons influencing the screening process. On the other hand, too much agreement amongst the team may reflect some degree of group-think, which can be dangerous if no critical judgment is present.

3.2.1.1 Conflict between Marketing and Research and Development

A frequent source of friction invariably arises between members of the research and development and marketing groups. A policy of fluidity of personnel (allowing members of both to move with the project) can assist in avoiding conflict by letting members of these two groups work more closely. Marketing people live in a world of optimism, chutzpah, hyperbole, and persuasion where sooner rather than later is more appropriate. Technical people, by contrast, prefer a world of logical methodology where organized skepticism is the rule. Technical people live in a world of verifiable facts. They keep perfecting and testing; they are inveterate tinkerers seeking protection and solace behind irrefutable data; this latter tendency is much to the annoyance of marketing personnel. Technologists are devil's advocates, doubting Thomases. They never want to let their pet project go—they become very attached to them.

Marketing personnel complain that research and development people are intractable and inflexible and do not, or cannot, respond to the rapidly changing environment in the marketplace. Marketing personnel see the marketplace as highly volatile and requiring rapid about-faces that technologists cannot keep abreast of and that technologists react negatively to changing ideas. Technical people become too absorbed in the science of a project and not absorbed enough in the needs of customers and consumers. Marketing people complain that technologists do not understand that the introduction of a new product must be timed precisely and that speed is essential.

Technical people see the development process only as a narrow field incorporating their particular scientific disciplines. Ideas are opportunities for resolving technological challenges. A personal incident explains this better:

While working at the Mellon Institute on the H.J. Heinz Fellowship, I was asked by a sister company to identify some red specks in a rival's product. My forte at the time was gas liquid chromatography. Thrilled at the prospect of this challenge, I promptly gathered samples of the red specks, extracted them and prepared them for chromatographic determination. I got nowhere. Solvent extraction with subsequent concentration only confused the issue as all the contaminating material, including trace impurities in the solvent, was itself concentrated. It took time. Finally we received a call from the sender of the sample. They had looked at it under a light microscope borrowed from a local high school. The specks were red pepper pieces easily identifiable under a light microscope! In my defense, I can only say I was young and wedded to technology.

Technologists fail to appreciate that ideas must lead to products that satisfy the demands of both customers and consumers on whom the company depends for its survival. They are frequently reluctant to accept the ideas from sales and marketing personnel that they consider not only impractical but—sin of sins!—unscientific. As French physiologist Claude Bernard has said, "Science repulses the indefinite."

Marketing and technical people speak different languages and use different measuring tools in their trades. Each is skeptical of the merits of the others' tools of the trade. Technologists' tools are objective; those of marketers' subjective. This challenge of understanding different language issue can be a very real one. For example:

I used the words *rheological properties* in a product development review meeting to describe the flow properties of a sauce that was under development. This engendered hoots of laughter from the marketing personnel amidst pleas to speak English. Yet they felt no discomfort about dropping such terms as *perceptual mapping* or *non-metric multi-dimensional scaling analysis* during the same meeting. I had no idea what they were talking about.

The vagueness ("airy-fairyness" as I heard it put) of terms used in quantitative scaling techniques in consumer research and concept testing disturbs the technical person used to logical methodology and verifiable data with statistically significant results.

3.2.1.2 Conflict between Production and Marketing

Confrontations are not confined to research and development and marketing personnel. Production personnel and marketing personnel find cause for friction. Production personnel live in an ordered world ruled by planned scheduling of labor, supplies, and produce in order to manufacture uniform, safe product at the end of the processing line. Their operations are run to rigid tolerance standards. They must meet delivery dates for finished product. Any disruption to this ordered existence affects their bottom line. New product development, especially the sudden scheduling of trial plant runs, disrupts their orderly production schedules. The importance of new product development, if it is not properly communicated, can be strongly resented. The production department's belief is that they could have produced more quality product more efficiently with more modern equipment purchased with the money spent on product development.

Tensions arise during the push for new product development. They arise between all segments of the new product development team; between marketing, production, and technical personnel. What one does not expect, and does not want, are problems at the interface between each of these groups that result in the breakdown of communications. As James Humes aptly put it, "The art of communication is the language of leadership."

This, then, is the environment of problems and constraints in which new product development must be managed productively and efficiently. Whether the company is large or small, the same problems exist. They differ only in size and the complexity that size brings.

3.2.2 Applying the Criteria

Applying the following criteria to ideas must be done as objectively as possible. Certain bitter facts must be faced; not all ideas will succeed. Those that do must

- Satisfy the goals for new product success set by senior management
- Lead to profitable products according to criteria established by management
- Satisfy needs and wants of targeted customers and consumers
- Respect certain financial impositions set by management (i.e., be developable within certain budgetary and time constraints)
- Be within the marketability and sales skills of the company

Ideas garnered from all sources and screened on the basis of initial market research will range from the exciting to the seemingly illogical or bizarre. As screening progresses with the accumulation of more refined marketing and technical data gained as development goes forward, there is advancement of those product ideas most fitted to meet the needs and expectations of customers and consumers and to satisfy the goals of the company for growth if successfully introduced. The wildest ideas if shaped to customer and consumer needs have an element of brilliance if shaped by skilled technical and marketing personnel and daring management dedicated to the growth of the company.

Table 3.11 must be seen as suggestions to refine the process that moves product concepts through to new products with promise to fulfill consumer needs to the company's advantage. Screening does not eliminate ideas; ideas that do not make the cut need to be catalogued along with any development work done on them and filed. Reasons for having abandoned those ideas may be no longer valid in the future as markets change and new technologies emerge, and if they are valid, there will be no need to reinvent the wheel.

Several elements are encompassed in screening new product ideas. First, the new product development teams are guided by clearly stated company objectives establishing the goals to be reached within a specific time frame and within stated costs. These objectives are the backdrop against which subsequent phases of screening and development are assessed by how closely these objectives are met. The team cannot operate divorced from management's financial, strategic, and tactical planning and objectives that dictate the direction for development to follow.

Second, there must be an organization with a leader to coordinate the tasks involved with development and to manage a team to carry out these tasks. This leader will review progress with responsible management personnel and have the responsibility of deciding, based on the collective advice of the team, whether to advance any product in development for further exploration.

Finally, the new product development team requires physical facilities, ideally a laboratory, test kitchen, and pilot plant facilities or access to these: contract research companies or university-based research institutes can complement a company's facilities. As well, there must be marketing skills to explore ideas for products with qualities that meet the needs and expectations of consumers. If a small food company has no such in-house capabilities, new product development companies and market research companies are available who will research markets and develop products for a fee.

Screening is a continuing activity throughout development; it does not stop with the selection of an idea but continues as development progresses. Development takes months or even years during which markets and marketplaces are changing, the originally targeted consumers are getting older and are exposed to other market stimuli. Screening proceeds throughout the entire development process based on the collation and interpretation of new research findings about consumers, markets, and technology.

3.2.2.1 Reality of New Product Development Ideas

The preceding has attempted to show that the means to develop ideas for new product development are well within the capability of all food companies. The rub is, of course, do they exploit all these avenues? The disparity that personal experience has shown me between what idea generation and subsequent product development could and should be and what it is in reality is enormous and somewhat terrifying. Ideas, in my experience, have come from the following:

- The personal whim of a president or other senior executive ("I think it would be a good idea if we had …"). This is the gut-feel approach.
- The competition had come in with one, so we should, too.

- A consultant had got the ear of a senior executive and convinced the executive it was a good idea based on the consultant's research.

There is a constant stream of "ideas" from consultants, academics, and research and development companies.

Goldman (1983) surveyed 47 food companies in southern Ontario ranging in size from those with fewer than 100 employees to those with more than 500 employees for their management practices respecting food product development. The companies represented a wide range of product and processing abilities.

The method most used for idea generation was imitation of competitors' products already in the marketplace. This leads to products of the me-too variety with little innovation or originality and no thought for each company's individual customer's or consumer's needs and expectations. Techniques employing focus group discussions (Marlow, 1987; Cohen, 1990a) and brainstorming sessions were next in frequency of use. (Focus groups will be discussed in greater depth in Chapter 4.) Surprisingly, lowest for frequency of use were those techniques most easily performed or most readily available; attribute testing, recipe books, company personnel suggestion box (asking the sales force what new products they would like to sell), and the patent literature.

Goldman (1983) noted that as the level of formalized organization of new product development increased, there was a greater tendency to make use of a wider variety of techniques for idea generation. This might also suggest that where companies employed a more disciplined approach to new product development, formal techniques for idea generation were considered a more valuable tool.

3.2.2.2 Caution about Copy-Cat Products

It is risky business at best for a company to follow a competitor who has introduced a new product onto the store shelves with a me-too product, for example:

- The originator of the new product has researched the market to obtain a clear picture of the needs and expectations of their customers and consumers. The copy-cat manufacturer does not have this picture. Does the product satisfy the imitator's customers and consumers?

- The originator is into the market first with a carefully planned marketing program. It will cost the copy-cat producer more marketing dollars to get market introduction into the competitor's already established market.

- The originator has the processing know-how and has established distribution channels. The manufacturer of the me-too product must learn the technology, and time is not on this company's side.

The analysis of a competitor's product should be used by a company, not for copy-cat products, but for ideas that will compete with the next generation of products aimed at the constantly changing consumer. That is, today's products on today's shelves are the prelude to tomorrow's products. This is where the developer should be looking. No company can spend too much time generating ideas based on all the information that can be gleaned from all available sources. Shrewd screening will weed out bad and unprofitable ones later in the process of new product development.

4

Strategy and the Strategists

You can never plan the future from the past.

Edmund Burke, Letter to a member of the National Assembly (1791)

4.1 Strategy

Strategy is a simple word but one with a complex meaning: generalship and the art of war. This chapter discusses the art of war, a war in the various food marketplaces where a company finds its customers. All companies vie with one another to defeat competing companies with products similar to their own. Defeat is not too strong a word to describe activities in the marketplace: a more appropriate and euphemistic expression is to gain greater or more market share than their rivals.

Before strategy can be plotted, companies need to know who they are and what business they are in; that is, what their side stands for? Who are they? When, in the course of my consulting, I asked senior executives and owners of companies this question, I received a wide, and wild, variety of answers. None were satisfactory or correct. There was a disbelieving "What the hell are you on about?" to "We're in the food business" to "We're in the business of making money." They indicate a lack of awareness of what business the companies they represent are in. Weren't they in the business of satisfying the needs and wants of customers with unique products to fill those wants? "Knowing who and what you are" is an essential first step to doing business and developing new products. This establishes direction, policy, and philosophy.

Once a company has an understanding of what business it is in—"in the food business" is not a sufficient understanding—and has defined who they are and what their company philosophy is, then they can plan goals and strategy. Strategy is planning to get to those goals.

There are several routes to reaching goals; a company must decide which to take. If a goal is expansion into new markets, then acquisitions or new products to suit that market or both are means to that goal. To increase a company's market share in its local targeted markets, a company must increase its marketing, promotional activities, and advertising with its existing array of products and go head-to-head with its competitors in these marketplaces. More dangerous is to enter into a price war in the hopes of gaining market share or eliminating a competitor. Another route could be to increase its expenditures on product development to come up with products far superior in quality, taste, nutrition, etc., and challenge its competitors in this manner. There is another route, a financial one: eliminate a competitor by buying the competitor. With goals, strategy can be plotted to attack or to defend against the competition.

4.1.1 Defining the Company

If a company classifies itself as a food canning company or a frozen food plant or an ingredient manufacturing company, it knows itself only as a category. It cannot have substantive goals. They must define themselves in terms of what values, services, or assistance they provide their customers and consumers with (Levitt, 1960, 1975). It is this knowledge that gives direction to the new product development program and clarifies the way for marketing strategy. In short, senior management must clearly provide an answer to "what business are we in?" Levitt (1975) provides the now classic example of the railroad industry, which declined because of a lack of definition of what they were. However, the need for passenger traffic and freight transportation did not decline. This need was filled by other means of transportation and not by the railroads. Levitt believes the reason for the decline of the railroads was that the railroads defined themselves as being in the railroad business exclusively when they should have looked upon themselves as being in the transportation business, the business of moving people and goods. In short, the railroads became product oriented when they should have been customer oriented.

Some companies have spent considerable effort on exploiting the properties of nutraceuticals (i.e., phytochemicals, prebiotics, probiotics, and functional foods) by fortifying their food products. Some have added mood-altering phytochemicals to foods. Are these companies fogging up their identities? Are they a snack food company or a pharmaceutical company dispensing medications for real or presumed health problems of self-medicating customers? Getting into what has been called the "wellness market" presents some very definite strategic changes in direction.

I worked with a company whose products naturally contained phytochemicals that had been determined to have significant health benefits. I suggested to the president using this feature in promotions. He and his management were adamant that they were a leisure food company and neither a pharmaceutical nor a health food company. Consumers used their product because it added pleasure and sophistication to their food; they did not use their product because it aided in warding off some disease. They also had no desire to wrestle with the legal problems involving health claims. They knew who they were.

Food companies are not in the canning or freezing of products: they are in the business of serving their customers and consumers with products that meet defined needs. Definition is important.

Defining the company is the role of senior management, and it is their duty to protect and build on this identity with sound business programs that attend to the broad aims and policies of the company. This definition of what the company is and its communication and understanding throughout the organization are essential for effective new product development. It is only by understanding what the company is can true direction be given to product development.

4.2 The Strategists

One group within the company plots the company's strategy: senior management. However, within senior management are two other resource groups almost inseparable from senior management that they are usually incorporated as part of, but subservient to, senior management. Thus, three groups within the company develop the strategy for growth, and if that strategy involves new product development, they provide its direction and hence the nature of the screening tools. These groups are as follows:

- Senior management: that is, the boss, the owner in small companies, corporate management in large companies; or the CEO (chief executive officer) and COO (chief operating officer) either alone or together with the board of directors. Senior management provides direction and defines what the company is.

- The head of finance: that is, the chief financial officer (CFO and usually an integral part of senior management), the accountant, cynically the "bean counter" or the bookkeeper in smaller companies who may be either an internal resource or an external resource such as a consulting chartered accountant. These advise management on the financial health of the company and monitor the progress of all aspects of the business.

- The vice president of marketing: this position may have several subvice presidents divided by geography, for example, vice president of marketing for Europe, or by product category, vice president of consumer goods. They are the visionaries, a desirable quality when visions are based on technical and practical feasibility, financial practicality and company goals; herein, one sees a nascent clash with other resources within the company. The head of marketing in small companies is often also the sales manager—although sales and marketing are worlds apart in philosophy. They often work in consultation with outside market research and public relations firms. Their function is to research customers and consumers in whatever marketplaces the company's products are sold and to provide information on these customers, consumers, and marketplaces to discover hidden wants and unexplored market niches as well as develop promotional materials.

The latter two are information-gathering elements that contribute to policy by feeding information that is the basis for policy making into senior management; senior management always has the final decision-making responsibility. All that follows from idea gathering to the finished product can reasonably be expected to satisfy the goals and expectations established by senior management.

This triumvirate is usually considered as one element in the company, but here each will be examined separately.

4.2.1 An Involved Senior Management

Management is part of the product development team if for no other reason than to see and be seen as having a deep interest in, and providing support for, new product development. In new product development, management performs several functions:

- Management establishes early and clearly its interest in and commitment to new product development. Senior management's interest and commitment play an important role in encouraging team spirit.

- Management ascertains that the company's business objectives are strictly adhered to and that divergent paths of endeavor are not dissipating the energies of the team.

- Management ensures that ideas selected for development fit the corporate (or brand) image of the company (i.e., they answer to what business the company is in).

- Management's presence acts as a check to rivalries or abrasiveness that may arise among the disparate members of the development team as pressures and deadlines take their toll of even the most integrated venture team.

- Management has an opportunity to assess the strengths and weaknesses of individuals possessing different skills as they cooperate under pressure with other members of the team. This allows justly rewarding good work—in itself a morale booster—and lets management earmark a cadre of future leaders.

- Management removes obstacles from the development process, thus facilitating development and thereby confirming their commitment to the objectives of the project.

In small companies, senior management (often owner-presidents) are understandably very intimately involved with their companies' growth and especially with new product development plans. They are often micromanagers.

I consulted with a plant manufacturing meat delicatessen items and smoked salmon whose owner often worked on the line going from one work station to another assisting wherever he saw a need or an occasion to instruct workers in how he wanted things done at that particular moment. There were no written job descriptions or procedural protocols and confusion usually followed in his wake. He, alone, handled all aspects of new product development in a seat-of-the-pants manner, informally and haphazardly. There were no formulations approved and signed off on by either management or the quality control department (there wasn't one). There was no market research or consumer surveys: if the competition did it the company followed. I had been called in to correct an instability occurring in a newly introduced smoked salmon paté. No stability testing had been done on this product before its introduction. Both the formulation and manufacturing procedures were secure in the head of the owner and were related to me on the plant floor. Some compositional and microbiological analyses done by an outside laboratory were provided to me by the owner. I made suggestions on reformulation based on this data and what the owner told me. My suggested changes were conducted on the line. I also indicated sanitary and good manufacturing practices (GMP) violations that would certainly contribute to lowering microbial loads. I strongly urged written sanitation and GMP protocols as well as approved and written formulations.

Lack of a clear definition of what the product concept is can cause terrible friction. This lack of clarity was brought glaringly home to me on one assignment.

The president of a small company was not pleased with progress on the development of a vegetarian spread. When I arrived, the project had gone from a product spreadable at room temperature to one spreadable at refrigerator temperature; then from one with a smooth buttery texture to one with a coarse egg salad-like texture, eventually to one which could be grilled on toast pieces and finally to a REPFED (refrigerated processed foods of extended durability). There was no written product description. As test samples were developed, the president/owner would sample them and usually comment negatively on color, flavor, or texture. The concept was always changing direction—all to the frustration of the technical director. My final, and terminal, report to the president placed blame solely at his door for the lack of clarity and progress in product development and suggested strongly he put a description of what he wanted the product to be in writing.

I found myself in a similar Catch 22 situation on my first new product assignment when I was a greenhorn fresh out of university. This time it was not a small company where the lack of communication occurred but a large multinational, multiproduct company. The company's American headquarters were in New York City where the product manager was also stationed. I worked at a branch confectionery plant in upper New York State.

My assignment (verbally given!) was to make a stable icing product – that was it, simply a shelf-stable icing. The product manager visited regularly on Fridays to review what had been done and taste samples. (I never knew who the other members of the team were, if indeed there was a team.) After weeks of frustration—my samples were either too sweet, too sticky, too dry, too runny, too chocolaty, too smooth, were not freeze stable (a characteristic I had not been asked for), and so on—I complained to our plant's research director, a capable veteran of many years of experience, that I was going nowhere. The next Friday visit, I met my director and the product manager. The director said only, "Give us the complete product concept and its desired quality attributes." This was done rather reluctantly and the product quickly accomplished after I was told it was to be identical to a competitor's product!

Goldman (1983) in her survey of the product development management habits of food companies in southern Ontario found that in just under half of the smaller (less than 99 personnel) companies surveyed, the president was largely responsible for product development. In confirmation of Goldman's findings, from my own experience, in all the small plants under 200 employees, the president/owner has had a major, often sole, impact on the new products in development, usually with no market research guidance. Unfortunately, there was often a disastrous consequence.

As companies increase in size, the president's involvement in product development becomes, of necessity, more remote. More junior managers such as product managers take his place. Only 12.5% of presidents of companies with more than 500 employees affirmed they had the main responsibility for product development (Goldman, 1983). In large companies, the product manager serves this function and maintains a close liaison with senior management, usually at the vice presidential level.

Larger companies had a more organized approach to product development (Goldman, 1983). In these companies, 37.5% of survey respondents indicated their job function as one of product development. On the other hand, no person from a small company who responded to Goldman's survey described their job function solely as product development. In smaller companies, staff wear several hats.

The intensity of the involvement of senior management obviously varies widely with the size of the company. Their greatest involvement is often in the initial screening phases when they decide whether the screened ideas submitted by the team fit the goals of the company and fit the brand or the company's image. This is not necessarily so in large multinational companies where this chore is left to product or brand managers. The involvement of upper management tapers off until the time when "go" or "no-go" decisions are to be made for introductions.

(In this text, any reference to management is to be broadly interpreted as those persons with authority over product development whether they be product managers or the company CEO or COO.)

4.2.2 Shaping the Company's Objectives

When senior management are clear on who they are, what business they are in, and where they want to go, then a company's product development objectives can be stated clearly, be understood by all, and be seen to focus on attaining these company objectives.

Any lack of communication about these objectives leads to confusion within and between the marketing and technical development team members as they try to second guess management with their own interpretations. The team does not know what they are developing or for whom they are developing it.

An understanding of what the company is provides direction to the ultimate course of new product development. In Figure 4.1, a generalized picture of company's identity as seen by the community it lives in and the greater community its products serve is composed of several elements:

- The objectives established by the owners (senior management) cannot be divorced from how manufacturing is conducted or how promotional activities are undertaken in the communities where the company is situated or serves. The company is part of the several communities it serves and has altruistic and humanitarian feelings that lead it to indulge in various charities, to support artistic endeavors, to become a patron of the arts, and to participate in or promote sports activities in the community where it lives. These indulgences may be entirely personal or may be well-orchestrated

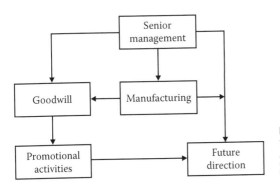

FIGURE 4.1
Understanding the company's corporate identity in new product development.

financial moves for the company's publicity or product promotion. These activities usually provide networking opportunities with the potential for future endeavors.

- The manufacture of quality products in an environmentally friendly, and socially responsible, progressive company leads directly to a goodwill factor that enhances the value of the company and provides opportunities for promotional activities. A poor labor or pollution record dampens goodwill and any activity the company undertakes. In the social networking society, such ill will can go viral very quickly.

- The above have both a direct and indirect impact on the company's future—its continued existence, hence its new product development and its growth plans. A simplistic example might explain this better: participation by a community-spirited confectionery company in local sporting, artistic, or social events might suggest to developers to work toward a nutritional bar, adult candies with a sophisticated taste and for marketers to use such events for promotional purposes.

This, then, reflects the company with knowledge of itself and its objectives that influence its direction within its several various communities.

4.2.2.1 Company Objectives That Shape Product Development

Companies are rarely so single- and bloody-minded as to want to make only more money. They do want profit, of course, and management decides whether short-term or long-term profits are required. If short-term profits are needed, then activities are directed to product development ideas that will return quick profits; this usually means products requiring little cost and a short development time. Where management does not feel so constrained financially, they can opt for pursuing ideas that require a less restricted time horizon, more effort, and a greater amount of money to accomplish.

Attaining financial objectives, always desirable, is not the only force driving the development team. Some other objectives are as follows:

- Management's desire to grow (such growth depends heavily on new product development)
- Management's desire to expand geographically to lessen dependence on the vagaries of the economic climate of local or regional marketplaces or caused by a major competitor
- Management's need to reduce dependence on commodity-type products and increase profitability and competitiveness with added value products
- Management's desire to expand its product base and thereby reduce the company's dependency on a narrowly focused product base of one or two bell-ringer products

- Management's desire for greater local or regional market penetration with a broader range of products to maintain and protect a competitive position

The above compress into three broad ambitions:

- Financial objectives such as obtaining a greater return on investment
- Strategic objectives, often defensive in nature, to protect the company's market position against inroads by the competition or offensive to gain market share from the competition
- Tactical objectives by which the company charts its path toward its goals that are derived from its strategic objectives

With clearly stated and communicated aims, the development team has guidelines for creating criteria for screening and selecting suitable ideas for further development. For example, if going toe-to-toe to gain greater market share is the goal, then the company wants new products that are competitive or more so in price and quality than those already in the marketplace. If the need is to defend an already established market being threatened by a competitor, then a faster more responsive action is required with more promotion, perhaps even price wars, and new products as soon as possible to compete with the invader. Here, the two objectives require different strategies, criteria for ideas for development and need for urgency.

4.2.2.2 Sanctioned Espionage or Competitive Intelligence?

Senior management cannot lead or provide direction in a vacuum. They need vast amounts of information and careful assimilation of this information to provide the intelligence required to move advantageously in their marketplaces or into new markets.

A recent cartoon depicted a candidate for a job in his interview with a personnel manager. The manager says, "Your several convictions for computer hacking will help you fit into our corporate plans." Humor or cynicism?

There is a saying—anonymous as far as I know: "Know your enemy." It is best, however, to know what the competitor is doing before the competitor does it. Knowing that they have done it is far too late for defensive action. That is the key in competitive intelligence gathering. Knowing what competitors are doing is important for planning counter-strategies that help a company to survive and to protect its market position. Yet many companies have no idea what their competitor is doing until they see the results in the marketplace when retaliatory action is ill-thought-out reaction.

Competitive intelligence is generally the domain of senior management—there are vice presidents of competitive intelligence—but the function has also been considered a marketing duty. Companies use many techniques to spy on their competition. Espionage is a dirty word that suggests the unethical collection of information done by secret agents lurking in trench coats in darkened laneways. Competitive intelligence is gathered in ethical and respectable ways.

Intelligence gathering masquerades under many innocuous names in corporate directories: corporate intelligence, competitive intelligence, strategic business development, business planning, or as very innocent market research. Such activities, if carried out by trade officials assigned to embassies in foreign countries, would have been cause for the officials to be accused of spying and declared persona non grata. Yet, it is nothing more than the systematic gathering of information about the competition from as many legal and public sources as possible and the intelligent interpretation and use of this information to plan strategy against the competition or to counter the competition's strategy.

Corporate intelligence gathering has become very professional. There is even a Society of Competitive Intelligence Professionals in the United States: its Web site provides interesting information about its activities. Ten years ago, the industry was expected to become an industry exceeding the $110 billions (Lewandowski, 1999). The many companies specializing in this activity have dossiers on particular industries as well as on specific companies or even products in these industries. An Internet search of competitive intelligence reveals numerous companies with specific company or specific industry data to sell.

The first line of developing an internal competitive intelligence network involves having a presence in the marketplace to see what is going on and a center to collate this knowledge. Traditionally, information gathering has been carried on innocuously by salespeople interfacing with the retailers selling their products, but any information gathered found its way back to management in a haphazard fashion. When this review of marketplace shelves is done by second party distributors, knowledge is lost. A company's sales force has the greatest advantage of gathering information from the marketplaces the company services and feeding this back to management systematically, information such as the following:

- What product facings the competition has, how extensive these are, and where these products are located in the store. Product location in a store can be a rough idea of how sales are progressing.

- What new product introductions the competition is launching, in what geographical areas these new products are being launched, and whom they are targeting. Additionally, salespeople can get a general idea of the success of the products.

- The extent of their competitors' advertising and promotional materials in the marketplaces and again, whom these are aimed at.

- How the pricing of similar competitive products compares to the company's own products.

- What deals the competition may be making with the outlets carrying their products.

In short, sales people on site in the marketplaces are able to gather valuable information about their competition in these marketplaces. If the sales force does not stock shelves then some responsible company personnel needs to make the company aware of activity in the marketplaces.

This information—indeed all information about the competition gleaned from any source by the company's staff—is fed back to a central repository, the war room. The term *war room* is found in a glossary used by one competitive intelligence firm (Crone, 1999). The war room is defined as a central location in a company where competitive intelligence obtained from all sources is sifted and analyzed to reach strategic and tactical decisions.

An ongoing activity in any food company is the examination and analysis of products by grading, tasting, and chemical and physical analysis (i.e., the cutting of competitive products). The ingredients found and identified are costed and based on the amounts found a crude formulation, and costing could be established to provide a rough estimate of the competitor's profit margins.

To get value from competitive information gathering requires that this information find its way back in a very focused manner to those who are able to analyze it and get information to their management in a timely fashion. Many large companies require that their staff be aware of any actions by their competitors and report such information back to the war room.

Companies can gather competitive information ethically (Table 4.1) without hiring outside resources. Many of these activities are broadly classified as networking, that is, building up of a broad association of contacts and using these as conduits for information. As such, they are part and parcel of every business person's activities. Networking can be a source of critical information that analytical wizards of the war room utilize.

All organizations from chambers of commerce to professional associations meet regularly for business and for pleasure. At these meetings, invited business leaders may reveal in their talks some of their company's business activities. During informal sessions, cocktail hours, or receptions, they may discuss or let fall some unguarded words describing their company's programs. Companies with active competitive information programs use such settings to ferret out snippets of information.

New names or the absence of old names of executives in company literature hints strongly at a corporate policy change. Executives have a management style, and their policies in previous positions indicate how they will operate in their new role. For example, a company hires a new executive CEO. This appointment is widely circulated in the business news media. A check of the electronic databases allows profiling of the new executive. Executives have established patterns of behavior (their previous record was most likely the reason they were chosen for their new role); it is likely the new CEO will act as profiled. Is the new man known as an expansionist, a hatchet man, or an aggressive marketing person? This information allows a rival company to be

TABLE 4.1

Commonly Used Sources of Competitive Information Gathering by Which
the Activities of Competitors Can Be Gathered for Analysis to Prepare
Counter-Defensive Movements

Source of Information	Specific Activity or Information
Print media	Executive moves and removals (reported in business newspapers and trade journals) and mergers suggest policy changes within companies
	Press releases issued by companies describing activities
	Help wanted advertisements, especially those specifying the need for particular skills or training
	Articles in scientific and technical journals where researchers indicate source of funding by companies and hence reveal direction of research interests of these companies
	Major equipment purchases announced by companies or major equipment sales contracts made by companies
	Listings/registrations of land purchases by companies
Corporate publications	Annual and quarterly reports record personnel changes, policy directions, and financial status (purchases and sales of properties and assets) of companies
	Corporate Web sites with product information
	Company blogs and other social networking sites
Conferences and trade shows	Exhibition booths at trade shows demonstrate latest equipment and products
	Suppliers often inadvertently or purposely reveal buying activities of other companies as sales ploys
	Speeches or panel presentations made by senior management especially in question and answer periods
	Research papers delivered by technical staff or presented by research groups supported by companies
Pro bono activities	Dinner speeches by senior executives at charity events, Chambers of Commerce, fraternal organizations, Young Presidents' Associations, etc.
	Volunteer participation in activities of professional associations, e.g., Institute of Food Technologists
	Participation in alumni association activities
Access to government information	Information and forms filed with various government agencies
	Grants applications requesting cooperative research ventures
	Searches of patent notices, patents granted and pending which provide information of company's research direction
Company sponsored social events	Cocktail or other sponsored receptions or dinners
	Sports outings such as golf tournaments

on the qui vive and be prepared to act accordingly against the anticipated actions the new executive might put in place.

The technical and patent literature that is publicly available reveals research projects the competition had supported, in what institutes it had been performed and what patents have been granted or are pending. From the time lag in such publications, intelligence analysts can prepare time lines of the developmental activities of its competitors. The least that is determined is knowledge of the competitor's technical interests.

More surreptitious activity is arranging to seat oneself at a conference dinner beside a competitor's director of research or one of its senior technologists. I personally have always found the conversation around the breakfast, luncheon, and dinner tables at, for example, the Institute for Food Technologists' annual and quarterly meetings or at other conferences to be very useful opportunities for glimpses of other company's technical activities. The information came either from academics to whom the work was contracted, the competitors themselves in unguarded, cross-table talk, or suppliers looking for customers ("Didn't you know that so and so company was using our ..."). Our company's practice was to spread out our technical staff at the luncheons and dinners at conventions such that at each table where a competitor was recognized, one of our staff attempted to be seated there also. We learned nothing sitting together.

Companies plan social outings for their suppliers, their clients, and potential clients at trade shows or expositions; these are carefully orchestrated affairs. Seating arrangements at dinner tables, foursomes for golf, and outings for wives are arranged so that maximum benefits to obtaining information are had. My wife has provided me with many interesting bits of information innocently dropped as she attended spousal programs at various food conferences.

Through access to information regulations researchers can explore grant applications, land rezoning applications, and temporary import permits or any of a multitude of forms their competitors may have filed with their governments. Each piece of information has significance and may be pieced together to form a bigger picture: they provide awareness of one's competition.

Another avenue to competitive intelligence is through social networking systems, the most common of which are Facebook®, blogs, and other online platforms. These have given rise to what has become known as citizen journalism. Often what is read here on these sites must be accepted "with a grain of salt" coming as it may from disgruntled employees, cranks, and amateur news reporters. A comment made on the reliability of these sources was "I would trust citizen journalists as much as I would citizen surgery." Nevertheless, it profits a company to know what is on social networking sites about themselves and their rivals. A nasty rumor may need to be squelched.

4.2.2.3 Benchmarking

Much of the foregoing may be recognized as a form of benchmarking, a comparatively new management technique. Benchmarking is derivative of intelligence gathering and as such is a tool wielded by management to attain its goals (usually those related to management style and quality of products) for the company. A company undertaking the practice of benchmarking closely examines its own operations in all aspects. Then it compares itself with the successful leaders in the same field. The company then applies what it learned from the leaders to its own operations. The hope is that this imitation will help the company as it did the leader. A comparison might be made between the Japanese auto industry and the American auto industry. Japan emulated American styles of quality control and manufacturing, even to hiring American engineers, and eventually refining the techniques. Unfortunately, there was a lapse in business practices beginning in 2009.

The principle of benchmarking is a process of constant incremental improvement of services, of key quality characteristics of products, and of all areas of operations by comparison with the successful leaders in their industry or product category, a form of imitative strategy to become, it is hoped, more like or as successful as one's competitor. Companies must study the strengths of their rivals while at the same time acknowledge their own weaknesses. Such knowledge involves researching their own customers as well as those of their competition. With this knowledge, management strategies are developed with which companies move quickly in the marketplace to gain superiority with products and services.

There is a side issue, competitive cost benchmarking. Knowing the competitor's costs for a product leads directly to being able to estimate profit margins. Such knowledge of a competitor's product's profit margins can be used in the marketplace to wreak havoc to the competitor's product launch.

Benchmarking is more a means of enabling a company's self-improvement and is only indirectly a tool leading to new product ideas.

4.3 Finance Department: The Cautionary Hand in Development

> The golden rule: Whoever has the gold makes the rules.
>
> **Anonymous**

4.3.1 Finance's Not So Passive Role in Development

A very blunt fact that must be learned by all in any company is that to survive, a company must ultimately be successful, and making money helps its survival. A company is certainly not in business to lose money or simply to keep its employees happy and busy. Only if the company is making money can it enjoy the luxury of new product development, and new products represent one way to make money. The financial department's duties include the following:

- Monitoring the day-to-day financial affairs of the company
- Alerting management to the financial health of the company and the financial consequences of any action, precipitate or otherwise, that the company undertakes
- Advising management of the company's financial ability to undertake new ventures and evaluating the financial consequences of such ventures
- Tracking the costs of any venture the company undertakes and providing a cost/benefit analysis

Another fact of life to be learned by the novice product developer is this: the finance department (as does every other department in the company) has its own plans for how the company's objectives can be met. They view with disdain spending money on uncertain high risk projects, which new product development is. Financial people usually are, by nature, conservative people. They have their own ideas of how to make money for the company in the financial instruments with which they are familiar.

As senior food scientist I attended a food product development meeting with senior management, marketing and production personnel to discuss the new product development program. That meeting was demolished when the senior vice president (corporate) of finance bluntly pointed out to the assembled staff, that by transferring company funds into various foreign currencies and by investing in bonds or stocks that his department would make more profit more surely for the company than our food division would!

In many ways, the financial department shapes much of the activities for new product development. They have the advantage of knowing intimately what the financial health of the company is and hence how much financial exposure the company can withstand. They know how much money is readily available for development and, hence, how much development can be afforded. Their criteria for screening any venture hinge on many intangibles such as expected profitability, probability of success of any ventures undertaken (i.e., risk assessment), and financial stability of the company in the current economic times. These three intangibles can be wielded by an uncooperative, unconvinced, or conservative financial department very facilely to make their presence known.

The financial member of the team monitors the costs of the project as it progresses, and it is only costs that accrue until well after launching when it is hoped costs can be recouped. This tracking of costs alerts all to whether the project is within budgetary limits, is in keeping with expected probability of profit generation, and has minimal financial risk exposure. Marketing personnel, with this cost information, can compare the projected introduction costs with sales predictions. The result, it is hoped, is a reasonably accurate estimate of the net profits that the new product is expected to earn. As development progresses, these cost figures are refined as more data are accumulated and financial projections become clearer. Management must assess all inputs to determine whether company objectives are striven toward in a prudent manner.

4.3.2 Financial Realities of Product Development

The relationship between developmental progress and costs can be more readily seen in Figure 4.2. In the upper graph, the number of ideas under consideration or in development is charted against time. The number of ideas decreases during preliminary marketing and technical evaluation and recipe development as screening takes its toll. Sensory testing, further consumer researching, and production scale-up weed out the unpromising ones until few are left, from which finally one is selected for introduction. The upper graph (Figure 4.2b) continues, but now sales volume of the newly

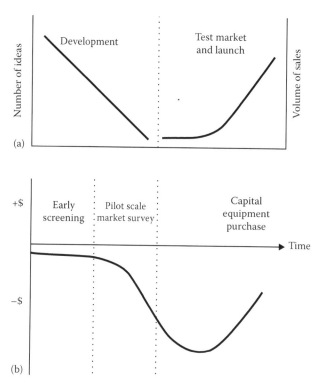

FIGURE 4.2
The relationship between (a) the course of development and (b) the costs of development.

launched product is plotted against time. The lower graph plots money flow (costs are −$: profit is +$) against the same timescale as the upper graph.

Costs are minimal in the early phases of screening as many ideas can be evaluated and eliminated over discussions with readily available marketing, production, technical, and financial data. When preliminary work begins on evaluating and tasting formulations and when outside marketing, consumer research, and development companies are hired for more sophisticated studies, then expenses rapidly mount. Sensory testing eventually goes beyond small in-house panels onto a larger scale that requires full production line runs of product. These are used for home-use tests, mini-tastings at trade fairs or county fairs, in supermarkets, or free sample distribution with follow-up questionnaires or interviews.

Plant trials are costly in two ways to a company:

1. They require labor, and perhaps extra labor, and they use up raw materials, all for product that brings in no financial returns. That is, they produce an overhead for which there is no recompense except the gathering of consumer information.

2. They are also disruptive of routine plant production from which the company makes its money. Disruption of routine plant production can be a great annoyance to production personnel. Production targets are compromised and merely add to costs.

Frequent disruptions of production routines introduce an intangible cost, that of irritation between manufacturing and research and development departments.

Costs increase as development proceeds and research techniques become more sophisticated. New equipment may have to be purchased, specialized equipment designed de novo, or pilot plant equipment leased from equipment manufacturers or research institutions. If new equipment has to be designed and fabricated, a steep increase in costs can be expected. To circumvent these costs, outside packers or co-packers may be required. Costs of having products packaged by a specialty packer or co-packed by another manufacturer will impact on the profit picture. Costs for mini-test markets and their attendant market research also increase the amount of red ink.

A successful test market justifies a wider initial launch area. But the costs of a wider introduction can be astronomical. Eventually, sales volume and with this profit, it is hoped, overcomes the red ink (compare Figure 4.2 and Figure 1.3a and b).

4.3.2.1 Slotting Fees

Retailers demand monies, slotting fees (or slotting allowances) that are a cost of doing business. They are not the result of the development process itself.

Slotting fees are monies that retailers require to be paid by manufacturers in order that the retailers allow them shelf space, obviously an important requirement for new products.

These expenses are major deterrents to product introductions for both large and small food companies and are an impediment to new product introductions for all companies. Retailers justify these fees as necessary to compensate them for the costs and risks they face by putting new products on their shelves. If a new product goes on the shelves, the retailers argue, some established product must be removed or be given reduced facings. This represents a loss of income for retailers. If the new products should fail, retailers lose in two ways. They lose good returns because of the newly introduced product's poor return, and they lose the income that would have been generated by the established products that were displaced. Manufacturers object to these fees as too expensive, unfair, and punitive. "Criticism for slotting fees is rooted less in the need to recover the cost of new product introductions and more in potential abuse" (Hollingsworth, 2000). In-depth discussions of slotting fees are presented by Hollingsworth (2000) and Stanton (undated).

Large food companies do have some clout that is not available to small companies, yet even the giant food companies must face giant retailers. Big companies can pay more for space and squeeze the small manufacturers off the shelves. This is the eternal complaint of small companies. As a result, smaller companies are forced into smaller markets (niches) and marketplaces for introductions of their new products. Another tactic of big companies that I was informed of but have never personally encountered is the following: large companies would deliberately short a bell-ringer product from a retailer who did not allow shelf space for the introduction of a new product. Competing retailers who cooperate with the large companies got the shorted item. When noncooperating retailers allotted shelf space on the large company's terms, then the shortages disappeared and the shelves were filled with the bell-ringer product.

4.3.2.2 Financial Criteria

Financial criteria for screening depend very heavily on what the company's objectives are, how vigorously senior management pursues these objectives, and how financial criteria are applied. Objectives vary widely with the economic environment a company finds itself in. To illustrate, two hypothetical situations are presented with financial criteria that are very different:

1. A food manufacturer is dependent on seasonal processing of locally grown produce. Management chooses to broaden their product base with the development of added value products and thereby reduce dependence on seasonally grown commodities. They believe this would allow them two benefits: first, to keep a trained work force year-round with some community goodwill generated; and, second,

to be more productive and increase their return on investment. Financial success is measured by

- Less reliance on seasonal manufacturing and its correlative, developing a year-round manufacturing operation
- Gaining a foothold in a new business in new marketplaces with added value products

Their investment is for the longer term; it is expensive. They can be more patient in their expectations of a return on their investment.

2. A company is fighting intrusive action by competitors into its marketplace. It needs to protect its market share or to regain lost market share. This company's time frame is shorter. Its investment will be concentrated on aggressive marketing programs as well as new products such as line extensions or the closely related "new, improved product" with reduced risk, less development time, and less costly research. They require a more aggressive product maintenance program. The company's resources will be directed toward projects consistent with the restraint it faces. Success for this company is

- Maintaining—certainly not losing—market share or even improving market share
- Thwarting inroads of a competitor

The finance departments of both companies need, nevertheless, to monitor expenses and advise management of the financial health of the company.

As development progresses, the direct costs of development and the indirect costs associated with the new product's impact on the company's existing infrastructure become more apparent and more critical. These direct and indirect costs are figured into their development expenses that must be recovered. For example, the costs of development are treated as a loan; the current cost of money, the loan, is added to the development costs. The finance department argues that this money could have been earning interest for the company: this lost interest should be an added expense of development.

Reliable forecasting of sales volumes based on consumer research and firmer cost estimates give clearer information of when or if expenses will be recovered in a timely fashion. If these projections do not fit the company's interpretation of *timeliness* of financial success and expected rate of investment, then those products in development in the pipeline are terminated; the data and research work to date are catalogued and filed for future reference.

Financial criteria must be applied fairly (see Section 4.3.1). Company controllers can, by using accepted accounting techniques, assess certain expenses as assets (investments) and can regulate the rate of depreciation. How the accounting is carried on the company's books can be made to reflect badly on the progress of product development according to the attitude of the financial department to the project. Management must recognize any bias that financial people introduce into their financial statements to the progress of development. Where short-term financial gain is favored, company philosophy and policies are usually directed to cutting development budgets and reducing their staffing, and thus, companies risk losing technical skills that could be their salvation in the future.

Projects lost for financial reasons with attendant technical staff reductions demoralize younger staff who fear for their future and are disappointed to see challenging work terminated. Projects can prove a training ground for younger staff and allow senior management opportunities to evaluate personnel for those who will be the company's leaders tomorrow.

Other departments within the company, besides the financial department, would be pointing out how their pet projects would have been making money for the company. These departments can produce cogent arguments to convince, and perhaps bias, the financial department to their merit. When Imperial Tobacco Company of Canada diversified to create a food division, the tobacco group—a highly profitable group—resented the foods group as it saw its profits used to prop up what it saw as a weak sister. They felt they could use their monies to better purposes. There is always competition for money within companies and no end of advice on how best to spend it. The financial department must be aware of and fairly assess the financial viability of all avenues to the company's goals.

4.3.3 Financial Tools

A number of rough rules of thumb have been generated to estimate the potential profitability of projects. They are all crude tools based on the best available reliable data the strategists can get; in short, they are guesses used to estimate a guess. As arbitrary estimates of the economic viability of a project, they can be useful to counterbalance the intuition and "gut feeling" that is frequently behind the unwarranted, continued support of pet projects with questionable probability of success. Nevertheless, these tools for project cost estimations are themselves based on assumptions (that may not be true), on estimates (that may be biased by those deeply involved in them), and on faulty interpretation of customer research.

4.3.3.1 Comparing Costs with Anticipated Revenues

The simplest and crudest financial criterion is to compare total projected costs against the projected gross sales for the period within which the company wants its payback. Table 4.2 describes projected costs. It rapidly becomes apparent that this index is based on uncertain costs in combination with equally uncertain income derived from projections of hypothetical sales figures derived from marketplaces that are influenced from many sources. The potentially misleading nature of this criterion highlights its danger if it is

TABLE 4.2

An Analysis of Projected Costs of New Product Development and Introduction

Breakdown of Projected Costs

Production costs (largely?)

Overhead costs: these can be assumed to be the same as those for similar in-house product (but the novelty of the new product may confound this assumption)

Co-packer fees and costs (if product cannot be produced in-house)

Capital expenditures for new equipment or other plant facilities (special handling equipment; storage or warehousing facilities)

Packaging and labeling materials (new containers, packaging and container design, label design)

Raw material and ingredients: seasonality, quality standard demands, availability may challenge financial assumptions

Process and product quality control (can be assumed to be part of overhead costs but additional impact on facilities cannot be assessed fully)(?)

Development costs (largely?)

Market research (focus groups, taste testings [mini-market tests, in-home placements, etc.], surveys, etc.)

Technical research (laboratory research, test kitchen development, shelf life testing, pilot plant trials, toxicological and environmental testing)(?)

Consulting fees for outside market or technical research

Marketing, launch, and sales costs (largely?)

Test market and follow-up(?)

Advertising materials (brochures, handbills, coupons, etc.)(?)

Advertising preparation, and promotional costs, media costs(?)

Slotting fees and other retail trade promotion practices (Stanton, undated)(?)

Note: (?) indicates costs that cannot be projected accurately.

wielded too vigorously to screen out products that are rightfully worthy of development. (Admittedly as development progresses, some of the cost figures emerge more clearly.)

The very crudeness of this technique underlines its shortcomings. This measure focuses only on direct costs associated with new products. The impact new products have in other business areas is ignored. For example,

- Were the production department's additional costs respecting added downtime for line changeovers, extra labor costs, and costs due to regular production disruptions included?
- What extra costs occurred due to handling, warehousing, or transportation?
- Were the extra costs for sales calls included, whether calls made by the company's own sales force or an agent's, or the costs of extra salespeople? A broader product line will be more difficult to sell in the few minutes sales personnel are allotted with the retailer's purchasing agent.

The comparison of total projected development costs and total projected sales alone is not a reliable index. Promotional costs to educate consumers to innovation are unpredictable but are certainly highest when attempting to get market penetration until retailers welcome the new addition to their shelves. Where novelty of product is such that safety or toxicological testing is required or its processing may have an undesirable environmental impact, these concerns require large expenditures of money to allay (foods derived from genetically modified organisms). A high level of novelty or innovation brings its own initiation fee as it were, that is, too much novelty frightens customers, and there may be a high price to pay to convince of or educate people to its value (cf., irradiation and genetically modified foods).

What is management's time horizon for return on investment? Does senior management consider products as failures if they become impatient? They wanted quick paybacks. The failure then was not due to lack of customer response to the product or problems inherent in the product but to senior management's disinterest in continuing with marketing if they will not get their money back in a short time frame. Management had unrealistic expectations of a satisfactory payback period; given time, the product may have had a good success. Most companies want rewards as soon as possible. No new product is ever introduced to the market with the implied intention of withdrawing the product in any foreseeable time frame. Therefore, management's unreal time horizon introduces a real conflict within the development team and with retailers who made space for the product.

Not getting sufficient share of market is akin to not quick enough return on investment: it is also a failed ego trip in that the project did not succeed as management had planned. They failed. However, if some market share is obtained and this can be maintained with sustained profitability, is this not success? Why must the greatest market share need to be wrested from the competition? One company, with the most popular line of Italian style products (middle shelf at eye level facings), wanted its competitor there to fill the upper price and quality range product category. The popular brand had the greater market share and the other brand a smaller but very viable market share. A company need not aim for market dominance. There was a tag line to the effect; we are second best, so we try harder. There is nothing wrong with having the second or even third largest share if the profit is growing in a growing market.

4.3.3.2 Probability Index

Attempts using all possible combinations of market and research data have been made to improve the predictability of success. One such attempt, the profitability index (Holmes, 1968), compares the expected return to the total cost, that is, return/total cost, and multiplies this by the probability of success:

$$\left(\frac{\text{return}}{\text{total cost}} \right) \times (\text{probability of success}) = \text{profitability index}$$

This is hardly a great improvement over the tool described in (Section 4.3.3.1) above: a guess (expected return) is divided by another guess (total cost) and the answer multiplied by a guess.

The shortcomings of all arbitrary indices to predict success or profitability or market penetration are due to the imprecise nature upon which the indices are based. Predictions of sales made by marketing for new products that have no proven sales record are imprecise. The indices make no allowance for the retaliatory action of competitors. Estimates for the probability of success, time for completion, and costs are exactly that, estimates. They are only as good as the information that went into their estimating. Garbage in, garbage out, as the saying goes. The indices remain tools to assist decision making not tools to replace decision making.

4.3.3.3 Other Tools

Malpas (1977) discussed the use of Boston experience (or learning) curves for what could be another criterion for determining whether research and development dollars should be spent. In general, when volume units of a product double, costs usually fall by approximately 20%–25%. If this generalization does not hold, then it is time to seek a new process. Argote and Epple (1990) discuss the value of learning curves in nonfood manufacturing and cite their value in pricing, marketing, and predicting competitors' costs.

The competition's retaliatory action in the marketplace will confound all the above indices and will have an enormous impact on cost calculations as the competition fights back. How will retailers react? All predictors fail to give due consideration to the multitude occurring in the marketplaces. The activity of the competition is the least of the development team's worries: they, too, are indulging in competitive intelligence gathering and may have prepared elaborate counter strategies based on more accurate market information.

4.4 Strategy: Marketing's Perspective

> Failing to see relevance when it is present is a form of ignorance inadvertently encouraged by traditional practice in science.
>
> **Zaltman (2000b)**

> ...one of the more pernicious scientific fallacies: assuming that the absence of evidence amounted to evidence of absence.
>
> **Pearce (1996)**

Zaltman's and Pearce's statements above give fair warning to the scientific arm of the development team: do not ignore one's observations or assume too much from the absence of same (to which I would draw attention to A. Conan Doyle's Sherlock Holme's story *The Hound of the Baskervilles* and the dog in the night). Marketing personnel must tread a fine line in what is, as yet, an imperfect science, often termed a soft science, marked with imprecision and inadequate tools that provide more subjective rather than objective evidence. Researching people and their likes and dislikes and what motivates them to buy and understanding what influences them is slowly gaining in improved technology and understanding. But as a science, it is, as yet, looked down upon by practitioners of the hard sciences. Soft sciences with their subjectivity are anathema to scientists and technologists.

> Science earns its reputation for objectivity by treating the perils of subjectivity with the greatest respect.
>
> **Cole (1985)**

4.4.1 Marketing's Functions

Marketing people have three primary functions:

1. To understand their targeted customer's and consumer's actual and perceived needs and from this understanding develop strategies for the marketplaces they service that communicate clear messages of these needs.

2. To develop sales and promotional materials suitable for use in the various marketplaces where they have found their customers and consumers. "Marketing [focuses] on the needs of the buyer" (Levitt, 1975). Levitt does not distinguish between customer and consumer as used throughout this book. He comments further that marketing "[is preoccupied] with the idea of satisfying the needs of the customer by means of the product…and finally consuming it." These sales and promotional materials must also reflect the wants of the retailer in the targeted marketplace.

3. To monitor the reactions of customers and consumers to their strategies to ensure that both are pleased with the product at the time of its purchase and with its first use and remain satisfied and happy for repeat purchases. From their research, they develop product maintenance strategies to ensure a continued product life.

Marketing must know and understand the various marketplaces and, in particular, understand the marketplaces they service and their customers and consumers in these. They must also search new markets as the environment surrounding their customers and consumers changes. This requires extensive customer, consumer, and market research with, but not limited to, a physical inspection of all the marketplaces and of all products in those marketplaces, a vigorous product integrity program within the food plant, good relationships with retailers in the marketplace to obtain customer feedback from the retail scene, and research into consumer usage habits of the product.

Attendance to these functions results in a marketing plan that reflects the nature of the company and the types of products to fit this uniqueness. The marketing department's plan must respect the company's past and present historical brand positioning *if the past image fits the current management's objectives and new product positioning.* Brand image cannot be changed abruptly: it must be done smoothly, but the plan must also reflect new directions that the company wishes to go. An example fits here:

Imasco Foods Co. purchased S&W Fine Foods, whose high quality products served the carriage trade of canned foods. The S and the W were originally the initials of the founders Sussman and Wormser. Imasco Foods tried to lighten its image by using the S and the W for Swinging and Wonderful to move the brand from a stodgy carriage trade image to a more vibrant youthful image. It was not the smoothest transition encountering as it did resistance from within the company's culture and from the trade.

All successful product development starts with the consumer just as all successful selling starts with the customer. Coincidentally, selling must include the needs of the retailer in the marketplaces in which the customer, consumer, and retailer are to be found. Senior management in small companies often start with preconceived ideas of what they think they intuitively know consumers will want. This is flawed thinking. The misconception (not unknown in large companies) of what consumers want frequently has its genesis in a feeling ("a gut feeling") a member of senior management has for some pet product idea: "It will sell. I know it. I have this gut feeling." They may very well be correct, but they also have a very good chance of being wrong.

Marketing strategists have a misconception that is somewhat more prevalent in technology-dominated food companies that technologically advanced products will sell. Technologists have convinced the rest of the development team that gimmickry in packaging or product disguised as novelty or innovation will draw customers. Wrong! Customers determine what will sell if and when their and their consumers' needs are satisfied. Marketing department task is to discover, through market research, what these needs are. That is their strategy.

4.4.2 Market Research

The difficulty is to find the common denominator that governs the actions of men.

W. Somerset Maugham, *A Writer's Notebook.*

Market research is an organized and unbiased investigation to

- Measure qualitatively and quantitatively all factors influencing the marketplace in which a company's market niches, targeted customers, and consumers are found
- Provide information from the data gathered to guide the strategy and tactics of the development team in decision making

A subtle oxymoron, "organized and unbiased" stands out like a sore thumb; herein lies the potential weakness for companies in the interpretation of market research data. If any investigation is organized, a priori, some bias is introduced by the researcher. It is the market researcher who

- Selects the populations to be researched
- Selects the research techniques used in studying the selected populations
- Prepares the wording of questionnaires and other survey materials
- Analyzes and decides how the data are to be weighted and interpreted

> In the preconscious process of converting primary data of our experience step by step into structures, information is necessarily lost, because the creation of structures, or the recognition of patterns, is nothing else than the selective destruction of information.
>
> **Stent (1987)**

Biases are unconsciously (sometimes consciously) introduced by market researchers, whether internal or hired. Market research companies do not want to antagonize their clients and lose them, especially those clients who are determined they know best. When clients show great affection for pet projects, the research company might feel obliged to shape research methods or to interpret findings to massage their clients to continue a liaison with them. This is not a condemnation of market research firms or of techniques for market research, the latter of which are becoming very sophisticated. However, an objective assessment of the reality of the marketplace is obtained only with careful and close liaison between the company's own marketing and sales personnel and the other members of the development team and the outside market research company. Interpretation of consumer data is highly subjective.

A further danger respecting market research arises when expansion into foreign markets or even new geographic areas within the same country is being considered. Other marketplaces are not extensions of the domestic or local market and cannot be considered as anything else but a new and unknown market with equally new and unknown customers and consumers. The buying habits of customers are different no matter how close the countries may be geographically, culturally, and linguistically. Countless times my American clients have plaintively asked me, "But why are Canadian food laws and regulations different from ours?" or "Why is that flavor popular in Canada? It isn't popular in the States." Or more commonly, "Why do Canadians celebrate Thanksgiving at a different time than we do?" (this said in response to discussions on timing of promotions or arranging visits to their clients). Developers must recognize that food laws, customer and consumer habits, and their tastes differ from country to country. Habits differ from region to region within the same country (e.g., both what constitutes the classic pizza and chili sauce are highly variable across the United States). A classic example of these differences was demonstrated when Snapple® appeared first in glass in Japan, where most such drinks are dispensed from vending machines for which cans are more customary for the Japanese.

Some cultures are reluctant to offend and may distort answers to questionnaires used in customer research and convey the wrong message to survey takers. Telephone surveys may be skewed when it is found that in some countries there is not a heavy concentration of telephones and, hence, fewer land line telephone users or that cell phones dominate.

There is an interesting aside to population and hence cultural differences that confounds market research: we are taught how we choose. People from communist countries have been found to choose differently from those brought up in freer societies. Parents nearly always choose what we eat or what is acceptable to eat. They impose cultural norms on what constitutes "food." While we may grow up to dislike certain foods that we were made to eat, we nevertheless, accept them as edible. In some cultures, adults would not think otherwise than to choose foods sanctioned by their parents.

4.4.3 Time: A Critical Element in Marketing Planning and Development

The time necessary for development confounds market researchers in two ways:

1. Volatility of the needs or wants of those originally surveyed: their opinions about product concepts may have changed subject to local, regional, or national happenings.
2. Aging of the targeted population: people have different needs, preferences, and opinions—for example, tweens, if targeted, can have very different preferences as teenagers in just 1 year. All ages display this change with the passage of time and experience.

Suppose, for example, that an analysis of demographic data predicts that the population in a given market area will drop by a million people over a period of 4–5 years; the consequences of that decline are that a million meals or meal items will be lost each day of each year in that market area. Such losses are not uncommon as the economies of communities change. A company has less than 4 years to find new products or develop new markets to fill the void caused by the gradual loss of those meal items as people move away. Market research must track these changes to keep development aware of any necessary changes. Four or five years is not an unusually long one for the development of some product concepts. In this time frame, new markets and marketplaces could evolve for which a total reengineering of the company and its objectives may be necessary.

When a new product is introduced, the originally surveyed consumers are figuratively and literally gone. They are older with, perhaps, different needs. Consider the tweens: if surveyed a year ago, then a year later, many have gone from their grade school into a high school that is fed from many diverse grade schools. There will be a new set of stimuli. The old surveyed consumers have been subjected to a totally different set of stimuli and were never surveyed. Those still in grade school are a year older with different values.

Market research must continue throughout the development process feeding new information to the team with which to shape the product to fit any necessary changes. Development takes time during which time customers' and consumers' needs, desires, and tastes change. What were once thought good ideas for products based on today's consumers can be short-lived fads and dead by tomorrow when they are introduced into the marketplace. Information about the consumers' volatility is best learned early in the development process through a continuing market research feedback before time and money are wasted developing the wrong products.

Developers are not aiming for today's customers and consumers in development. They are always developing products for tomorrow's targeted customer or consumer. Therefore, when products for development are conceived, they must be based on the best available demographic and psychographic data describing customers and consumers at the time of the product's introduction. The data only a few months old are old in marketing. Consider, for example, that the electronic communication market and its cognate social networking can change public opinion and tastes almost overnight—a trend or fad goes viral: this has implications of all consumer product industries including the food industry. The data, today's data, must be extrapolated to describe the needs of the consumer of the future; extrapolation of market data over time takes no account for the unexpected.

The marketplace is replete with change, not all of which can be extrapolated forward with certainty to define trends despite all the surveys, for example, the sudden emergence of nutraceuticals on the health food market and the fortification by food companies of their products with these. Then, equally suddenly, came the warnings over safety and overconsumption and government and medical concerns over their use. At best, one can only say there is observable confusion among customers and consumers about nutraceuticals and in the marketplaces in which companies are promoting nutraceuticals. There are very few facts to be found in the marketplace; observations, yes, but precious few facts. Claude Bernard, a renowned French physiologist wrote, "Observation is a passive science, experimentation an active science."

4.4.4 Nature of Market Information

Entities to be averaged must have homogeneity, a commonality. This was discussed in Chapter 3. Thus, the average customer or consumer does not exist. Any casual observation in any food marketplaces reveals that customers and consumers have little in common except the fact that they were there at the time the observation was made. This is a not very useful piece of information. One must look for common traits before averaging becomes meaningful.

Nevertheless, there are attempts made to categorize customers, to determine what traits consumers have, or do not have, in common, and whether these can be classified to advantage by marketers. The observation, for example, that "the average consumer is a white, female between 23 and 35 years of age, etc...." conveys very little useful information for either market researchers or product developers.

Market research companies categorize customers and consumers in their own way, and there are many such classifications. MacNulty (1989) arrived at three main groups in one such categorization. People, particularly those in the Western world, were classified relative to their personal values and motivations:

- Inner directeds: These are "highly confident, self-determined individuals." They are best described as individualistic and as leaders.
- Outer directeds: According to MacNulty, "they want others to see how well they are doing." They need to be seen as having status. They are less confident and inclined to be followers.
- Sustenance driven: These are more safety and security conscious. They like the comfort of the traditional and avoid risk taking. They are very conservative in their buying habits.

Market research is, therefore, directed toward the further characterization of the needs of the inner directeds, the leaders, for new product ideas. MacNulty describes at length as the pacesetters and "will provide us a picture of what the outer directeds will be doing in the next year or two." In a little more time the sustenance driven will catch up to the outer directeds who will in turn be trying to catch up to the inner directeds.

Bass Taverns, a British brewer and food service giant, used a combination of fuzzy logic and search algorithms to locate their taverns and restaurants according to their consumer classification (Davidson, 1998). Eight consumer classifications were apparent:

1. First tasters: young, affluent, and adventurous (perhaps identified with MacNulty's inner directeds).
2. Blue-collar hunters: often unskilled laborers who prefer a local pub with arcade games, bright lights, and canned music. Simple tastes run to draught ciders and ordinary lagers.
3. Premium wanderers: these are singles on the prowl and out for enjoyment. They usually belong to the bottled beer drinkers.
4. Pints and pensioners: an older, limited-income crowd who stick close to home for a quiet social evening.
5. Student crowd: a younger pint-and-pension crowd on a limited income and stay close to campus.
6. Quality diners: these have good incomes and a love for good food and are willing to travel for it.
7. Cards-and-dominoes set: a group that wants only the quiet sociality of the local pub.
8. Nightclubbers: these are entertainment-lovers with the money to afford it.

Yankelovich Partners (2000) broke U.S. consumers into eight categories:

1. Up and comers: young, upbeat, and childless "yuppies" (young, urban professionals) who lead the market and are not led by it. They tell marketers what to do.
2. Young materialists: these are single people who cynically equate money with happiness.
3. Stressed by lifers: people, often parents, with heavy burdens.
4. New traditionalists: forward-looking, family-oriented upscale people who "set the agenda for boomers."
5. Family limiteds: for these, families are central to their interests.
6. Detached introverts: these are successful, moneyed but lonely "geeks."
7. Renaissance elders: seniors who are wealthy and have a zest for life.
8. Retired from life: uninvolved seniors who "hear you knocking but you can't come in."

These three categorizations demonstrate the diversity of classifications that market research companies will use and the distinctiveness of consumers.

The diversity of potential market niches becomes very apparent. Currie (2008) defines two categories of customer found in the singular niche of supermarket private label brand buyers:

1. Primary: the decision makers of the purchase who control the purse strings (cf., the gatekeeper concept).
2. Secondary: the users or consumers, but not the purchasers, of the product.

In this book, the Currie's secondary customer was more clearly distinguished as the consumer and the customer–consumer (one who was both purchaser and user) introduced. Knowing the characteristics of the many niches to be found provides direction for screening and developing ideas that are suitable for the consumers that marketing people have categorized.

> Culture is smitten with counting and measuring; it feels out of place and uncomfortable with the innumerable.
>
> **Jean Dubuffet**

4.4.5 Qualitative and Quantitative Market Research Information

Two methods of studying markets have already been discussed in Chapter 3: the Delphi model and the simple extrapolation of current trends. The dangers of following trends have been discussed. In the previous discussions, these were used as techniques for getting new product ideas. They can also be used to study markets and market trends.

Another novel way to study markets starts with the future, that is, a likely development to foretell the purchasing of food according to one's genomic type for optimal health and work backward thinking what must be in place in each successive step to have reached that future goal. A more amusing example comes from the Harry Potter book series by J.K. Rowling. In this, Harry gets a newspaper, *The Daily Prophet*, in which pictures move about as in a video. Now comes the iPad™ and the suggestion that the venerable newspaper, *The New York Times*, might be available on iPad with pictures moving à la *The Daily Prophet*. An idea of the future is presented and then figuring out how to get there from here. This is called normative forecasting (Jantsch, 1967).

Market research information can be qualitative or quantitative, or both. Qualitative market information is based on the interpretation of focus groups and interviews. These tools require subjective interpretation, a technique prone to subjective errors of interpretation. Quantitative market information is based on the statistical analysis of data obtained from surveys and questionnaires and direct measurements of what customers buy, when and where they buy, and what combinations of purchases are made. They record historical data. Quantitative research techniques aided by computer-assisted analyses of the data are the norm. Techniques employed in researching the market (and there are many) require that the researcher clearly understand what information is wanted and users of this knowledge understand its limitations.

4.4.5.1 Focus Groups

Focus groups are a commonly used research tool for developing product statements or concepts. A focus group is an assemblage of consumers sequestered in a specially designed and equipped room. Typically between 8 and 12 people, selected as representative of the target consumer that the new product development company wishes to reach, participate. Market research companies keep lists of consumers whose backgrounds are well documented. With little effort, they can enlist consumers with any desired profile that the client wishes to have participate. The samplings of respondents are not randomly chosen and represent those the market research company has on file that meet the description of the targeted audience.

The group is led by a moderator—a professionally trained discussion leader. This moderator leads the group through a discussion aided by a prop to stimulate a more focused discussion. The prop is either a simple description of the client's proposed product or descriptive artwork depicting the proposed product, or a prototype product for the group to see. The moderator elicits comments from the group about this proposed product always probing their reactions and always gently pushing to get more focused attitudes to the prop. Sessions usually last 2–3 hours. Clients can investigate the impact of a variety of product concepts on several different consumer profiles through these groups.

The rooms are equipped with cameras and voice recorders and are often paneled with one-way glass allowing outsiders to note the reactions of the group without themselves being seen. Proceedings are audio taped and filmed to enable consumer research companies to analyze the group's oral and body language for hidden clues in the responses. Usually, no more than three or four focus groups with different individuals are required before a clear, concise concept statement for a new product emerges that embodies what the product is and how it will meet the needs and expectations of consumers. Often, representatives of the client company are present as observers.

There are as many variants of the above as there are market research companies. The technique masquerades under various, sometimes quite obscure, esoteric names such as real-time knowledge elicitation groups or simulated test markets (a form of focus group). They all have in common a trained discussion leader who focuses the discussion of a group of targeted consumers with some food concept embodied in a prop. The leader later collates the reactions of the group to the prop and reports the findings to the client.

Some consumer research companies use the same consumers over a 2 day session of discussions to get a good definition of a concept statement.

The participants are paid and thus, for some participants, a focus is a source of income. This introduces a potential danger of bias in the results as participants may tend to please or agree with their moderator.

After a lecture at Concordia University in Montreal on focus groups, a marketing student related that she participated in focus groups regularly because they provided extra pocket money for her. As for her comments in these sessions, she said whatever it was obvious the moderator wanted to be said. It kept her on the recall list. This did not surprise me greatly after my personal experiences with focus groups conducted by a market research company in the San Francisco area.

Because of these experiences, I strongly urge companies to use any information derived from focus groups cautiously.

If done well, focus groups provide qualitative (never quantitative) information concerning customers' and consumers' interactions with the product concept. Done poorly, focus groups can be a waste of both time and money because they are expensive. The results can be distorted by "group-think" with one dominant person swaying other participants' opinions. A market research firm can manipulate and interpret the results to jolly a client along. Both Marlow (1987) and Cohen (1990) discuss the value of focus groups, how best to use them, and the mechanics of using them in product development. A caution is in order: a food company interested in using the focus group technique for market research should seek out a professional market research company with an established reputation. As Marlow remarks, focus groups have value for suggesting direction for ideas, but they cannot be the basis for business decisions. The main function of the focus group is to determine consumer reaction to product concepts and from this reaction to redesign the concept for products to be on target. Interpretation of the data is highly subjective.

4.4.5.2 Beyond Focus Groups: Neuromarketing—Invading the Consumer's Inner Space

Marketing research has changed drastically. Much of the previous has discussed the traditional techniques, surveys, closed circuit observation of customers moving through stores, and demographic and psychrographic data collection, that is, field work. It involved groups of customers. Market research has moved forward (or backward depending on one's point of view) to neurological studies on individuals to determine what pleases or attracts them.

My opinions of the frailty of focus groups in providing reliable customer or consumer data are probably obvious. Olson Zaltman Associates (undated) describe research methods such as mini-market tests, structured surveys, taste test comparisons, including focus groups, etc., which Olson Zaltman Associates describe as Type I market research, as providing only descriptive marketing information. These methods introduce a bias originating with the market researcher since the researcher is eliciting the reactions of consumers to a prop (a stimulus) presented to them. That is, the researcher is providing the stimulus (this is a critical complaint running through surveys—the questioner supplies the stimulus, the question). Therein is the fault. The researcher is forcing the consumer to focus on what the researcher wants information about. That researcher conducting the test is directing consumer's responses: this bias faults the results of the research according to Olson Zaltman Associates.

Type II research attempts to discover and understand the customers' and the consumers' needs, desires, and goals. The questions used and their responses are more complex but have the bias of being, still, directed questions, leading the interviewee. Some typical questions asked by companies are as follows: What do my customers and consumers want and how and where do my products answer these desires? What does our brand or product mean to our customers and consumers? Or this more unusual question "why do consumers use our product differently in different situations?"

Market researchers have attempted to overcome this bias introduced by the researcher. One of these methods is Zaltman Metaphor Elicitation Technique (ZMET); another is neuromarketing. Both are described here from literature available from their developers.

4.4.5.2.1 *Zaltman Metaphor Elicitation Technique*

Zaltman's technique, ZMET (Zaltman, 2000a), explores consumers' relationships with a brand or product. These are complex relationships that involve a multitude of constructs forming a mental model. This model Zaltman refers to as a "consensus map." To understand how consumers value a brand (or a product), this consensus map must be uncovered by an understanding of these various constructs making it up.

In ZMET testing the subject brings their own stimuli (e.g., a picture or pictures) to the researcher in response to a prompt. Consequently, the subject's responses are not couched in the nuances of meaning of the researcher's language nor in the need to develop such a language (Bone, 1987).

ZMET methodology is based on several established assumptions (Zaltman, 2000a). In practice, a group of 20–30 individuals are first selected. They have expressed an interest in, intimacy with, knowledge of, or recognition of, the product or brand under test. The participants are told what the subject matter is and are required to bring a number of pictures (from magazines or newspapers) or to take photos (at least a dozen; camera supplied) that characterize their good or bad feelings and thinking about the topic.

The participants are given individual appointments for one-on-one interviews with skilled interviewers backed up with computer imaging specialists. The interview consists of several steps in which subjects relate their ideas and uncover hidden thoughts (metaphors) that describe very closely how the subjects relate themselves through the pictures to the product or brand. The constructs so developed are measures of this relationship. The example in the reference, using data from an actual test, provides greater understanding of the process (Zaltman, 2000a).

Two further studies (devoid of promotional advertising to be found in Zaltman, 2000a) may give further light on this technique. One by van Dessel (2005) demonstrates the technique with a sample of three students. The other study is by Vorell and Shulman (2004) in a study of racial profiling.

4.4.5.2.2 *Neuromarketing*

The BrightHouse Institute for Thought Sciences in Atlanta, Georgia, has developed a technique that they call *neuromarketing* for analyzing people's reactions to consumer products. They have used functional magnetic resonance imaging (fMRI) to study activity regions within the brain of subjects as they look at, study, or think of products, services, and advertising (Lovel, 2002). Researchers claim the fMR images reveal brain activity indicative of how the subject is actually evaluating a product, a service, or any advertising or promotion when viewing such object. Thought science researchers further claim that this knowledge is a more accurate measure of consumer preferences than focus groups or surveys.

Prior to a test scanning, participants are surveyed to evaluate their responses (likes and dislikes) to a variety of food products, promotional materials, and other consumer goods. In this preliminary survey, the participants have identified their preferences, and the researchers have identified activity regions for liking, disliking, aversion, etc., in their brains. The subjects are placed under an fMRI scanner and then shown objects on a screen whereupon a brain picture is taken. Comparison of the picture displaying brain activity and the survey results enables researchers to find the preference center of that participant's brain. With such data, researchers believe they can help clients develop better products, services, and marketing campaigns (Brighthouse, 2002).

Brat (2010) describes the Campbell Soup Company's use of biometrics to determine people's reactions to soup and determine how to sell more soup. They measured such biometrics as skin moisture, heart rate, depth and pace of breathing, eye movement, and pupil width. Critics of these biometrics claim these reactions are too slow compared to what has happened in the brain. Proponents of fMRI claim their readings are immediate, but the biometrics tell only that a person has responded and do not measure a person's emotions.

4.4.5.2.3 *Summary*

Both ZMET and fMRI scans are new tools to measure consumer valuations of products, brands, or advertising. With no experience of either process, I cannot describe advantages or disadvantages for either ZMET or neuromarketing from a personal stance except to parrot what critics of these techniques say; criticism seems to center on the reluctance of many market researchers to let measurements on individuals, which ZMET and fMRI are, be extrapolated to represent the integrated thinking of groups of individuals. I have leveled criticism at taking several observations and averaging these down to an "average consumer." I must then be critical with the reverse procedure, taking observations of individuals and assuming by grouping them this is what the masses would do or feel.

4.4.6 Marketing's War Room

Earlier in this chapter, management's role in competitive intelligence gathering and its war room were described. It is in this war room that management gathers business intelligence about the general business climate it is in and about its competitor's activities and policies in particular, then formulates its own policies and counter activities.

Marketing, in addition to its market research, also conducts its own war room activities, perhaps in a small corner of management's facilities. It gathers market intelligence about customers, retailers, and selling locations.

Customer shopping habits have changed due in part to the computer age with its access to the Internet and telecommunication with its cognate social network systems providing instant messaging. Prior to social networking, advertising had to tell a story about the product. The message was all about the product. Social networking is killing that; advertising is now becoming a conversation between marketing and customer and consumer. Selling, buying, advertising, and customer service (product maintenance) have all been influenced by these tools (Table 4.3).

TABLE 4.3

Getting to Know Customers and Consumers in the Computer Age

Tools	Description and Use
Cookies	A note (small text file) on user's computer dropped by a Web server. It allows third-party advertisers to track the user via any site the third party advertises on. Hence they know the user's interests.
Web bugs	These are invisible surveillance devices used by governments and by big business to monitor Internet activity. They are 1-pixel GIFs (graphic image also referred to as 1-by-1 GIFs and invisible GIFS) acting like a banner advertisement that "talks" to cookies and reports back. The privacy concern is that they can track the user, but the user does not know they are present because the user does not see a banner ad.
"Spyware"	Software that is installed on a surfer's computer without their consent or knowledge. It is often part of free software downloaded from the Internet. They report back on the surfer's use of the Internet. Food companies may not use spyware, but they might possibly buy market information from companies that do.

A cartoon, *PC and Pixel*©, by Thach Bui and Geoff Johnson, depicted a customer reading a sign at the entrance to a supermarket. The sign read, "Here at MegaFoodCorp, you're more than just a customer... You're a completely predictable compilation of spending habits and product data." Privacy is rapidly becoming a thing of the past, and vast data banks of information are available for companies to target promotions, advertising, and products.

A dramatic and personal example of customer profiling or purchase profiling occurred to me:

I received a telephone call from a person with my bank's credit card fraud department (I verified name and source). He asked first if I had my credit card (I had it); whether the night before I had visited a particular sushi bar where I must have been quite a bon vivant and spent over $700 on a meal; and if I had that very morning spent over $300 on sunglasses. I had not done either. I was told to destroy my card at once and a new one would be in the mail immediately. The fraud department was alerted by the fact that *neither purchase fit my pattern of card use for purchases.* (I hate sushi.) In addition, after the second purchase my card had been used, unsuccessfully, for a pay telephone call. The latter action was a sure giveaway that the credit card thief was trying to see if the card was still active. It wasn't: my bank had cancelled it after two non-pattern fitting purchases.

Courtesy cards, privilege cards, and client cards (e.g., Air Miles® cards) provide the card companies with tremendous customer information: what purchases were made, where they were made, when they were made, the addresses of purchasers, and their credit rating. All this information was requested on the form when customer registered for the card. One value of these cards is the discouragement of comparison shopping; customers are often held by their desire to accumulate points for redemption. With data-mining software, a very complete picture of customers and their purchases can be obtained with very little effort. Service stations, food chains, and box stores requiring membership all can gain valuable customer purchasing data.

Web sites are doing this profiling for a profit, to sell a product or to target an audience. They maintain that it is actually to serve the interests of their customers, the surfers. They can tailor advertisements to suit the interests of their customers. Thus, they claim that they serve the interests of Web surfers and, at the same time, they better target potential customers for advertisers on the Web. This online profiling is justified as the ability to deliver the right message to the right people. It is no longer broadcasting a message but *narrowcasting* one.

Hot lines (1-800 numbers) and help lines provide companies with valuable customer and consumer information. As the last number is dialed and even before it rings at the company's reception area, reverse directories have identified the caller's location and name while search engines pull up as much information on the caller as possible (e.g., previous calls and subject matter, other contact interfaces). The receptionist is then armed with extensive information about the caller to provide any product service, suggest new products, or reroute the call to others more able to help. Even if the caller routinely calls from an office telephone, the software can eventually associate that business number with the client caller.

4.4.7 Marketing and Sales Departments

Monitoring the marketplaces for changes that would influence the course of development is the important responsibility for marketing and especially for sales departments. Their vigilance in noting activities of the competition or within the retail outlets in the marketplace alerts the team to reconsider how these changes might influence new market opportunities or alter products under development.

Marketing and sales staff develop the marketing and advertising strategies for products as they are still under development. With cooperation of technologists and the legal department (or an outside resource), they prepare label statements; print material for promotions; advertising copy for newspapers, radio, and television; and any product claims as well as recipe and usage suggestions. They oversee artwork for labels and copy for advertisements and media campaigns that will be used in promotions.

Marketing personnel consider what impact new products may have on a company's established branded products: these, after all, bring in the money and the positioning of these products in the marketplace and would be dangerous to lose. Will there be fragmentation of the market? If so, is this fragmentation good or bad? A milder version of a hot sauce caused a drop in sales of the bell-ringer hot sauce and flavored versions of the sauce brought even more fragmentation and competition from rival sauces. Fragmentation was bad, but the company no longer relied on a single bell-ringer product. Within the larger market, there may be several smaller markets—niches—that might be profitably exploited making fragmentation a good thing. The instant coffee market was profitably fragmented with the development and introduction of flavored instant coffees and to a lesser degree the use of instant coffee as a flavoring ingredient for baking and roasting.

Marketing and, in particular, sales members of the team, must evaluate what impact products may have on the retailer. The following example illustrates this concern:

> A micro-brewery (craft brewer) introduced a prestige beer with an old fashion, wired-on cork stopper. Mini-tests were conducted in near-by campus pubs. Complaints from bartenders and waitresses poured in: opening bottles was inconvenient; service was delayed; hand injuries resulted. Wires proved a hazard underfoot and the corks made good missiles. The beer was fine. The container was redesigned.

Much of the preceding has been devoted to getting to know customers, consumers, and retailers in order to be able to satisfy those needs and expectations of all parties involved for a new product and its successful launch. Data generate information when interpreted. Information, in turn, generates ideas. Quantity of ideas, not quality, is important at the initial stage of the process. Critical screening will eliminate bad ideas later.

The generation of product ideas based on needs and desires, that is, the "I want," must come first from within either the customers' or the consumers' psyches. This requires an intimate knowledge of consumers and their gatekeepers, the customers. The first cast of the net for ideas is cast wide. As both demographic and psychographic data about consumers are gathered and converted into information, the net is pulled tighter and tighter still with neuromarketing. A picture of the needs and expectations of a very specific group of consumers will emerge.

4.4.8 Marketability and Marketing Skills

Does the marketing department have the ability to market the product to the targeted consumer? Will the product require unique marketing skills to reach the intended consumer? These are self-examination questions. Not all consumer food product companies possess the marketing skills to introduce new products into, for example, the food service or into food ingredient markets that they are unfamiliar with. It may prove more profitable to sell a by-product of manufacturing to an ingredient supplier than to develop and manufacture it into an ingredient oneself. The contacts and experience with these markets are not available. Unfortunately, companies often are not cognizant of their weaknesses respecting these very basic marketing considerations. All have been implicated, with the benefit of hindsight, as causes for new product failures (Anon., 1971). Companies need awareness of dangers in unknown markets and should recognize their own shortcomings in their abilities in markets they are unfamiliar with.

Products that present major difficulties in marketing for a company fall rather loosely into two categories: (1) They are so far ahead of their time that they cannot be marketed (perhaps in time they will be) without an excessive amount of consumer education (an enormous initiation fee) or major marketplace reform or (2) they possess a negligible point of difference from existing products that consumers can neither perceive nor appreciate the difference or they are inferior to products already on the market or cost appreciably more than superior products already available. These products obviously should be screened out early in development.

4.4.9 Summary

Any of the criteria used to screen, that is, marketability, technical feasibility, manufacturing capability, and financial criteria, could be reason enough to abort development at any stage. Two of these criteria, marketability and technical feasibility, must be applied carefully and dispassionately. Probabilities of success in both these areas are based on subjective assessments. Those responsible for them can, and do, get emotionally attached to pet projects and are reluctant to accept the need to drop the project.

Being objective can be difficult for the product development team. Leadership must be enlightened and compassionate. It is, after all, in the "people business." At the same time, the leadership must be dispassionate in applying criteria in screening. It is here that leadership must be demonstrated.

5

The Tacticians: Their Influence in Product Development

5.1 Science and Technology in Action

> There is something fascinating about science. One gets such wholesale returns of conjecture out of a trifling amount of fact.

> **Mark Twain**

The departments of research, engineering, and manufacturing and their subdivisions are the tacticians. They carry out, among many other duties, the product development strategy of the company through a winnowing of product ideas as development progresses until only a select few that promise the greatest chance of success remain.

These same departments in small companies are amorphous, with less-defined communication lines. Often only the technically trained individual heads the quality control group and usually heads up research and development when any such work is in progress. The head mechanic stands in for the engineering department, and the plant manager becomes the manufacturing department with duties of raw product purchasing, packaging, and traffic departments. At Trappey and Sons, at one time part of the McIlhenny Company, the quality control manager performed new product development duties, handled purchasing of raw produce, and supervised the fermentation operations. Small companies usually do not have an active, ongoing tactical group with a new product development program in continuous operation. They are more reactive, responding when new product needs are apparent or asked for by management.

Large companies, on the other hand, may have wondrously complex infrastructures for organizing the tactical groupings of product development for their brands and different product categories.

5.1.1 Research and Development: Meeting the Challenges

The development process begins when ideas from all sources are gathered and noted. Next, these ideas are explored for their appropriateness for the company, their capability to bring in needed revenues, their marketability based on more detailed market research findings, and their technical and manufacturing feasibility. Then, product concept statements are prepared—there may be many concept statements for different product ideas developed, especially in large companies—for the ideas surviving to this point. Technical research and development process begins when culinologists (food technologists trained as chefs or vice versa) have rough concepts of products-to-be; now, basic kitchen-top recipes for these products can be formulated. Their sources are cookbooks, ingredient suppliers, or analyses of similar products in the marketplace. These basic recipes serve several purposes:

- These prototype recipes become props used by marketing personnel to refine concept statements for focus groups and serve as test samples for in-plant tastings.
- Rough financial cost figures are estimated through the purchasing department's efforts to find suppliers and cost materials.
- Engineering personnel get preliminary data of raw material properties and can anticipate possible problems in handling or processing these materials in plant.
- Food technologists research the problems associated with maintaining quality and safety of the product that might be encountered during subsequent processing, warehousing, and distribution and throughout the anticipated shelf life of the product.

Consumer opinions uncovered in market and consumer research, preliminary cost estimates, and foreseeable engineering difficulties of these prototype products eliminate some ideas and send other ideas back for recipe tinkering and rethinking. Eventually, clear, concise statements of several possible product ideas emerge. Usually, several products reach this stage as companies often keep several products in development stages.

5.1.1.1 Recipe Development and Recipe Scale-Up: Meeting the Challenge

The first recipes may come from home recipes, cookbooks, analyses of similar products on the shelves, or, more frequently, from the de novo creation by chefs based on consumers' preferred tastes. These are used for stove-top samples only. They are inadequate at this preliminary stage for use in pilot plant runs or for large-scale plant production. Family-owned small food companies have difficulty understanding why their wife's spaghetti sauce recipe or Mama's chicken soup recipe is not readily adaptable to commercial production techniques without extensive ingredient or processing or both modifications.

Two instances of this lack of understanding of problems associated with scale-up of home recipes come to mind:

In the first instance, a president of a medium-sized west coast food processor insisted a particular canapé spread made by his wife for home entertaining could be processed as is – just increase the quantities. Against the advice of all including the plant manager, a trial was made. Much production time was lost cleaning and chipping out enamel-like gunk from the walls of the steam-jacketed kettle and getting the pump functional after the first disastrous trial run.

To commercialize a cherished family recipe and have it meet the needs and expectations of targeted customers and consumers (who are not family) and at a price the customer is willing to pay require much research and development for scale-up by food technologists.

In the second instance, before the days of culinologists:

A large, east-coast food company had hired a prominent T.V. chef to formulate several frozen main course items suitable for factory scale production. The items were impossible to prepare in quantity on commercial equipment with unskilled manpower, with readily available commercial ingredients, and within the cost parameters laid down for the products by management.

This fault was not the chef's. On the factory floor, one does not have a multitude of trained sous-chefs to assist in the preparation of dishes. The chef's dish preparation required more culinary skills than line people had. Nor can a company use the quality of ingredients chefs would use in their restaurants; they personally select produce from markets and often deal with growers they know. Plants buy produce according to grade standards. Factory operations must be such as to produce a consistently high-quality product rapidly and repeatedly at a price customers are willing to pay.

The need to modify family or cook book recipes arises for several reasons (Table 5.1). Two points in this table deserve the technologist's attention. First, some manufacturers are beginning to realize that consumers believe products with uniform quality are synonymous with mass production and therefore are associated with being highly processed foods:

I worked with a manufacturer of home-style Italian food products who credited the wide popularity of their products to their variable, but always high, quality. They did not want uniformity in their pour-on sauces with its attendant glossy, gloopy smooth appearance. They cultivated non-uniformity by allowing a degree of syneresis and lack of gloss in their product.

TABLE 5.1

Comparison of Difficulties in Using Home Recipes for Commercial Production Runs

Commercial Recipes	Home-Cooked from Family Recipes
Safety and stability	
Products must be safe with respect to all hazards of public health significance. Products must be stable with respect to their high quality attributes from factory to table.	Home-cooked product is cooked and eaten within minutes of preparation with no, or few, hazards associated with packaging, storage, warehousing, distribution, or retailing.
Costs	
Commercially prepared product must be kept within well-defined cost limitations (labor, plant overheads, labels, packaging, advertising, and promotions).	Home-prepared recipes have flexible budget allowances (with the exception of low income and poor families). Labor is free, and there are no overheads or packaging and labeling costs.
Ingredients	
Ingredients must be adaptable to mass production technologies for uniformity of processing characteristics (density, viscosity, particle size, thermal properties, etc.) and final product characteristics.	Ingredients and raw materials do not always have uniform characteristics. Home-cooked products often have a high degree of variability of size (e.g., spread of cookies), thickness and viscosity of sauces and gravies, and textures.
Volume of production	
Large-volume requirements of commerce require high-speed food-processing equipment.	Equipment and processing technology used in the home could never produce the volume of product demanded in commerce.
Expectations	
Commercial products must consistently meet the needs and expectations of a wide cross section of customers and consumers.	Home-prepared products need only meet the expectations of family members and guests who have accepted that mama's cooking is the best and that the family's way of preparing anything is the proper, traditional, and correct way.

These imperfections that gave their products the appearance of being home-made made perfection in their opinion. Lightbody (1990) makes the point that "there is evidence to indicate that uniformity of appearance, texture and taste within some manufactured food products can be judged by some consumers as unattractive and, in some cases, as a sign of "excessive" processing." Many bakeries now try for this nonuniform appearance in cookies.

Second, Worsfold and Griffith (1997) and Daniels (1998) have seriously questioned the safe and experienced hands of home preparers of food. In their studies, they have noted abysmal home food preparation techniques. It demonstrates the need for developers to take this consideration of a home hazard potential into their safety design for new products. Today's home food preparers are more rightly called assemblers of meal items and are not skilful or knowledgeable in handling sensitive food products.

Basic recipes can be obtained:

- By deconstructing similar products that exist in the marketplace with such simple techniques as sieving, filtering, and microscopy to estimate quantities and identify plant material used. Such analyses can give an idea of the competitor's product costs. Consulting laboratories can provide this service for a fee.
- By exploring the vast amount of historical, classical, and scientific literature on foods that is readily available in public and academic libraries.
- By consulting with chefs whose experience with foods and flavors can provide working formulations.
- Commercial flavor houses usually have commercial recipes, using their flavor preparations, and will supply information about flavors for a fee. They are able to duplicate most flavors presented to them.

As development progresses, the limitations of the individual skills available to the company or the limitations of the available technology or both cause a modification of the product statement as certain product characteristics cannot be attained. The product appears to be diverging from the statement much to the consternation of the strategists. This differing between strategy and tactics (marketing and technology usually) is often unavoidable but must be contained before it disrupts cooperation between team members.

5.1.2 Spoilage and Public Health Concerns

With recipes based on the product concept in hand, food technologists and food engineers experiment with ingredients and processes to duplicate the desired attributes described in the product concept and required in the final product. The nature of the product dictates failures (sensitivity to spoilage or hazards of public health significance) such products are prone to. With awareness of this sensitivity, a suitable stabilizing system for a safe product of acceptable quality throughout the expected shelf life can be designed. Literature searches and food stability databases provide safe processing information on similar products. Safety and quality are designed into the product at the very beginning.

Products must be safe: they must not cause injury or disease to those consuming them or handling them. They must remain with their high quality attributes intact throughout their shelf life for economic and esthetic reasons. Spoiled products in the marketplace are a serious economic problem, but products unsafe with respect to hazards of public health significance are a far more important and serious concern than spoilage.

5.1.2.1 Food Spoilage Concerns

Margaret Hungerford's well-known comment "beauty is in the eye of the beholder" could very easily be rewritten by food technologists: "spoilage is in the eye of the consumer." The consumer decides whether a product is acceptable or spoiled. For consumers, spoilage is the failure of products to meet their expectations as promised by promotional material that convinced the customer to purchase that product. Spoilage in the consumer's mind is simply: "you promised me this, but the product I got didn't fulfill this promise. It …." The manufacturer can finish the previous sentence of the consumer with "was bad smelling," "was bad textured," "was unsightly," "was moldy," "didn't function as promised," "tasted nothing like…," "wouldn't set," "was cooked unevenly," and so forth to complete the consumer's reaction.

A novel example of the consumers' idea of spoilage occurred in southern California. Tortilla chips spoil either by losing their crispness or by going rancid. Moisture-proof packaging will stop the former and elimination of oxygen the latter.

I worked with a manufacturer of an up-scale tortilla chip snack. They had ascertained that rancidity was the limiting factor in the shelf life of their product, a determination based on an in-plant, expert tasting panel. Accordingly, the product was packed using a gas flush of nitrogen to combat oxidation. Rancidity was stopped. But loss of crispness became the limiting quality factor, the product lasted longer. As well, surprisingly, consumer complaints regarding flavor poured in. Apparently consumers liked the hint of rancidity in the product. With that gone, they stopped purchases, leaving product sitting longer on the shelves and losing crispness.

Technology misapplied or too little reliance on what consumers considered "quality" and too much on what expert panels considered quality was our downfall.

McGinn (1982) classified new products into three categories based on their stability:

1. Highly perishable foods: Acceptable quality shelf life is short and measured in days. Typical examples are refrigerated fluid and flavored dairy and soy milks and yogurts; fresh meats, fresh sausages, poultry, and seafood; delicatessen meats, fresh peeled and cutup fruits and vegetables, mixed salad greens, and delicatessen salads (slaws); fresh bread; and cream- or custard-filled baked goods.

2. Semi-perishable foods: An acceptable quality shelf life is calculated in weeks. Examples of this category are conserved meat products such as bacon, hams, and some fermented and semidry sausages; some bakery goods (fruit cakes); dairy products (natural cheeses); and potato and tortilla chips and other snack foods.

3. Highly stable foods: Acceptable shelf life of these products is measured in months or years. Dehydrated foods and food mixes, canned foods, many cereal products (flour, pastas, breakfast cereals), confectionery products (toffees, hard candies, and some chocolate products), jams, and jellies typify this group.

TABLE 5.2

Categories of Changes in Foods with Examples
of Such Changes

The Reaction Mechanism		
Physical	**Chemical**	**Biological**
Phase changes	Oxidation	Respiration
Evaporation	Reduction	Oxidation
Concentration	Hydrolysis	Autolysis
Crystallization	Condensation	Fermentation
Mass migration	Decarboxylation	Putrefaction
Irradiation	Deamination	
	Browning reaction (e.g., Maillard reaction)	
The sensible changes in foods		
Exudation	Off-flavors	Wilting
Separation	Off-odors	Softening
Precipitation	Discoloration	Discoloration
Clumping	Browning	Slime formation
Clotting	Fading	Clotting
Textural changes	Exudation	Exudation
Grittiness	Textural changes	Off-flavors
Staling	Container interactions	Off-odors
Toughening	Rusting	Toxin formation
Discoloration	Delamination	Senescence
Fading		Excessive cfu/g
Opacity		

A food product's shelf life and its safety are compromised if adverse storage conditions for the product pertain, if the integrity of the product's container is broken, or if abusive handling has occurred.

Defective product that is found in the marketplace can be devastating. Bad news spreads quickly, especially via the Internet and its associated social networking systems, and may put the economic damage of spoiled or defective product much higher with the growth of gripe sites (see Chapter 2).

Spoilage in food products (see "the sensible changes in foods" under Table 5.2) is a result of any, or all, of the following three causes:

1. Microbiological changes resulting from growth of microorganisms characteristic of the particular foods, of the ingredients added to these foods, or present in the environment in which foods were prepared
2. Biological changes caused by enzyme systems naturally occurring within foods that cause unacceptable quality changes

3. Nonbiological causes resulting from chemical and physical reactions between the various chemical species within a food matrix that are promoted by processing, storage conditions, and packaging materials

When products under development are similar in compositions to products presently in the marketplace, then it can be anticipated the new product will spoil in a similar manner as these known products. This knowledge clarifies what preservative systems will possibly be necessary to maintain the quality characteristics desired in the final product. Now, with this knowledge, suitable packaging to prevent recontamination in handling and distribution can be similarly guessed at. There is a constant juggling of activities and refinements to produce a product as close to the product concept as possible. Any changes in the product statement must be reported to marketers for their assessment of acceptability.

Obviously, marketing's desires for their product set constraints for technologists who must devise ways to accommodate the product description. How sensitive is the product to temperature fluctuations? There will be extra costs—and extra time—incurred by the tacticians to accommodate any unusual and innovative characteristics market people want in their product. Extra costs alter projected profit pictures; all the tacticians' extra cost efforts must be reported back to the strategists for their consideration. The criteria dictated by the product concept are not immutable; they must be adjusted for by the existing technology, and compromises may need to be made.

Not all the changes noted in Table 5.2 are undesirable. Desirable changes occur in the aging of whiskey; fermentation and lipolysis in cheese making, the maturation of wine; or the fermentation of vegetables (krauts), the hanging of meat for tenderization, etc. These reactions are carefully controlled to allow the product to reach a desired characteristic when it is at its best.

Some measures used to control or prevent the changes (Table 5.2) may be unacceptable to customers and consumers as well as marketing personnel or senior management. For example, the use of irradiation or chemical additives or biogenetically derived foods can arouse passions in customers and consumers that marketing people or senior management wish to avoid. If marketing personnel want to project a natural brand image for a product with a "green" label, then the use of additives would not be wanted.

Where a product as natural appearing and tasting as if freshly picked is desired but if that product is unstable and requires rigorous stabilizing processes, then some give-and-take between marketing personnel's product demands and the existing technology's capabilities is required. Each compromise affecting the product statement requires further marketing research assessment; is it still the product that will meet the targeted market's expectations?

There will be conflict marked by compromises and concessions between marketing personnel and technologists, strategists vs. tacticians, visionaries vs. pragmatists, as development progresses concerning the product concept.

In general, as processing technologies applied to a product increase its shelf life, the causes of instability alter from biological (largely microbiological) to physical or chemical causes. Short shelf life products (e.g., hamburger patties or pasteurized flavored milks) primarily spoil from microbiological causes, not usually for chemical or physical reasons.

On the other hand, UHT (ultrahigh-temperature-processed) milk has a long shelf life and is more likely to fail for flavor and grittiness (lactose crystallization): beef jerky, a product with a long shelf life, fails due to fat oxidation causing off-flavors. These unacceptable changes are accelerated by temperature fluctuations, light, and presence of oxygen and compounded by abusive handling throughout storage and retailing. Biological concerns as factors in spoilage are minimal but always present, for example, in the long-term storage of flour, lipases and lipoxidases cause losses of "oven spring" (McWeeny, 1980).

Chemical and physical changes that make foods unacceptable are somewhat more complex to treat. Some processing techniques used to delay or prevent chemical and physical changes are as follows:

- Chemical and biological reactions are slowed by cool storage temperatures: the result will be more costly refrigerated handling and storage, a factor affecting economic projections. Refrigeration can cause deleterious changes as solutes crystallize out and gels and emulsions destabilize.

- Homogenization prevents separation of oils: changes in mouth feel and color may or may not be desirable.

- Agglomeration and crystallization prevent caking of powders and improve their solubility.

- A judicious choice of ingredients—for example, choosing non-crystallizing sugars over crystallizing sugars—and the use of doctoring agents (additives) will control staling (e.g., moist cookies) and grittiness in products, but there will be a more complex ingredient list.

- Encapsulation separates and protects labile components from reacting during processing stresses and minimizes flavor losses.

The quality attributes of any food product deteriorate with time, but each attribute often deteriorates at a different rate according to environmental changes occurring during storage. That is, the spoilage mechanisms for flavor or color or texture, for example, can have different rates of reaction (Q_{10} values) (Labuza and Riboh, 1982; Labuza and Schmidl, 1985). At one set of environmental conditions, the color of a beverage may deteriorate faster than its flavor. At another set of conditions, off-flavors may develop more quickly. Stresses likely to influence rates of deterioration are temperature changes, light (especially the wavelength and heat of incident light), vibration, and moisture transfer within plastic packaged goods.

Colored glass bottles or the use of opaque plastic/paperboard/foil-laminated containers will protect beverages from light damage but not from the radiant heat that incident light might cause. Laminated containers, while providing excellent protection, are regarded as environmentally unfriendly and banned in some jurisdictions. Glass bottles are often part of recycling or reuse programs that retailers are not happy with since such programs require a deposit and return-to-retailer system.

The desire for naturalness can be thwarted by the desire for stability of color, texture, flavor, shelf life duration, and, most importantly, safety. It is now obvious that the product description and its packaging may change how marketing personnel wanted to present the product.

Others factors besides packaging that influence cost and profitability projections are as follows:

- Ingredients and raw materials that meet the demands of marketers and technologists for quality and functionality
- Availability of raw materials, their seasonality, and their variable costs
- The preservative techniques necessary to sustain the quality and safety of the product from the factory floor to the consumers' tables
- Confounding these is the ever-growing demand both by governments and the public for more hazard protection from tampering

All are subjects of discussion between marketing personnel and technologists to choose the most acceptable compromises that still fit the needs and desires of the targeted customers and consumers.

5.1.2.2 Microbial Spoilage

The determinants of microbial spoilage of foods can be classified (Mossel and Ingram, 1955) as follows:

- Intrinsic factors that are characteristics of the food, properties of components in its composition, its processing, pH, a_w, antimicrobial constituents, colloidal state or biological structure of the matrix, etc.
- Extrinsic factors that include all environmental factors surrounding and in contact with the food or its package
- Implicit factors that characterize the microorganisms involved, their population dynamics, with all the synergistic and antagonistic pressures they are affected by, and their nutrition and metabolism

The implicit factors are those developers must always be on guard against; does one know one's microbial enemies? There always exists the danger of the introduction of a *"harbinger of change in food safety"* (Wachsmuth, 1997). For example, the emergence of the microorganism *Escherichia coli* 0157:H7 and the *Listeria* spp. (Warriner and Namvar, 2009) has challenged traditional concepts of food safety and techniques to control them. Buchanan and Doyle (1997) discuss virulent strains of *E. coli* in detail.

There are three broad approaches for stabilizing a food are based on the above classification. They are

- Knowledge of the chemical and biological properties of the food, its ingredients, the processing system used, and the susceptibility of these to microbial challenge.
- Handling of hazards (temperature variations, relative humidity, sensitivity to light, abusive treatment of the packaging material, etc.) that may be expected throughout processing, filling, packaging, casing, palletizing, warehousing, distribution, and retailing.
- Experiential knowledge with similar food systems and from the technical literature of microbial hazards associated with similar foods and the growth characteristics of microorganisms implicated with these hazards. The growth characteristics used to control microorganisms are
  - Optimal growth temperature
  - pH requirement for its growth
  - Reducing environment (Eh) requirements for its growth, water activity requirements for its growth
  - Its specific nutrient requirements
  - Its sensitivity to competitive pressures of other microorganisms (Helander et al., 1997; Jay, 1997)
  - Sensitivity to antimicrobial agents (Roller, 1995; Helander et al., 1997; particularly interesting for the new antimicrobials derived from bacteria)

That is, an understanding of the properties of the product under development and knowledge of the most likely microorganisms to spoil the food allow one to prescribe the preservative techniques to be applied; in a nutshell, one simply changes conditions for its growth to conditions that are inimical to the spoilage microorganisms. This is done in a limited fashion with formulation changes, the use of permitted preservatives, and temperature and water control.

An example illustrates the application of this knowledge. Food technologists have determined that the expected cause of deterioration of a hypothetical new food product is an equally hypothetical microorganism that has the following characteristics:

- It grows poorly in pH conditions below 4.8.
- It does not tolerate water activity conditions below 0.9.
- It is a mesophile.

- It grows best in aerobic conditions.
- It is sensitive to sorbate as a preservative.

Food technologists can delay spoilage and increase the shelf life by

- Acidification of the product to below pH 4.8. This requires a reformulation that will affect taste. Acceptable or not to marketers and their targeted consumers?
- Reformulation with ingredients to lower the water activity of the product to below 0.9. This will alter texture, mouthfeel, and flavor. This represents a possible challenge to the product description.
- Control of extrinsic factors by presenting the product as a chilled product with refrigerated distribution or by vacuum packaging or by increasing the reducing potential of the food matrix or by considering modified or controlled atmosphere packaging or by employing either the first two or the last two options. These all challenge cost and profit projections.
- Addition of sorbic acid or its salts as a preservative. The preservative may not be permitted in the type of product. Marketing personnel's desire for naturalness is compromised.
- Heat treatment, irradiation, or ultrahigh pressure plus mild heat techniques to preserve the product. These will cause texture, color, and integrity changes, and irradiation may be unacceptable to the marketing department.

If suitable combinations of techniques are used, then several multiparameter methods of preservation are possible.

It is readily apparent that not all the methods will be suitable for conformation to the product statement or acceptable to marketing personnel for reasons outlined above.

5.1.2.3 Naturalness: Minimal Processing

Today's customers and consumers want safe, healthful, fresh-looking food products that keep their fresh characteristics of nutrition and flavor throughout their shelf life that, paradoxically, must be as long as, if not exceed, that of the natural, unprocessed product. They also want added value and convenience. Technologists attempt to meet this challenge with "minimally processed" foods. Meeting this challenge brings problems of cost increases and concerns about safety. Minimally processed foods actually require more technology input and perhaps more processing.

The processes (Table 5.3) associated with minimally processed foods produce as little damage as possible to the quality attributes of a food and extend the product's shelf life. Often, the techniques for minimal

TABLE 5.3

Techniques Associated with the Production and Manufacture of Minimally Processed Foods

Techniques For Minimally Processed Foods
Controlled atmosphere storage
Postharvest treatments
Clean-room technology[a]
Protective microbiological treatment (Helander et al., 1997; Jay, 1997)
"Hurdle" technology (see references authored by Leistner)
Nonthermal processing
Mild thermal processing
Packaging
Controlled atmosphere and modified atmosphere packaging
Active packaging
Edible coatings

Source: Ohlsson, T., *Trends Food Sci. Technol.*, 5(11), 341, 1994.

[a] Clean-room technology has not been discussed in this book as it is a somewhat limited, esoteric use and is mentioned only for reference sake. Further information can be found in the reference.

processing are used in combinations. Ohlsson (1994) has reviewed minimal processing from storage and postharvest treatments of foods. The challenge for food technologists is stated deftly by Ohlsson: "...very short shelf-life products require preservation methods that will prolong their shelf life, while long shelf life products require methods that reduce shelf life but improve quality."

Marechal et al. (1999) reviewed the application of thermal and water potential stresses mainly with some references to pH lowering, pressure increase, and temperature decrease on the viability of microorganisms in order to develop optimal kinetics for minimal processing. Their goal is to optimize minimal processes to obtain maximum preservation of quality characteristics of foods while maintaining adequate safety of the minimally processed foods.

5.1.3 Maintaining Safety and Product Integrity

Safety and quality are designed into new food products. For simplicity, the expression "stabilizing system" will be used for processing technologies that ensure safety of foods with respect to hazards of public health significance and that preserve their quality characteristics throughout their expected shelf life.

5.1.3.1 General Methods and Constraints to Their Use

The techniques to stabilize foods (Table 5.4) can be used singly or in combinations designed uniquely for a particular product's quality characteristics and based on the characteristics of the microorganisms of concern.

TABLE 5.4

Traditional Techniques to Stabilize Foods

Stabilizing Stress	Possible Mechanisms
Thermal processing	
Temperature > 100°C	Spore inactivation; vegetative cell destruction
Temperature < 100°C	Vegetative cell destruction
	Enzyme inactivation
Chilling	
Refrigeration	Slowing of microbial metabolic pathways
	Slowing of chemical and enzyme-mediated reactions
Freezing	Immobilization of water
	(Some microbial destruction)
Fermentation	Alteration or removal of a substrate
	Acidification
	Production of antimicrobial agents
	Overgrowth by beneficial or benign microorganisms
Control of water	(Partial) removal of water
	Humectants (control water activity)
	(Freezing)
Acidification	Hostile pH for microorganisms
	Suboptimal pH for enzymic reactions
	Preservative (specific property of acid)
	(Fermentation)
Chemicals	Specific preservative action
	Modification of a substrate
	Enzyme antagonist
	Specific chemical action (e.g., antioxidant)
	Control metabolic pathway
	(Control redox potential)
Control redox potential	Prevent senescence
	Prevent growth of some harmful microorganisms
Irradiation	Inactivates microorganisms
	Inactivates stages in metabolism
Pressure	Protein denaturation
	Disruption of cell organization

Some limits to which techniques can be used have been referred to in previous sections. Freezing may well be impossible as a procedure for preservation if a company has no experience in, and no desire for, a venture into frozen foods. Many packaging materials are restricted for use because they are not recyclable, and other packaging materials may simply be unpopular with environmental activists (Akre, 1991). Customer or consumer resistance to the use of irradiation, chemical preservatives, or genetically modified ingredients can deter the use of these technologies, and senior management may be sensitive to upsetting customers. This reluctance effectively screens what ideas and products can be followed up.

Many new developments for stabilizing food products have emerged, and technical advances in older, traditional technologies have improved their usefulness in making high-quality products. Some of these advances will be reviewed here, not because they are the newest or most novel but because they show promise in providing consumers with exciting new products in new market niches.

5.1.3.1.1 Thermal Processing of Foods

Technologists have long known that transferring heat rapidly into food and removing it rapidly results in less heat damage to the quality and nutritional characteristics of the food during pasteurization or commercial sterilization. Higher temperatures to inactivate heat resistant microorganisms or spores led to shorter processing times—if the thermal properties (conductivity and diffusivity) of the food permitted the rapid movement of heat into and out of the product.

5.1.3.1.1.1 Continuous Flow and Swept Surface Heat Exchangers The use of continuous flow plate or swept surface heat exchangers allows HTST (high temperature short time) or UHT (ultrahigh-temperature) processes to rapidly heat foods to high temperatures. Rapid cooling in the heat exchangers followed by aseptic filling into sterile containers completes the process. There is a problem; residence time in the heat exchanger for the convection heating fluid component of the food is much less than for the thicker, conduction heating particulate component of the food. Therefore, foods with large, irregular-shaped particles do not have a sufficiently long residence time in the hot portion of the exchanger and may not be adequately heat processed.

5.1.3.1.1.2 Agitating Retorts and Thin Profile Containers Thermal processes for foods are drastically reduced if heat penetration is speeded up by agitating the food while it is in its container in the retort. Many commercial retorts can now agitate can contents by an end-over-end fashion, axially or in a rocking motion. As the container is agitated in the retort, the headspace as well as food particles in the container moves through the food, mixing the contents and thereby assisting heat penetration to cold spots. A shorter thermal process is obtained with improved product quality. With the faster rate of heat transfer, higher processing temperatures can be used with still shorter process times.

In general, agitating retorts employing end-over-end can rotation or a rocking motion are batch-type retorts. Retorts that provide axial rotation to agitate the container's contents are usually continuous-type retorts. Production speeds and manpower requirements depend, therefore, on the type of retort used.

Altering the geometry of the conventional cylindrical can to speed up heat penetration will also minimize heat damage to the container contents still further. This altered geometry is accomplished with thin-profile (or low-profile) containers such as the (flexible) retort pouch, the semirigid container, or the larger institutional half-steam table tray. These containers present two broad surfaces separated by a shallow width for rapid heat penetration and cooling. Consequently, the contents of thin-profile contents are subjected to less heat damage (Brody, 2003; Chapman and McKernan, 1963; Rizvi and Acton, 1982; Tung et al., 1975). Quality of the product with respect to color, flavor, nutrition, and integrity of particulates is greatly improved.

The retort pouch first became a very popular container in Japan for thermally processed sauce-based products such as curries or spaghetti sauces or stews (Saito, 1983) and is certainly well-known in military rations (Mermelstein, 1978; Tuomy and Young, 1982; Lingle, 1989; Brody, 2003). It has not received great success in consumer markets in North America despite its many advantages (Mermelstein, 1978; but see Brody, 2003). These packaging containers cannot be used on conventional metal can or glass lines.

Developments in thin-profile containers and retorts able to agitate the can contents have given thermal processing a new appeal for the production of added value, high-quality gourmet products that are gaining acceptance as main course items, particularly in the food service industry (Adams et al., 1983; Eisner, 1988).

5.1.3.1.1.3 Ohmic Heating   Ohmic heating occurs by passing a low frequency alternating current through an electrically conductive food. The resultant heating depends on the electrical conductivity of the food. Since neither convection nor conduction heating plays major roles in ohmic heating, there are no large temperature gradients between the fluid portion of the food and any large particles in it; all heat at roughly the same time. The size or other physical properties (e.g., thermal diffusivity) of particulate food have less effect on heat penetration (Biss et al., 1989; Halden et al., 1990; Selman, 1991). Packaging is, of course, done aseptically. Biss et al. (1989) have described the development of ohmic heating, and Sastry and Palaniappan (1992) have discussed applications for liquid and particulate mixtures.

Effective ohmic heating requires that the product be an electrical conductor; this is not a problem since most foods have water contents in the 30%–40% range with dissolved ionic constituents present. Nonionized food components, for example, fats and oils, sugar syrups, alcohols and nonconducting solids (bone and cellulosic material), are heated only indirectly by ohmic heating. Thus, not all products are suitable for ohmic heating.

Ohmic heating promises to rival plate heat exchangers as a stabilizing system for rapidly heating foods containing nonuniform particulate material and liquid. Ohmic heating overcomes many of the problems (burn-on, for one) encountered with continuous flow through plate heat exchangers. There is minimal damage to temperature-sensitive quality characteristics.

Several observations suggest that developers should proceed cautiously in establishing safe processes for products using ohmic heating (Halden et al., 1990; Parrott, 1992; Sastry and Palaniappan, 1992):

- The electrical conductivity of meats varied only slightly with temperature, but preheating increased the electrical conductivity of some foods.
- Processes must be designed to sufficiently heat-treat the slowest-heating, that is, nonconducting, food component.
- There were very specific plant tissue responses to ohmic heating as Halden et al. (1990) observed with aubergines and strawberries. Such unexpected responses should be a caution to those wishing to explore ohmic heating.
- Starch gelatinization caused a change in electrical conductivity.

Halden et al. (1990) caution that "electrical conductivity data from sources other than ohmic heating must thus be treated with care when designing an ohmic process."

5.1.3.1.1.4 Microwave Heating   Microwave heating, like ohmic heating, is an internal heating process; that is, food heats from within. The two differ in that microwave heating is by both conductive and dielectric heating while ohmic heating requires that the food be (electrically) conductive.

There have been suggestions that microwaves have a sterilizing effect by themselves and quite apart from their heating power. Mertens and Knorr (1992) leave the impression that they believe with others that the inactivation of microorganisms is primarily the result of the thermal effects of microwaves. Nevertheless, they do state, "If we assume (deleterious cellular effects) are real, it is difficult to imagine how these sub-lethal and long-term effects can be upgraded to a useful food preservation method."

Four very interesting presentations on the use of microwaves for pasteurization and sterilization were given at IFT's Food Engineering Division symposium at the Annual Meeting of the Institute of Food Technologists in 1992. Schiffman (1992) described the early history of microwave processing and posed reasons for its less than enthusiastic reception by the food industry. Harlfinger (1992) and Schlegel (1992), both representatives of commercial equipment manufacturers, described the basics of their respective companies' equipment, but it must be remembered that this describes equipment designed two decades and more ago although not much in equipment design has changed since then. Datta and Hu (1992) reviewed quality characteristics of foods processed with microwaves. More recently, Clark (2002) describes microwave equipment displayed at IFT's Food Expo.

5.1.3.1.2 Nonthermal Processing of Foods

5.1.3.1.2.1 Stabilizing with High Pressure The history of, and developments in, the use of high pressures to stabilize food systems has been reviewed by Farr (1990), Hoover et al. (1989), and Hayashi (1989). The technique, nearly 100 years old, was first reported by Hite in 1899 for the preservation of milk (reported in Hoover et al., 1989). The pressures used are in the order of 3500 atmospheres to nearly 10,000 atmospheres (1 atmosphere = 14.696 lb/sq in. = 1.033 kg/sq cm).

Changes occur in isolated proteins from 1000 atmospheres and up, but Dörnenburg and Knorr (1998) in their review discuss pressure-induced responses in plant tissue with pressures as low as 50 and 100 MPa (MPa is million Pascals; this range is roughly the range 500–1000 atmospheres). My experiences with high pressure processing to stabilize a fresh-pack salsa produced undesirable, glassy appearance in onion tissue and a faint but distinct flavor change that my client found undesirable.

Okamoto et al. (1990) studied the effect of combinations of different pressures and temperature and duration of pressure application on egg and soy protein solutions to produce gels. They found considerable changes in the gels formed respecting softness, adhesiveness, and cohesiveness; these varied with the amount of pressure applied.

Based on these observations, Okamoto et al. (1990) suggested that the mechanisms of gelation caused by the two techniques were different: pressure-produced gels were softer than heat-produced gels. Taste and color were unchanged in the pressurized foods.

Messens et al. (1997) reviewed the effect of high pressure on milk, meat, egg, and soy proteins with process parameters such as pressure, time, temperature, protein concentration, pH, and the presence of salts to produce new textures and tastes.

The changes observed in protein foods with high pressure techniques should also produce alterations in the proteins of microorganisms. Metrick et al. (1989) studied the sensitivity of *Salmonella senftenberg* and *Salmonella typhimurium* to high pressures using two different media. Cell injury and death occurred in the pressure ranges studied and increased with increasing pressure. Inactivation was greater in the phosphate buffer medium than in the strained chicken baby food medium. The more heat resistant strain of *S. senftenberg* was more pressure sensitive than was *S. typhimurium*.

Several caveats emerge for developers using high pressure to develop safe food products:

- High pressures significantly affect different proteins in different ways.
- Properties of pressure-coagulated protein are different from those of heated proteins.

- Food matrices influence the survival of microorganisms subjected to high pressures. High pressures appear to be more efficient in inactivating cells in acid pH than in neutral pH (reported in Hoover et al., 1989), but other factors in the food matrix may prevail (Metrick et al., 1989).
- Different strains of the same bacterium (Metrick et al., 1989) may have different sensitivities to the same lethal doses of pressure. That is, pressure lethalities (cf., z values in temperature studies) for different microorganisms vary.

Hayashi (1989) claims the advantages of high pressure for stabilization are the avoidance of heat damage and the preservation of natural flavor, taste, and nutrients (but see my observations earlier in this section). Hydrostatic pressure is transmitted instantaneously and uniformly into food, unlike heat transmission where thermal conducting properties influence the rate of transfer. High pressure has been used to kill insects as well. High pressure is also used in conjunction with other stabilizing methods such as acidification, antimicrobial agents, or mild heat to stabilize foods.

In Japan, jams, with their high acidity and solid content, are stabilized with high pressure, and surimi from different fish sources is gelled with high hydrostatic pressure (Farr, 1990).

Vidacek et al. (2009) used high pressures to determine its effect on *Anisakis simplex L3*, a parasite of hake. They found that the larvae of this parasite were killed, but there was still a danger for allergenic reactions for some consumers. They also noted small color and textural changes on the hake flesh.

Tiwari et al. (2009) studied various nonthermal processing techniques such as high hydrostatic pressure on the anthocyanin content of fruit juices. They discuss the mechanism of destruction of anthocyanin and possible ways to enhance its stability and hence improve color and phytochemical content of juices.

Knorr et al. (1998) improved the quality characteristics (flavor, lessened amount of thaw loss, color) by using pressure to freeze and thaw products. They describe the need for much more accurate instrumentation to verify the kinetic data on crystal formation as well as the effects of pressure freezing and thawing on the food systems to which it is applied.

In other work, Dörnenburg et al. (1998) reviewed the effect of high pressure processing on the production of a plant stress response hydrogen peroxide on the triggering of anthocyanin synthesis, on enzymatic browning, on the presence of polyphenol oxidases, on cell membrane integrity, and on texture as a function of polygalacturonase and pectin methylesterase activation. Their substrates for these model systems were plant cell cultures of grape, potato, and tomato. This article is part of a series begun by Knorr (1994) to demonstrate the feasibility of using plant cultures to study the effects of processing stresses.

Information on an equipment supplier is available in Clark (2002).

5.1.3.1.2.2 Control of Water: Water Relationships in Stabilization Water is necessary for enzymic reactions. It brings chemically reactive components together, brings nutrients to microorganisms, and permits the movement of motile bacterial species. Consequently, controlling the movement and availability of water in a food is a tool to lengthen its safe (with respect to hazards of public health significance), high quality shelf life.

Water is controlled by

- Removing it from the food with dehydration by heat, filtering, pressing, or reverse osmotic techniques
- Immobilizing it within foods by chilling, freezing, or adding solutes that bind water

To control water, it is obvious that naturalness will be compromised. Both technologies risk altering shape, form, and flavor in some manner.

Jezek and Smyrl (1980) describe the dehydration of apple slices by osmotic dehydration using a sucrose solution to remove water, followed by a further drying by vacuum. Advantages were an increase in apple volatiles and improved appearance of the slices. Silveira et al. (1996) studied the kinetics of osmotic dehydration of pineapple wedges followed by air or vacuum drying.

Tregunno and Goff (1996) used apple slices to study changes in the microstructure of apple tissue in a process they called osmodehydrofreezing. Apple slices were dehydrated with different sugar solutions, after which the slices were rinsed and frozen. They found the different sugars used in osmotic dehydration did cause different microstructural changes to tissues—an effect developers should note.

Trehalose is a sugar found in high concentrations in cryptobiotic organisms. Cryptobionts have a remarkable ability to survive harsh conditions, for example, the cryptobiotic resurrection plant, which, when fully dry, can withstand heating to 100°C and megarad doses of irradiation (Roser, 1991). Roser reviewed the use of trehalose in the drying of foods; the secret would appear to be the ability of trehalose (and other compounds with this ability) to dry as glasses rather than as crystals. Trehalose shows promise as an aid in producing superior dehydrated products.

MacDonald and Lanier (1991) reviewed the use of low and high-molecular-weight materials as cryoprotectants for meats and surimi and the mechanisms by which protection was obtained. Katz (1997) described and discussed innovations and application in water-binding technology and the water-binding activities of minerals, carbohydrates, and plant-derived products such as raisin paste and oat bran.

As awareness grew of how the control of water in foods could stabilize foods a new food form, semimoist foods also known as intermediate-moisture foods (IMF) was made possible. Early developments in IMF can be found in Davies et al. (1976), an excellent basic book on the subject.

The concept of water activity (a_w) and its measurement as a tool is comparatively new to many in food manufacturing despite the technology's availability for two decades or more. For developers, water activity control is, as Best (1992) aptly described it two decades ago, a minimal processing concept which is undergoing a major upheaval; it is still awaiting full acceptance as a tool.

Slade and Levine (1991) revised thinking on water relationships in foods when they introduced phase transitions as a factor in food quality and stability. Labuza and Hyman (1998) discuss the importance of moisture content and moisture migration to quality and safety in multidomain (multicomponent) foods at the macromolecular level (examples are dry cereals with raisins or a frozen pizza crust with sauce) and at the micromolecular level (examples are water in a starch granule or water in baked goods). Moisture migration causing the loss of crispness in chips, staling of bread with loss of crustiness, or crystallization is a function of the thermodynamics (water activity equilibrium) and dynamics of mass transfer (rate of diffusion of water) of the food system. The former is dependent on the water activity of the components in the multidomain food; the latter is dependent among other factors on pore dimensions within the components of the food.

Suggestions arising from Labuza and Hyman (1998) for controlling moisture to improve quality of foods are as follows:

- The components (domains) should be chosen with very similar water activities to avoid moisture transfer between them.
- Processing technology is required that assures that pore size of components is as small as possible and that pore size distribution is as narrow as possible.
- Viscosity within the components should be increased to inhibit diffusion and mobility of moisture.
- An edible barrier placed between components will deter diffusion.
- "Use ingredients with a high monolayer moisture content and high excess surface binding energy."

Practical examples of the application of these control measures are given by Taylor (1996) for the manufacture and packaging of sandwiches for the vending machine and retail trade and Cauvain (1998) for frozen bakery products.

5.1.3.1.2.3 Controlled Atmosphere/Modified Atmosphere Packaging Gases, used either singly or in precisely defined mixtures, provide a longer fresh shelf life to many foods in bulk storage and in unit packages. The gases, usually mixtures of nitrogen, oxygen, and carbon dioxide, control plant metabolic pathways that lead to deleterious flavor and texture changes within the tissues and slow or stop the growth of some microorganisms on plant material.

This property has given rise to controlled-atmosphere and modified-atmosphere unit packaging as technologies to prolong the shelf life of many fruits, vegetables, and meats. An overview of Controlled Atmosphere/Modified Atmosphere Packaging (CA/MAP) for fresh produce in Western Europe is provided by Day (1990).

Church (1994) reviews the different gases used in CA/MAP and various techniques of CA/MAP with their applications. He describes new development in what has been termed *intelligent* packaging. The new developments and their applications that are described are mixtures of oxygen, water, ethylene, and taint removal; edible films; oxygen barriers; gas indicators; carbon dioxide release; and time–temperature indicators.

Despite its many advantages, CA/MAP has not been as popular in North America as it has been in Europe. Reasons for this as suggested by Day (1990) highlight shortcomings not only of CA/MAP foods but also of chilled or frozen foods in general. CA/MAP foods are only as successful as the care and control in the distribution systems that are in place nationally and locally.

Anthony (1989) expressed concern about the adequacy of the then available gas-packaging technology, the distribution system to maintain proper cool temperature control, and customer and consumer acceptance of these products. Anthony also cautioned that product liability due to any abuse in the chilled food CA/MAP chain could result in serious product losses and the potential for risks of public health significance. Product developers should be cognizant of these concerns. Day (1990) echoed these concerns. Day (1990) and Anthony (1989) both stressed the need to control (see also Mossel and Ingram, 1955):

- Implicit factors such as the numbers of microorganisms present
- Extrinsic factors including the gaseous atmosphere and humidity in the package, storage, and distribution temperature, as well as the film composition
- Intrinsic factors that are characteristic of the food itself

Geeson et al. (1987, 1991) successfully extended the shelf life of some varieties of apples, but the same technology failed to extend the shelf life of Conference pears. This demonstrates that a preservative system successful for one product cannot, with certainty, be applied holus-bolus to another. Each product may have its own unique CA/MAP preservation system (intrinsic factors).

CA/MAP stabilizing systems are generally used in combination with other preservative techniques, for example, refrigeration. Fresh, sweet cherries, like many other soft fruits, would benefit greatly from extended shelf life both for marketing and for commercial processing. Meheriuk et al. (1995) stored Lapins sweet cherries gas flushed with a mixture of oxygen, carbon dioxide, and nitrogen and obtained acceptable quality characteristics for up to 6 weeks of refrigerated storage.

The greatest strength for CA/MAP might be for premium, added-value products.

5.1.3.1.2.4 Irradiation

"We cannot control atomic energy to an extent which would be of any value commercially, and I believe we are not likely ever to be able to do so." Baron Rutherford, the acknowledged father of the study of the atom, made this statement in a speech delivered to the British Association for the Advancement of Science in 1933. Rutherford's remarks notwithstanding, Lieber (1905) had earlier taken out a patent for the preservation of "canned foods, meat, beef extracts, and other manufactured or prepared foods, milk, cheese cream, and the compounds thereof, fruits, jams, juices, jellies, and preserves generally."

He preserved the foods by impregnating them with emanations from "thorium oxide," thus rendering the substances radioactive! Irradiation has come a long way since 1905 and 1933.

The mechanism for the effectiveness of irradiation (at the levels used in food processing) in inactivating enzymes and reducing the counts of microorganisms is discussed by Robinson (1985). The effect irradiation (for some levels used see Table 5.5) has on plant or animal tissue is dose dependent (Giddings, 1984; Gaunt, 1985; AIC/CIFST, 1989).

The irradiation of food at doses up to 10 kGy presents no hazards, toxicologically, nutritionally, or microbiologically, of public health significance (AIC/CIFST, 1989). A steadily growing number of countries have accepted irradiation as a food process (Loaharanu, 1989). Developers using irradiation for increasing the shelf life of new food products for the export market should be aware of regulations respecting irradiation in the importing country.

TABLE 5.5

Dose Dependency of Irradiation on Plant and Animal Tissue

Food Product	Dose
Sprout inhibition	0.15–0.2 kGy
Flour disinfestation	Up to 1 kGy
Spice cleaning	Up to 5 kGy for 2–3 log cycles of count reduction
Parasite elimination	Up to 6 kGy
Reduction of bacteria	Up to 5 kGy depending on level of reduction desired
Non-sporing pathogens	Up to 10 kGy

Sources: AIC/CIFST joint statement on food irradiation, Ottawa, 1989; Gaunt, I.F., *Inst. Food Sci. and Technol. Proc.*, 19, 171, 1985; Giddings, G.G., *Activities Rep. Res. Dev. Assoc.*, 36(2), 20, 1984.

Irradiation, for several years, has replaced fumigation to control infestations in fruit and vegetables; to prevent sprouting in potatoes; to sterilize sewage sludge for conversion into fertilizer; and to sterilize surgical equipment (Anon., 1981). Tsuji (1983) describes the use of low-dose irradiation to reduce microbial counts in fish-protein concentrate used as flavoring in vitamin/mineral supplements in veterinary products.

The advantages of irradiation for product development have been glowingly summarized by Josephson (1984) and AIC/CIFST (1989) as follows:

- Irradiation can replace chemicals used for preservation and disinfestation of fruit, vegetables, and grains (see Giddings, 1989) allowing wider distribution of improved quality tropical products in more northerly market places. It effectively eliminates parasites such as trichina in meat products and cyclospora on imported raspberries [IFST(U.K.), 2003; Lund, 2002; Strauss, 1998]; it reduces or eliminates *Salmonella* and other pathogenic microorganisms in poultry and disinfests plant cuttings, fruits, and other agricultural products.

- Irradiation can be used to sanitize sensitive, labile products such as pharmaceuticals that cannot be done by other techniques.

- Food can be prepackaged prior to irradiation. Neither shape of the product nor the form (liquid, solid, frozen, or powder) is a major factor in irradiation as these are in thermal processing. Irradiation in the frozen state prevents formation of ionic species that cause chemical changes. Freezing immobilizes ionic species.

- Irradiation is cost competitive with other conventional techniques for food preservation and is economical of energy consumption.

- Irradiated foods provide convenience and versatility to meal components as well as snacks and reduce preparation and labor in the kitchen.

Unfortunately, there is opposition to its use that developers who wish to use the technology must be aware of; they must be prepared for physical opposition and be prepared for a difficult period of education of the public to its use.

Consumer reaction to and acceptance of irradiation in North America have been and still are ambivalent. For example, irradiated Puerto Rican mangoes were enthusiastically received by customers in Florida according to press reports (Puzo, 1986; Bruhn and Schutz, 1989; Loaharanu, 1989). A comparison test of irradiated Hawaiian papayas and traditionally processed papayas showed that customers had no aversion to irradiation (Bruhn and Schutz, 1989; Loaharanu, 1989). In January of 1992, irradiated strawberries were sold in Florida, United States (Marcotte, 1992). Despite protesters attempting to disrupt the market test, customers bought irradiated product and were in general favorably impressed by irradiated strawberries.

Deep polarization between proponents and antagonists of irradiation is well documented (see Giddings (1989) for the "pro" side and Colby and Savagian (1989) for the "con" side). Acceptance of irradiation is slow but customer resistance is falling (Pszczola, 1993; Demetrakakes, 1998a). Irradiation trials for test packs have kept the facilities for irradiation busy at Vindicator, Inc. (Lingle, 1992).

Acceptance of irradiation is conditional according to Bruhn and Schutz (1989) on favorable answers to the following questions:

- Do alternative technologies offer greater or less safety than irradiation for the food supply?
- How does industry weight the costs of failure vs. the potential rewards of success in the introduction of irradiated foods?
- Will responsible media coverage accompany the introduction of irradiated foods?
- Have consumers been adequately educated to the value of irradiation for them?

Answers to these questions are still not clear enough that compel company strategists to be staunch advocates of irradiation, and retailers share this reluctance to fully endorse irradiation. Best (1989a) makes an excellent point that customers and consumers do not see irradiation as an advantage to them and as satisfying their needs. It is a process they hadn't asked for. Unless consumers see an added value and safety of irradiated raw meat, poultry, and other products, it is of no value to them (Frenzen et al., 2000).

At present, the opponents of irradiation argue that irradiation is an advantage to processors who can ignore sanitary control features and "zap" product for sterility (Coghlan, 1998). Manufacturers fear that few customers will be willing to buy irradiated foods, despite favorable receptions in test markets, and they fear reprisals if they undertake a public educational program. A recent study of Brazilian consumer to irradiated foods (Behrens et al., 2009) demonstrated a reluctance to accept irradiated foods that was considered largely due to lack of knowledge of the possibility of nuclear power to benefit people. This test in Sao Paulo was conducted with three focus groups of consumers who were made up of the gatekeepers; when presented with irradiated and nonirradiated sample, the researchers found barely perceptible difference between the focus groups or between acceptance of irradiated and nonirradiated samples. Processors and retailers must learn to handle protesters and the "bad press" they can bring with sound crisis management techniques. Unfortunately, some journalists still think themselves quite funny by feeding the public's reservations with comments that by eating irradiated food, they will "glow in the dark."

Consumers are uninformed of the benefits of irradiation to them. They see irradiation as a great benefit to the food industry (combating insect infestation of cereal crops, elimination of parasites in products, sterilization of foods, etc.) but only indirectly to themselves as customers and consumers. Unfortunately, most of the answers to the questions posed by Bruhn and Schutz (1989) are provided by those in whom the general public has dwindling faith. Such an attitude is disappointing to the tacticians involved in its development as a tool for minimally processed foods.

Irradiation installations are capital-intense facilities, and, hence, irradiating facilities will, in future, be few and widely scattered geographically. Herein is the problem: food to be irradiated must be brought to the facility. The very high capital costs can be better understood by a description of an irradiation facility. The United States' first commercial facility for irradiating foods, Vindicator, Inc., located in Mulberry, Florida, uses cobalt 60 as its source of gamma rays rather than x-rays from a machine-made source (Lingle, 1992). It is a wet-cell irradiator; that is, the cobalt 60 is housed in and shielded by an 18,000-gallon pool of deionized water in a 28-feet deep well. Palletized food enters the chamber; the cobalt 60 is raised out of the water to activate the irradiation process. When submerged, the source of gamma rays is thoroughly shielded, and workers can enter the chamber safely. Walls of 6.5-feet-thick steel-reinforced concrete provide further shielding. Cobalt 60 does lose its ionizing strength with time and must be replaced. Other facilities use electron beams. This brief description provides some idea of the complex and expensive structures that are necessary to house these facilities. Demetrakakes (1998a) and Baird (1999) both describe other commercial gamma irradiation plants.

Costs for irradiation vary according to the radiation source used, electron beams (x-rays) or some radioactive source, most commonly cobalt 60. Tsuji (1983) estimated the cost of irradiation and associated handling to hover around the 5 cents/pound range for a product. Estimates made by Loaharanu (1994) and Frenzen et al. (2000) reported costs of irradiation of meat or poultry at 0.5–1.5 cents/pound in a plant with a throughput of 100 million pounds using an electron-beam system. Irradiated food appears to be very price sensitive—a factor of major concern to developers. Only added value, heat labile products capable of benefiting from irradiation can bear the added costs. Developers now engage a Catch 22 situation: added value products are often more sensitive, labile products. To be trucked any great distances to a radiation facility will surely damage them.

Irradiation doses are restricted to pasteurizing doses (10 kGy or less). Irradiation, therefore, is often combined with other stabilizing systems such as refrigeration and CA/MAP systems.

Designing safety and stability into products using irradiation requires a major caution. Microorganisms have different resistances to irradiation doses; Paster et al. (1985) found fungi to be more resistant than bacteria to low-dose irradiation of pomegranate kernels in combination with nitrogen gas flush packaging and refrigeration. The red flag for developers here was that the course of spoilage was altered. The dominant fungal contaminant common in nonirradiated kernels was normally *Penicillium frequens*.

In irradiated kernels, *Sporothrix cyanescens* became the dominant fungal contaminant. Thus, one spoilage microorganism was removed only to be substituted by a different one.

Dempster et al. (1985) irradiated raw beef burgers in combination with vacuum packaging and refrigerated storage. Shelf life was extended at refrigerator temperatures and counts of microorganisms were significantly reduced, but the irradiated burgers lost their redness, showed an increase in peroxide values in their fat, and developed a distinct, unpleasant odor during storage. Not only can the microbial path of spoilage be changed but also the chemical path of deterioration.

Grodner and Hinton (1986) also noted alterations of conventional spoilage paths after irradiation of crabmeat to study the interrelationship of storage duration vs. storage temperature vs. irradiation (up to 1 kGy) on the viability of *Vibrio cholerae*. Sterile and non-sterile samples of crabmeat were inoculated with *V. cholerae*. Three interesting findings are discussed in their work that should demonstrate the complexity of multicomponent stabilizing systems using irradiation:

1. No *V. cholerae* were found in crabmeat pretreated by sterilization and then inoculated, or in non-sterile, inoculated crabmeat at dosages of either 0.5 or 1.0 kGy after irradiation. Irradiation at 0.25 kGy reduced the counts by several log cycles.

2. Pretreatment by sterilization of the crabmeat removed competing microflora to *V. cholerae*'s advantage compared to nonirradiated crabmeat stored at each of the storage temperatures.

3. Survival of *V. cholerae* was greater in crabmeat that had not been sterilized prior to inoculation in both irradiated and nonirradiated crabmeat stored at −8°C. Irradiation was, however, effective in reducing the counts. Grodner and Hinton suggested that pretreatment of the crabmeat caused the loss of some protective factor (perhaps a protein) against freezing, which was contributory to the survival of *V. cholerae* in pretreated crabmeat at the coldest storage temperature.

In each of the above, irradiation plus other stabilizing systems produced an unexpected result either in the path of spoilage or altered the food's character in some manner. The interrelationships between the application of newer stabilizing systems, the product, and its microflora must be understood to ensure safety. Careless application of the newer technologies of stabilization can result in an alteration of the traditional courses of spoilage or of the microorganisms responsible for spoilage as these were recognized by food technologists. Unless this is understood, developers may find they have replaced the devil they know with the devil they don't know.

Irradiation if used with caution offers great promise for developers of new, minimally processed foods with desired, added value features for customers and consumers.

5.1.3.1.2.5 Hurdle Technology   By judiciously combining several stabilization techniques selected on the basis of a knowledge of the dominant path(s) of spoilage or of pathogenicity for a product, technologists can stabilize that product. Each of the individual systems selected inhibits or stops a spoilage vector in the food to which they are applied. They are singly not sufficient to wholly stabilize the food. But each, in combination, complements the stabilizing activities of other systems. In this way, the high quality shelf life and safety of food products can be maintained. The technique has been hinted at in previous sections when irradiation was combined with refrigeration and CA/MAP.

Hurdle technology (multiple stabilizing systems) has been practiced for many hundreds of years; it is recognized now as the preservative technology behind many traditional foods such as

- Sauerkraut: acidified through lactic fermentation; spoilable substrate removed through controlled fermentation; salt addition to control the course of fermentation and to encourage presence of highly competitive but benign lactic fermenters; lactic fermenters may also provide natural antibiotics (bacteriocins).

- Pemmican: sun-dried meat (lowered water activity), acidified by addition of high acid berries.

- Smoked, fermented, hard sausage: dehydrated by heat of smoking treatment (lowered water activity); stabilization by smoke constituents (chemical preservative); acidified by fermentation (lowered pH and removal of a labile substrate); preservative action of the lactic fermenters (biological preservative); meat additives such as salts, spices, and herbs, some of which have preservative action (Beuchat and Golden, 1989).

- Cheeses (very similar to sausages in their stabilization systems): low water activity; fermentation to remove labile substrate; the presence of bacteriocins from lactose and other fermenters; competitive microflora; some traditional cheeses are also smoked and some coated with ashes.

However, it was not until Leistner and Rödel (1976a) referred to this combining of stabilizing technique as *Hürdeneffekt* (hurdle effect) that the technology was formalized. Leister and Rödel (1976b) provide a further general discussion of the hurdle technology.

The application of hurdle technology has been likened to an obstacle race. The course is set with hurdles of varying heights representing different intensities of stabilizing factors. The hurdles are designed to slow or stop chemical, physical, and biological spoilage reactions. Some microorganisms (the "runners") fall at the first hurdle. Those microorganisms that pass that first hurdle are weakened for the next hurdle, where more runners are felled.

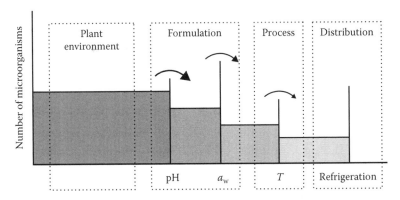

FIGURE 5.1
Pictorial representation of hurdle technology for a hypothetical food product formulated to have a low pH and low water activity.

Those that survive the first two hurdles are so further weakened when they face the third hurdle that they are in no condition to pass it or whatever hurdles that may remain.

Hurdles can be synergistic with one hurdle complementing the preservative action of another one such that the combined preservative effect of the two systems is greater than the sum of the two separately. The principles of hurdle technology are pictured in Figure 5.1 where a hypothetical food is shown preserved by four systems; that is,

1. Rigidly adhered to ingredient standards, purchasing from reputable suppliers, effective plant sanitation, maintenance, and HACCP programs, tight process control standards, and other plant support systems keep microorganisms in the plant environment and in the product low. These provide the first hurdle.

2. Product formulation introduced two built-in hurdles into the product: a lowered pH and an a_w of <0.90 that restricts the growth of spores of Clostridia and provides a further check on the growth of many other microorganisms.

3. Pasteurization eliminates vegetative microorganisms that can survive the lowered pH and lowered water activity and stops enzymic activities. The level of microorganisms is lowered even further.

4. Finally, the finished, packaged product is stored, distributed, and retailed at refrigerator temperatures to slow chemical and physical changes within the product.

The hypothetical product with the hypothetical microorganism contaminating it that was discussed under Section 5.1.2.2 above was stabilized with several hurdles: acidification; lowered water activity; chilled storage, vacuum packaging, or both; use of a chemical preservative; and heat treatment. If all hurdles were used, this would be overkill respecting stabilization.

The application of the concept of water activity, the enormous interest that developed for the hurdle concept, and its relation to food stability sparked the development of IMF. IMF have moisture contents between that of dried foods and fresh foods and whose water activities are above that of dehydrated foods but below that of natural foods. They can be consumed without the addition of water.

Davey and Daughtry (1995) developed a predictive model (a linear-Arrhenius predictive model—see reference for other predictive microbiological models) for the growth of bacteria using three hurdles—temperature, salt, and pH—based on a number of published growth curves of *Salmonella* spp. Cerf et al. (1996) developed a model based on published data on *E. coli* spp. also using three hurdles: temperature, pH, and water activity. They suggest this model has many advantages over previous ones including the property of additivity: that is, "the effect of environmental factors on bacterial inactivation can be summed."

Vega-Mercado et al. (1996) demonstrated the use of combinations of pulsed electric fields, pH, and ionic strength on the inactivation of *E. coli* in a simulated milk ultrafiltrate.

Sous-vide products, popular in the food service industry, use a multiple preservative system (series of hurdles) for their shelf life. They undergo a partial cooking, which both pasteurizes and blanches the product. Vacuum packaging inhibits aerobic microorganisms and protects the product from further contamination during distribution and retailing. These products are refrigerated or frozen, which provides further checks on biological deterioration. Product developers should, however, refer to the concerns of Livingston (1990) and Mason et al. (1990) regarding the safety of this food service process.

There are cautions to be aware of; Neaves et al. (1982) used a multicomponent system of stabilizers (sorbates and benzoates singly and in combinations thereof, salt or propylene glycol and combinations thereof, pH, storage temperature, and pasteurization) to inhibit the growth of *Clostridium botulinum* in a pork medium. Some combinations inhibited microbial growth and toxin production but required a lowered pH or reduced temperature. They found botulinal toxin could be present in a food when there was no obvious spoilage, and they feared inadequate preservation could result in the unwitting ingestion of toxic food by consumers. They suggested the possibility of inadequate preservation was far more dangerous than no preservation in a food system.

Webster et al. (1985) demonstrated the course of spoilage could be changed by the choice of hurdles. Their model system involved 72 hurdle interactions against a microbial cocktail consisting of *Staphylococcus aureus*, *Bacillus subtilis*, *Pseudomonas aeruginosa*, *Streptococcus faecalis*, *Lactobacillus casei* var.

rhamnosus, and *Clostridium perfringens.* The variables were four levels of a_w, four pH values plus the presence of sodium citrate and sodium benzoate. They observed that

- When a lowered a_w was the sole hurdle, the predominant microorganism was *S. aureus.*
- By adding another hurdle such as citrate and benzoate to that of reduced a_w, the dominant microorganism became *S. faecalis.*

In yet another example, Carlin et al. (1991) studied a range of controlled atmospheres using carbon dioxide and oxygen and storage at 10°C on the spoilage of commercially prepared, fresh, ready-to-use grated carrots (a product available on the French market). This storage temperature is not that recommended by the French government for such products but was the temperature at which such ready-to-use products were frequently sold. The growth of both lactic acid bacteria and yeasts was more rapid as the CO_2 concentration increased from 10% to 20% regardless of the oxygen concentration originally present. The changing composition of the atmosphere in CA/MAP products influenced the population dynamics of spoilage.

Chirife and Favetto (1992) found very specific solute and humectant effects in water activity, for example, (a) ethanol has an antimicrobial effect that is not due solely to its water activity-lowering ability and (b) sodium chloride and sucrose inhibit the growth of *S. aureus* around a value of 0.86, but when other solutes such as diols and polyols are used, growth is inhibited at a much higher water activity. Different combinations of hurdles may have complementary effects that alter the path of spoilage. Another disadvantage is that many humectants used to depress a food's water activity (see Table 2, in Chirife and Favetto, 1992) are precluded for flavor considerations.

Leistner (1985, 1986, 1992) and Leistner et al. (1981) emphasize that the hurdle effect requires that good manufacturing programs and HACCP programs, high levels of quality control, plant sanitation, and personnel hygiene systems must be in place and exercised scrupulously. Hurdles that are established for products are not designed to protect against high microbial contamination introduced from the plant environment prior to packaging.

Developers must understand fully the properties of the food product under development and understand the possible destabilization mechanisms that could lead to intoxication and spoilage. Any stabilizing system is a balance of stresses that is unique for that product composed of those particular ingredients and processed precisely in the manner it was.

The final caution is that the developer must understand the purpose and function of each ingredient or additive in the product and record these in the product description. Each serves a purpose; the safety and quality of the product were determined on the properties of these original materials used in its manufacture. Substitution of any original raw material, ingredient, or additive with another could seriously alter the safety of the product. Any reformulation needs to be undertaken carefully.

5.1.3.1.3 *Low-Temperature Stabilization*

Chilled, prepared foods (refrigerated processed foods of extended durability [REPFEDs]) were hinted at in the preceding discussion of hurdle technology and are discussed further in D.A.A. Mossel (Brackett, 1992; cited in Scott, 1987). Their growing acceptance by consumers justifies a separate discussion of their potential for use in new products. The major concern with these products is the growth of pathogenic psychotrophs coupled with weaknesses in the distribution and handling systems, that is, in transportation to, and conditions at, the retail receiving and display level or food service outlet. REPFEDs (e.g., sous-vide products) are used extensively in institutions such as nursing or convalescent homes or in restaurants where many people may eat the products, and, hence, controls to maintain their safety are of paramount importance.

REPFEDs are usually preserved using hurdle technology (Scott, 1987). In general, these products have four characteristics:

1. A sound HACCP program must be clearly identified to prevent contamination in their preparation and packaging. A very high level of quality control and sanitation is demanded in their manufacture to maintain as low a level of microorganisms as possible.

2. Packaging is usually done under vacuum (sous-vide) or under CA/ MAP conditions.

3. A pasteurizing level of heat is applied either during preparation of the product (partial cooking) or to the product in the packaged state. If the product is heated during its preparation, it is rapidly cooled and packaged as soon as possible, then chilled to prevent any microbial growth.

4. The products are stored, distributed, and displayed at refrigerator temperatures.

The design of stabilizing processes and their formulations for chilled food products must be such that potentially harmful microorganisms be adequately controlled. The realization that several pathogenic and spoilage microorganisms can survive and grow at good refrigerator temperatures reveals an inherent hazard in chilled foods. A recent outbreak of listeria intoxication in 2008 and 2009 in deli meat products produced in one plant but involving several brands resulted in the deaths of over 20 people across Canada.

The shelf life of REPFEDs is, as one would suspect, highly variable. It depends on the intrinsic properties of the food, the stabilizing systems used in the product's design and processing, and extrinsic factors. REPFEDs have anywhere from 12 to 60 days of shelf life with most in the 20–30 day range (Bristol, 1990). Fresh pasta at 40 days and their accompanying sauces at 60 days are among the most stable of REPFEDs. The Institute of Food Science & Technology (U.K.) (IFST(U.K.), 1990) classifies chilled foods into three categories:

1. Highly susceptible foods to be cooked immediately before consumption: fresh meat, fish, poultry, comminuted meats, sausage, etc. Recommended storage temperature −1°C to +2°C.

2. Products often eaten without further heating and may have no preservative properties. Chilling is primary preservation: pasteurized products, cured meats, soft cheeses, prepackaged sandwiches, fresh doughs, milk and soy milk products, etc. Recommended storage temperature between 0°C and +5°C.

3. Foods not relying entirely on chilling for stability and that have some additional stability system: hard cheeses, yoghurt, fermented sausages, etc. Storage is recommended to be between 0°C to +8°C.

Lechowich (1988) reported that low-acid foods (pH > 4.6) with mild heat treatment and vacuum packaging or CA/MAP had approximately 14 days of shelf life. Acidic foods were less of a problem. Sous-vide products had, in general, 2–3 weeks of shelf life if held at 2°C–4°C.

The short shelf life of chilled foods goes against their wider acceptance. Neither vacuum packaging nor MA/CAP alone plus refrigerated storage and distribution was sufficient to stabilize chilled foods adequately. This inadequacy has led Lechowich (1988) and Day (1989) to recommend that these stabilizing systems be supplemented with acidification, use of competitive microflora, addition of preservatives, or heat processing.

Waite-Wright (1990) recommended a five point program to ensure safe, high quality, chilled products:

- Raw materials of the highest quality. This requires purchasing standards for all raw materials from reliable and dependable suppliers plus audits of all supplies and suppliers. This is expensive but necessary.
- A safe controlled process employing all the plant's internal support systems. This involves active quality control and HACCP programs in place and close attention to good manufacturing programs.
- Strict vigilance against cross contamination caused by workers or by poor separation of cooked and raw materials and their associated equipment. (This and the following recommendations are echoes of the preceding one.)
- Close adherence to sound plant hygiene and sanitation programs.
- Employee training in sanitation and personal hygiene.

All are necessary for safe chilled foods, and all can contribute significantly to cost overheads. The development of REPFEDs as a new product venture requires a cadre of a skilled, well-trained, and disciplined workforce, something that unfortunately is often a rarity in food plants. Peck (1997) reviews concerns about the safety of REPFEDs with respect to *C. botulinum*.

5.1.3.1.3.1 Distribution and Handling: A Weakness in the Chain   Weaknesses in the distribution and handling system for REPFEDs are well documented. As early as 1980, Slight in a U.K. study of storage and transportation of chilled foods found that refrigeration systems for chilled foods were poor and none of the transport refrigeration systems studied were operating properly. In 1981, Bramsnaes discussed how the high quality shelf life of frozen foods was affected by poor storage temperature and the poor temperature control of retail cabinets. Similarly, while studying the shelf stability of a newly introduced chilled Mexican burritos in southern California in the mid-1980s, my colleagues and I found that chilled food display counters in supermarkets would at times reach temperatures of 9°C (48°F) and that they held that poor temperature for extensive periods of time. Light et al. (1987) in the United Kingdom found temperatures to fluctuate from –1°C to +10°C in chilled food vending machines. They also noted that some machines could not maintain the desired temperature range of 0°C–5°C for even 50% of their working cycle. Clarke (1990), for example, found multideck display units in retail outlets had day-to-day temperature variations as high as 15.8°C and as low as –1.2°C. The situation has apparently not improved with time. Audits International, reported in Brody (1997), found temperatures of delicatessen products in retail display cabinets across the United States to range from 14°F to 71°F (–10°C to 22°C) with a mean of 47.1°F (8.4°C).

The amorphous glassy state and the importance of the transition temperature at which a food passes into this state during freezing have been stressed by O'Donnell (1993) for its effect on the quality of foods (see also Section 5.1.3.1.2.2 and references therein).

5.1.3.1.4 Novel Alternative Stabilizing Systems

New nonthermal stabilizing systems include pulsed electric fields, oscillating magnetic field pulses, and intense light pulses. These and several other nonthermal processes are reviewed by Mertens and Knorr (1992), Institute of Food Technologists (IFT, 2000), Ohlsson and Bengtsson (2002), Leadley (2003), Green et al. (2003), Picart and Cheftel (2003), and Wan et al. (2009).

5.1.3.1.5 Pulsed Electric Fields

High-electric field pulses kill microorganisms by causing cell membrane rupture. Mertens and Knorr cite two industrial applications: one to improve fat recovery from animal slurries from slaughterhouses; and the other to stabilize pumpable foods. Gauri Mittal, at the University of Guelph, was reported in *The Gazette, Montreal*, (February 28, 1996) as having demonstrated the pasteurizing of milk products, fruit juices, and the brine used in smoke-houses to cool cooked meats and thus allow its reuse to prevent it becoming an environmental hazard. Vega-Mercado et al. (1997) discuss developments in pulsed electric field. Products successfully stabilized are orange juice, milk, yoghurt, and pea soup. All are pumpable, fluid foods. Details for the design and construction of a pulsed electric unit are provided.

Mañas et al. (2001) studied inactivation kinetics of *E. coli* in liquid, semi-solid, solid foods and model systems with different electrical parameters. Other applications have been studied: Sampedro et al. (2006) for egg and egg derivatives; Gachovska et al. (2009) as a pretreatment for carrots during drying and rehydration; and Marquez et al. (1997) for the destruction of *Bacillus* spores.

Álvarez et al. (2006) discuss microbial inactivation by pulsed electric fields. Design and other problems are described by Gerlach et al. (2008) in their review.

A description of pulsed electric fields with examples of many applications is written up in a book edited by Lilieveld et al. (2007).

5.1.3.1.5.1 Oscillating Magnetic Fields   Mertens and Knorr (1992) describe the principle of oscillating magnetic field thus: "when a large number of magnetic dipoles are present in one molecule, enough energy can be transferred to the molecule to break a covalent bond" but provide no commercial details about their use. Rupture of bonds in any essential enzyme, DNA, or protein within microorganisms would disrupt that microbe's metabolism and reproductive capability. Pothakamury et al. (1993) review the effects of this technique on microorganisms, cell membranes, and malignant cells in detail. They conclude from their review that (a) there is minimal heat damage to nutritional and organoleptic properties, (b) energy requirements are low, and (c) product can be treated already packaged in flexible film packaging. They caution that very little is known on the death kinetics of oscillating magnetic fields. Leadley (2003) echoes this view concluding that evidence for oscillating magnetic fields effectiveness in microbial inactivation is inconclusive.

5.1.3.1.5.2 Intense Light Pulses and Other Systems   Gómez-López et al. (2005) used broad spectrum intense light pulses: to examine the killing efficiency of intense light pulses in a food medium, to examine its decontamination effect on minimally processed vegetables, and to study its effect on minimally processes cabbage and lettuce during refrigerated storage along with MAP. They found the light pulses to be much less effective with high-fat or high-protein food material and sensory qualities were acceptable until storage was terminated (after 9 days). Mertens and Knorr (1992) suggest the inactivation effect of intense light pulses may be a combination of photochemical effects and photothermal effects.

Turtoi and Nicolau (2007) used intense light pulses for mold spore destruction on paper-polyethylene packaging material. Oms-Oliu et al. (2008)

are somewhat reserved about the use of pulsed light and indicate much more work needs to be done to determine its effects on the quality attributes of food.

Mertens and Knorr (1992) have also discussed carbon dioxide treatments and the use of chitosan (deacetylated chitin) as an antifungal agent, antimicrobial enzymes, and biological control systems that could be used in conjunction with the hurdle concept of food preservation. Roller (1995) also discusses the use of enzymes, bacteriocins, and microorganisms as preserving systems.

The main advantage of nonthermal techniques is the minimization of undesirable changes to flavor, color, texture, and nutritive value that thermal processing often brings about in foods (but see Oms-Oliu et al., 2008).

For a developer seeking information on minimal processing techniques for a delicate product, reference should be made to Ohlsson and Bengtsson (2002).

5.1.4 Summary and a Caution

The foregoing review of emerging technologies for stabilizing food products has demonstrated many opportunities for producing new products, in particular, for minimally processed food products. In the main, the references have been chosen to make developers, first, aware of these newer technologies and, second, aware of some cautions to be observed in choosing which of these techniques to use singly or in combinations. The danger in their indiscriminate use to produce novel foods is that they may alter generally recognized spoilage pathways of familiar products or expected (through experience with similar product formulations) spoilage pathways of new products. Something has changed in the food. Some hitherto unsuspected microorganism may emerge as a dominant cause of spoilage (Paster et al., 1985; Webster et al., 1985; Metrick et al., 1989), or a technique or complementary techniques applied to a product may cause physical changes in a product that are esthetically unacceptable to developers and consumers alike (Dempster et al., 1985; Grodner and Hinton, 1986). Great care must be used in selecting stabilizing systems for new products.

Food technologists are comfortable with their knowledge of thermal effects and moisture relations of microorganisms (but there are some surprises) but know comparatively little about the newer technologies for the stabilization of food systems or even how some microorganisms may behave when stressed by new techniques.

Many countries have introduced food laws regulating the introduction of novel foods. Their concern is with consumer safety. The observation that some combinations of novel techniques may hide or leave undetected by consumers previously overt and recognizable signs of spoilage (visible mold growth, loss of color, or an off-odor) is worrisome. No obvious signs of spoilage suggest to consumers that no spoilage has occurred. Nothing forewarns consumers of a problem. Rates of spoilage of some quality or safety attributes or both may proceed at different rates as different sets of stresses applied to a food are changed. However, unseen and unwelcome changes may have occurred that are economically damaging or more seriously are hazardous to public health.

5.2 Role of Engineering in the Development Process

The time element for completion of any new product development process would become impossibly long if it were a serial operation, with one step following after the other. Once prototype formulations meet the objectives of the company and reflect the perceived needs of consumers, then the team with their individual skills start working and interacting in earnest. When a suitable stabilizing system is determined, then the engineers begin planning processing line configurations, looking at necessary equipment changes that might be needed or new equipment needed. Information is shared about the availability of raw materials, their costs, their fragility; capabilities and limitations of in-house equipment; possible need for new equipments; special packaging requirements and handling problems, etc.

5.2.1 Engineers

Engineering personnel serve many functions in product development that vary in complexity with the product under development. For me-too products, their input is comparatively simple. For more exciting, novel products, their input may be very complex if they are required to investigate and design equipment for new processes for which standard off-the-shelf food equipment is not available. They identify how extensively existing plant must be modified to handle the new product. They design novel equipment for any innovative processes involved, modify existing equipment, investigate equipment suppliers, and work with these suppliers on specifications and standards. They, with the production department, identify for the team the process requirements on the basis of product throughput, manpower requirements, water usage, energy requirements, and any environmental concerns.

They conduct pilot trials where possible in the supplier's test facilities. They write equipment specifications for new equipment and bargain for the best delivery dates and prices for needed equipment. Delays occasioned by late delivery dates of equipment or the need to fabricate equipment to special standards must be communicated to the team early. Launch dates for product introductions and other marketing plans for promotions are determined by all the delays that are encountered.

5.2.1.1 Process Design

Engineers, with production personnel and food technologists, have two pathways when faced with the need to design a safe process de novo for plant-scale production of products needed for consumer taste and use studies:

- One is to use theoretical models, that is, mathematical modeling based on the best available heat and mass transfer data.
- The second option requires a vast amount of experimentation, analyzing data statistically and developing a process from this empirical information.

The former is prone to errors because most food processes are complex processes that cannot be easily modeled. Food is not uniform; pieces are not always regular in shape, density, or composition; raw material varies in composition by variety, by season, and by weather conditions. The second procedure is obviously time consuming, laborious, and, hence, expensive. De Vries et al. (1995) describe the design of a computer model for baking ovens; they were preparing biscuits of the Marie type.

First, a flow diagram of the expected process is prepared, identifying as closely as possible the unit operations and processes along with the materials, services, and energy required at each step. With this in hand, the team members identify critical points and the processes needed to control them. These critical operations can then be studied in isolation for developing design data. This is a combination of the two pathways above whereby empirical data are collected to build a practical model. The behavior of food material within these critical processes (changes in viscosity, in heat transfer properties, or phase changes) will be studied in order to describe the process as closely as possible. This permits mathematical simulation of changes that occur within the process.

5.2.1.2 Scale-Up

Test kitchen-prepared products on which the original consumer trials were based seldom resemble products prepared in larger batch pilot plant trials, and these latter never resemble entirely those prepared on production line equipment. These are the vicissitudes of scale-up. The mixing bowl of the test kitchen is only distantly related to the action of the 50 or 250 gal steam-jacketed kettle with attached mixer paddle on the food ingredients. The trials do, however, provide the preliminary data necessary to begin design and optimization of the process. Consumer reaction to the changes that scale-up brings must be assessed by consumers to determine whether satisfaction of their needs are still being met.

Taylor (1969) summarized causes of scale-up problems 40 years ago, and they are still pertinent. Taylor's summary follows:

1. Much of food technology lacks a sound scientific basis, or, as Taylor puts it, technology is running ahead of the science. There certainly does seem to be a gap between food technologists' understanding of the basic principles underlying many novel processes and their effect on microorganisms and food components and the equipment to take advantage of these with surety. When this occurs, "scale-up is not very soundly based." The caveats mentioned in the previous sections illustrate this; many of the newer processes lack a sound scientific basis. Although this situation is slowly improving, nevertheless, for many of the newer processes, there is no standard, off-the-shelf equipment, and, in some processes, there are no sound scientific principles to assess inactivation of microorganisms.

2. Raw materials used in food processing are highly variable from one variety to another variety of raw produce, from supplier to supplier, from season to season, and during the season from one geographical area to another. If there is one thing engineers dislike, it is variability as an element in designing new processes or new plant. Overcoming variability in processes requires operator skills, which, in turn, means relying on the judgment of operators.

3. Pressure mounts to cut corners in new product development to meet launch date targets or counter market intrusions by the competition. "There is regularly pressure by marketing to eliminate any intermediate scale-up on the grounds of time saving. Such elimination seldom saves time in the long run" (Taylor, 1969).

4. Engineers and technologists want pilot plant studies to get design data for scale-up. Product is secondary. Marketing personnel, however, see the need for product for test market or consumer research studies as paramount and the technologists' demand for ever more data as dithering around. These are the seeds for conflict. If potentially unsafe or unstable product is placed in the hands of consumers, repercussions could be severe.

5. Engineers tend to specify process designs based on the available data. Such designs require unique purpose-built equipment, which is expensive. Over the life cycle of the product, costs may not be recovered fully or may be a total loss if the development project is a failure. Consequently, standard, off-the-shelf equipment is used, and scale-up efficiencies are thereby compromised.

6. Echoing points 1, 5, and to an extent 2, the state of process control in the food industry is still not highly developed. Off-line controls and batch operations are the norm in many processing plants. There are time

delays getting crucial control information back to operators. Some have on-line controls providing faster control information; in-line controls delivering data in real time are only beginning to be developed to their full potential.

7. "The present food industry is still fairly labor intensive and in some parts of the country the labor force is unskilled with a high turnover rate" (Taylor, 1969). The conservatism of the food industry and its lack of skilled personnel that can cope with new technologies are discussed by Demetrakakes (1998b).

Taylor's remarks made four decades ago and intended for the U.K. food industry apply equally well today in North America and in most major food-processing areas. However, equipment technology and process control systems are rapidly improving in sophistication and versatility.

5.2.1.3 In-Process Specifications

Engineers need to know the sensitivity and reactivity of any raw materials and ingredients to any stresses to which products will be subjected, for example, heating, freezing, or size reduction, as they design processes. Processing must maintain the final product's quality characteristics as defined in the concept statement as closely as possible. Where a thermal process is required for a safe product, engineers work with the research and development group and professional thermal process authorities to develop safe thermal processes and the controls for these. All operations that alter the nature or state of the product or have an effect on quality must have tolerance limits specified (i.e., upper and lower temperature limits, mixer speeds, pump and flow rates), and equipment and their associated process control instrumentation must be designed to maintain these conditions. When this has been done, engineers will finalize their activities signing off on a product flow document identifying each and every unit operation and process, and the conditions under which these will be used. This includes the following:

- Temperatures of the product at each stage in this product flow; duration at these temperatures, flow rates, and tube diameters in heat exchangers need to be calculated to maintain the residency time; product viscosity changes require monitoring to prevent settling and turbulence in pipes, and amount and size of particulates present, etc.

- Cooling. Rapid cooling to reduce temperatures below critical values is a requirement as food is transferred within the plant. Heat balance studies are required. Recovery of heat is both an environmental and economic necessity and for worker comfort.

- Flow rates, pump speeds for minimum product damage, mixer speeds, pipe diameters, energy consumption of motors, etc. are all data that require specification. Shearing action as product is transported through the plant can be damaging.
- Pressure changes on the product plus its rheological properties are important characteristics in some processes, for example, extrusion cooking and supercritical carbon dioxide extraction.

Any operation within a process that might have an effect on a product's character must be described together with its safe (with respect to hazards of public health significance and maintenance of quality attributes) operating limits in the minutest detail.

5.3 Manufacturing Plant: A Stumbling Block or an Asset in Development?

5.3.1 The Plant

The manufacturing plant could be anthropomorphized as a Jeremiah crying in the wilderness; so often must it take a wet blanket approach to product development dampening the enthusiasm of the rest of the development team. Plant personnel see the need for new products, but they must also keep a plant running to earn the money for product development. They must tell the others either of their ability or inability to manufacture the product or to inform them of how disruptive the new operation will be. As engineers dislike variability so too do plant managers dislike disruptions in their orderly routines.

5.3.1.1 Concerns: Space, Facilities, Labor, and Disruptions

Manufacture of a new product puts a burden on existing processing lines of an established plant already working at capacity and on its labor force. Additional warehousing space is required for finished product, for packaging and labeling materials, and for raw materials and ingredients. This is disruptive of the profit-producing side of the company. If this disruption is not anticipated and accommodated early with planning, then the regular production and plant trials of new products can suffer. This can be disastrous for the timely launch of new and seasonal products. New products cannot kill the goose laying the golden eggs that are paying the bills.

As early as possible, the production unit must communicate their ability to produce the new product with their present labor force and physical plant within the cost and quality parameters required by marketing and without disruption of regular production. Together with engineers, they determine what modifications to existing lines may be required or what space is available for the installation, if necessary, of new production lines. They identify production costs and labor requirements.

New production lines mean finding space to put the new lines or the juggling of regular lines to accommodate the new production facilities. If accommodating the new product is too difficult, then alternate means of production must be considered such as going to co-packers or ultimately the construction of a new plant.

Production staff work with marketing staff to review their sales volume projections and to coordinate manning requirements for the processing lines. Their knowledge of local labor markets and the availability of skilled labor become factors in determining overheads to be associated with the new product's costs and ultimately profitability. If a high level of technically skilled is required, then these outside skills may have to be bought; this increases costs.

5.3.1.2 Co-Packers and Partnerships

Inability to manufacture new products is not sufficient reason to reject product ideas. It is the least consequential of the criteria, but it may have consequences for the anticipated financial success of the project. Manufacturing and packing capabilities can always be had through co-packers. There is, of course, a price to pay. Co-packers levy a fee per case packed. If new products are price-sensitive, co-packers' fees will obviously change profit projections.

A frequently overlooked cost associated with employing co-packers is the extra vigilance for quality control. Companies need quality control staff responsible to them to be resident in their co-packers' plants when their products are run to ensure that their products are manufactured according to specification: an exchange of quality control records is not enough watchfulness. This holds when the most amicable and trusting relationships exist between the two parties. Co-packed products require the same vigilance as self-manufactured products.

The use of a co-packer can be a very attractive route to new product development. Development costs and time are telescoped with a co-packer with experience with similar products. They have a more accurate assessment of their development and manufacturing costs and initially may be able to manufacture a product more efficiently with their experienced work force. If products should prove unsuccessful, no capital expenses or extra staff have been acquired. If products are successful by meeting or exceeding sales forecasts, then plans can be made to undertake self-manufacture either by plant expansion or acquisition.

Alternatives to co-packers are worth mentioning: partnerships and joint ventures. These can be fruitful if a suitable partner can be found. They can be dangerous. Where one of the partners is larger and hence economically stronger than the other, ultimately the larger may eat the smaller or simply leave the arrangement leaving the smaller helpless. Partnerships are possible but not recommended.

Joint ventures always present problems. Harvey (1977) hints at an analogy between joint ventures and marriage but points out the high incidence of divorce lawyers and the abundance of marriage counselors suggest the analogy is poor (see Chapter 9). He points out reasons for joint ventures often lead to conflict. To reduce competition is a bad reason because at once the partners are in conflict with their basic business objectives; that is they are in business to compete. Sharing a risk is an equally bad motive since managing risk requires strong undivided leadership. Harvey provides some interesting (and amusing) arguments against joint ventures. Houston and Johnson (2000), on the other hand, look at joint ventures (buyer–supplier contracts) and the variables necessary to make them successful. Their paper is a very technical contribution (see Chapter 9).

5.3.2 Roles of the Purchasing and Warehousing Departments

Two aspects that impinge heavily on the success of new product development fall under the umbrella of the manufacturing plant: the purchasing department and the warehousing and distribution department.

5.3.2.1 Purchasing Department's Activities

The purchasing department plays a role in product costs by having the task of finding inexpensive yet reliable sources of those raw materials, ingredients, and packaging materials specified by technologists. By reliability is meant availability, consistency of quality, and prompt and reliable delivery schedules. This often results in some buyer–supplier relationships that need to be developed. Obviously, if the purchasing staff can obtain materials and ingredients meeting specifications cheaply, product costs will be lower. If purchasing can negotiate delivery cycles from its suppliers and still maintain favorable terms, then the impact on warehousing will be lessened and warehousing costs reduced, that is just-in-time deliveries of supplies.

Food technologists identify and describe all *necessary* characteristics of all raw materials, ingredients, and additives in their written product standards. Where technologists have insisted on restrictive or unusual specifications, purchasing departments will require suppliers to submit samples of their products that most closely fit the requirements for technologists to assess. Purchasing agents research several sources to ensure that there will be a continuity of supplies in the future.

When specifications for ingredients or raw material are arbitrarily stringent and exotic, their costs are more than likely higher. I have always been bemused—and frustrated—to note how many technologists establish standards for ingredients. Suppliers send them samples to use in their formulations—standard practice this. But then, if that ingredient works satisfactorily, the technologists often use that supplier's specification sheet as the ingredient standard. This limits the source to one supplier and one cost. There is rarely an attempt by technologists to characterize what unique properties of an ingredient are essential to the product's quality—time constraints often prevent further study; these and only these essential characteristics are required in the purchasing standard. Better price, availability, delivery, servicing, and quality may be found with other suppliers if specifications are broader and less stringent while still providing all necessary properties.

Here is an example of what I mean respecting specifications:

I worked with a tortilla manufacturer which produced its own masa dough. The lime used was builders' grade and not stated food grade. The company had discovered that builders' grade lime met all the specifications of food grade lime (indeed surpassed them) and was considerably cheaper. For protection, every batch of lime purchased was sampled and analyzed for a wide range of chemical and microbiological components by the company's quality control department and by an outside U.S. government approved consulting laboratory. It was still cheaper, even after factoring in laboratory expenses, to buy builders' grade lime.

Purchasing personnel with the traffic department balance transportation costs with geographic availability, and the reliability of the source with the item's cost. This interplay between geography (transportation costs from supplier), availability and cost (they fluctuate inversely), reliability of supply and supplier, and quality (adherence to specification of ingredients) for material, ingredients, and additives is one in which great cooperation and interaction are required between suppliers and purchasers.

Reliable, quality suppliers often assist the development process for any application the developer wants. This is especially true with flavor houses; they own the flavor formulation they developed for the company. We asked a flavor house to develop a lemon flavored bread crumb flavoring. When we had the capability to manufacture in-house, we found we did not own the formulation. Developing and maintaining a working relationship with suppliers are discussed by Williams (2002). Practical advice and some cautions are described (see Chapter 8).

Ingredient, raw material, and packaging costs contribute significantly to any product's pricing structure; as such, they serve as a deterrent in some markets where price sensitivity is important. For example, for the military market, special packaging that meets military specifications may be required; for vending machines, both a different size and packaging material that are appropriate for machine dispensing are needed; for big box stores and warehouse clubs, still different forms of packaging are needed. Customers will reject a high unit cost if quality, novelty, and added value of the product do not justify that elevated price in their eyes. Purchasing personnel's success or failure in obtaining supplies at an economic price and meeting specifications is important to the success or failure of the project.

5.3.2.2 Activities in Warehousing and Distribution

Every square meter of warehouse space can cost several hundreds of dollars per year to maintain. The costs vary with the type of warehousing, dry warehousing being the cheapest but prices rise with refrigerated, frozen, controlled atmosphere or controlled temperature warehousing and especially for fresh produce warehousing, which can become very complex for controlling the ripening and maturing of produce. Popping corn and chipping potatoes, for example, require special temperature and humidity storage.

Warehousing considerations are usually overlooked or, at best, the warehouse as a structure is only considered to be a buffer between the end of the production line and the distribution system. If this latter attitude is taken, it will soon be realized by all that it is a very expensive buffer. Even an unheated dry warehouse cannot be considered a simple empty space: it is a space subject to the vagaries of summer and winter temperatures and with these temperature changes subject to varied humidities that can under appropriate conditions cause rusting of metal containers. They must also be guarded against vermin infestation that can soil and foul containers. They are not simple entities; they are complex structures.

New products require space for holding raw materials, its packaging, and the finished new product, and these items put an extra burden on a company's existing warehousing. It can be a chilled or frozen food warehouse and maintained at specific temperatures suited to the product—and expensive—or it can be a controlled atmosphere warehouse for storage of fresh fruits and vegetables or controlled to inhibit or accelerate the ripening of produce. Many fruits and vegetables require controlled atmosphere storage to keep adequate supplies of fresh produce on hand.

Warehouses can also be leased or put into a contracted third-party warehouse-cum-distributor.

Warehousing and distribution introduce such cost considerations as

- How many days of production are planned to be held in storage and for how long?
- Will production and distribution have seasonal peaks? Will it be necessary to store excess production off site? These are important considerations for potato chip and frozen French fry manufacturers who require year-around supplies and potatoes of particular properties.
- Are there special storage requirements, for example, refrigerated storage or heated storage in winter?

- Can production and distribution be phased to complement one another? Is it necessary to ship orders from stock in the warehouse or can it be shipped from production?
- Are special storage requirements for raw material and packaging material needed?

Products with short shelf lives often require a returns pickup system for product kept beyond their expiry dates; this is often possible to do with scheduled deliveries. Out-of-date, spoiled, or damaged product requires destruction under supervision if it cannot be recycled or reused by another means. Recyclable or reusable returned containers must have not only a pickup service but a cleaning operation as well before reuse.

5.3.3 IT Department's Contribution

Tacticians rely heavily on computers and specialized applications to accomplish their tasks. However, computerization of operations usually starts with the strategists in the company's accounts department for which there is an abundance of accounting applications available, then progresses to warehousing and distribution for use in stock-keeping and traceability of product (Mermelstein, 2000), into the manufacturing plant (Mermelstein, 2000), and eventually does find its way into technical departments and, at last, into new product development arenas (Gaisford, 1989). The ready availability of software programs capable of performing the myriad of tasks permits small companies the same computational ability as larger companies. These are all available, but how effectively they are used especially by small companies is not known.

Computers (and the applications to use with them) serve three basic needs in new product development:

1. Management of information: storage and retrieval of information. Management has real-time information on inventories and cash flow. Search engines coupled with external databases provide access to research technical literature as well as market information to assist product design.
2. "Number crunching": a capability that allows the use of sophisticated mathematical and statistical analytical programs to optimize experimental trials, thus reducing the number of trials and research costs, analyze sensory and consumer data, and provide least cost formulation techniques for optimization of ingredient usage (O'Donnell, 1991; Mermelstein, 2000).
3. Graphics programs that permit manipulation of three dimensional solids to complement statistical studies, to study the lay-out of processing lines, to assist package and packing designs to minimize waste, or to create and design food labels.

Developers, equipped with their computers and appropriate software, are now in a position to formulate a product to have a given nutritional value, meet a nutritional standard, and to reach these goals with the least number of experimental tests. These functions or variants of them provide developers with tools to reduce the work load and accelerate the development process to manipulate data for more efficient information retrieval from the data, to retrieve and classify data to see trends more clearly, to develop expert systems, and to communicate data and information more rapidly and efficiently.

5.3.3.1 Information Management and Retrieval

Technologists can tap many technical databases that are free or are available by subscription. The microbiology database, Combase (see Chapter 7), is available free and is a valuable tool to the study of microbial spoilage. Many food technology, food engineering and nutritional journals are available to technologists, and marketing personnel either in their entirety or in abstract form and the numerous search engines can direct researchers to original work on many topics.

The value of the collective memory (Chapter 3) cannot be overlooked as a valuable tool in product development. The archived memories of retirees may hold ideas worth pursuing that, once thought untimely, are now appropriate. Complaint files provide information submitted by consumers that highlight shortcomings or defects in products that must be avoided. Data mining (competitive information gathering) is discussed in Chapter 3.

5.3.3.2 Number Crunching

Mathematical and statistical software packages abound that are a boon to developers in the manipulation of numerical data in diverse fields from consumer preference studies to analyzing heat transfer data in processing trials to obtaining response surfaces from formulation trials. These programs (e.g., Origin® 80, manufactured by the OriginLab Corporation, Northampton, Massachusetts) permit rapid analysis of multivariables to extract information buried in the data.

An early use of computers for assistance in the collection and analysis of sensory data is described by McLellan et al. (1987). A review of this now very dated paper accentuates the progress that is currently available for the analysis of sensory data in real time. Software for analysis of data is expanding so rapidly that any mention of specific software applications is apt to find the company merged, out of business or the apps so improved, updated, or altered by newer versions as to be obsolete upon their publication here.

Nevertheless, a description of how the apps are used does have value. Software for generalized Procrustes analysis (*Procrustes PC©*) for sensory methods relying on consumers rather than on trained panelists and a program, *REST©* (Repertory Elicitation with Statistical Treatment), are described by Thomson (1989), with an explanatory example, as a structured method of qualitative market research based partly on the repertory grid method and generalized Procrustes analysis. A more detailed description of the software used in these studies with these same two techniques can be found in Scriven et al. (1989) who applied the technique to study the context (time, manner, place, or circumstances) under which consumers drank a variety of alcoholic beverages. Gains and Thomson (1990) also used generalized Procrustes analysis with the repertory grid method to study under what contexts a group of consumers used a range of canned lagers. Such data are invaluable in defining market niches for products and opening up new market opportunities. These earlier studies with their explanations and descriptions demystify the apps, and allow their users some degree of understanding.

By far the greatest use of computer applications by technologists is associated with statistically based experimental design software. These programs allow developers to take calculated shortcuts in the number of trials dictated by classical statistical experimental design. With a factorial design, the number of experiments mushroom rapidly as the number of variables and the levels at which each variable is to be tested increases; for example, the trials required in a factorial design is L^v where v is the number of variables and L is the number of levels at which each variable is to be tested. Astronomical, and costly, numbers of trials are required to test four ingredients of a product formulation at three concentrations. A pioneer in the field of experimental design is Dr. Genichi Taguchi. Dimou et al. (2009) used Taguchi methods to determine how to blend unifloral honeys to closely match Thyme honey. Any search engine using Taguchi method or "robust design" will bring up several references for those who wish more information.

Mullen and Ennis (1979a,b) describe the design for applying a linear equation process to a computer program to produce a six ingredient hypothetical product that supplied 10% of its calories from protein, 35% from fat (high by today's standard), and 55% from carbohydrate. Mullen and Ennis (1985) later refined their program to handle 15 variables but reduced the amount of experimentation by using fractional replication. The procedure is described in a detail that clarifies the assumptions used in the shortcuts, which underlie many of the statistical software programs available for experimental design.

Optimization designs are particularly useful as these permit the developer to investigate the optimum levels of ingredients to maximize a particular quality feature or to alter a process to get a maximum effect (or a minimum effect if the effect is undesired). Two techniques are used: response surface methodology designs and mixture designs. Henika (1972) used the example of improving the wettability and flavor of an instant breakfast cereal product to compare and explain classical testing versus the response surface methodology approach in getting answers more quickly and cost-effectively. Hsieh et al. (1980) developed a synthetic meat flavor using response surface methodology. Mixture designs treat the unique problem of formulations whose proportions of all the ingredients must equal 100%.

A history of the early work on experimental design, as well as an explanation of experimental design as an aid to product development, screening and optimization designs using response surface methodology and techniques based on mixture designs is described by Dziezak (1990). She lists then available software. Statistical software has proliferated and sources can be readily found in journals, technical and business magazines, and computer stores.

Further examples of response surface methodology in the optimization of processes can be found in work by Bastos et al. (1991) who upgraded offal processing by extrusion cooking to produce a finished product with good solubility and emulsifying capacity. King and Zall (1992) used a model system to study low-temperature vacuum drying using a design with three variables and three levels. Tsen et al. (2009) used response surface methodology for the optimization of process conditions of banana purée fermentation. They described a three variable by three-level design.

Skinner and Debling (1969) describe the application of linear programming techniques to classic management decisions with examples in food manufacturing:

- The allocation problem faced when several products use the same commodities that are available only in limited supply in their formulation. Which product to make? Their example was demonstrated with fruit salad versus fruit cocktail products.

- The blending problem that arises when ingredients of a particular product can be blended either to meet a quality standard or to achieve a cost standard (applicable to least cost formulations). Which proportions to use? A sausage formulation problem was used that is akin to a reformulation situation.

- A simultaneous blending and allocation problem exemplified with a pork versus beef sausage example.

The examples are worked through to demonstrate the principles, but, as the authors state, a statistical software package designed for linear programming could have been used.

5.3.3.3 Graphics

Computer graphics capability has grown immensely as anyone interested in the subject of virtual reality can attest. Architects now plan a building in their computer with suitable software. Then figuratively, they can stroll through the building noting the views from different aspects from within the simulated structure. This application has been a boon to plant engineers in the design of processing lines as well as to package designers to create new packages cutting and designing to minimize waste in packaging.

Dziezak (1990) describes some of the three-dimensional (3D) and contour plots that can be accomplished to view response surfaces to rapidly assess the most rewarding avenues of investigation. Rotation of these surfaces reveals information about areas of optimum and minimum values.

Bishop et al. (1981) make the case for the value of 3D graphics in food science applications. The programs they developed and used in their applications are discussed. Floros and Chinnan (1988), using optimization techniques based on response surface methodology, demonstrated the application of graphical optimization to the alkali peeling of tomatoes. Quality of the product was improved by studying response surface plots to select the optimum concentration of alkali and temperature.

Roberts (1990) demonstrated the use of computer-generated 3D graphics in predictive microbial growth modeling. While predictions may not be precise, he claimed that knowing the trend of growth would be highly important in designing stabilization systems for new products.

An obvious application of computer graphic techniques is in the design of labels. Lingle (1991) describes the use by several food companies of computer-based systems to control package and label design. The advantages cited are

- Ability to manipulate designs in any desired fashion for application to line extensions or to redesign packages
- Ability to store designs more easily on disk rather than as paper art
- Ease with which images can be communicated to others for decision making

Graphics software allows a company to bring package design in-house provided the company has talented graphic artists on the payroll.

5.4 Commercial Feasibility

A time comes in any development process when the tactical unit of the development team must ask itself, or will be asked by the strategists, what the chances are for the technical and (indirectly) the marketing success of the project.

5.4.1 The Loop: The Interconnectivity of Questions with Indefinite Answers

The technical and marketing success of any project is a function of several direct and indirect variables:

- Directly to the product's capability of development with the qualitative attributes, especially naturalness, demanded by marketing and indirectly to the time and costs of development to reach those goals successfully. Obviously, the longer the time the more costly; the more costly the less likelihood of success as customers might balk at the price.

- Directly to the product's successful launch and promotional campaigning with supportive advertising.
- Directly to the retailer's reaction to and acceptance of the product for prominent display (certainly not their passive resistance to the product).
- Directly to customers' and their consumers' acceptance and repeat purchases of the product and indirectly to ensuing "buzz" via social networks.
- Directly to the competitors' retaliatory actions.
- Directly to senior management's patience or impatience concerning ROI that determines success.

The foregoing is straightforward and uncomplicated. It is a guessing game.

Feasibility is different. Marketing personnel need to know, "Can it be done on schedule?" Marketers need firm time commitments so that promotional material, labels, and associated artwork for advertising are ready for the appropriate launch date, but, before any launch date can be decided, the distribution channels need to be filled by distribution personnel. And prior to the distribution channels being filled, manufacturing has to know whether it can produce the product. But even before this, any special equipment has either had to be designed or had to be modified by engineering or specifications had to be written for new equipment to be purchased. Financial interests should not be overlooked here. Accountants need to have financial estimates for the project in order to find funding for the project.

It is much like that children's nonsense song:

> There's a hole in my bucket, dear Liza, dear Liza,
>
> There's a hole in my bucket, dear Liza, there's a hole.
>
> With what shall I mend it? dear Liza, dear Liza,
>
> With what shall I mend it? dear Liza, with what?

Liza gives many suggestions to her companion, each one requiring a subsequent step until finally the last step requires a pail of water whereupon the entire song commences again.

All depends on the probability that the company's technologists or those contracted by the company can succeed in matching attributes demanded by marketing in its research with a safe, stable product. Development teams enter now into the realm of estimating probabilities and producing guesstimates. Products with little or no creativity, products that are imitations of existing products in the marketplace, or products that are simple line extensions can usually be brought on stream with a high probability of success in a comparatively short time. However, no development project is simple and without hitches.

Each step in the development of products can produce its own hitch and needs to be analyzed for its chance of success, its cost contribution to the whole, and its time to success. Any one step can present an insurmountable hurdle if its chance of success is impossibly slim or if the time to success is too long or is too costly; one step can thus stop a project's chance of moving forward.

5.4.1.1 The Art of Guesstimating

Determining the probability of certain events happening is familiar to every student of statistics, to every gambler, and to planners of outdoor events; we live with probability every day. For example, the probability of getting at least four heads when tossing seven pennies or the odds of picking a red ace from a deck of playing cards are common problems described in textbooks. Also familiar to students of statistics are problems associated with calculating the probability of a particular event occurring as the result of a sequence of events when the probability of each step of the sequence is known. (Readers unfamiliar with probability statistics should review Bender et al. (1982) and Parsons (1978) for concise readable accounts of probability statistics.)

If the chance of going from A to B in some sequence of events has a 9 out of 10 chance of success (.9), there is a high probability that B will be reached. If there is a third stage, C, and the probability of going from B to C is also .9, then the probability of going from A to C, as any student of statistics knows, is

$$.9 \times .9 = .8$$

The likelihood that C will be reached from A is still high but somewhat diminished. If more steps are added, even though each step has a high likelihood of success, the chance for success becomes less and less from the original starting point A. Instead of likelihood of success of an event or reaction, one could easily have substituted processing yields of some extraction procedure (Malpas, 1977). Thus, if, in this simple processing sequence, a 90% yield was anticipated at each step, the yield of C would be 80% conversion.

The a priori probabilities associated with tossing coins or picking playing cards from a deck of cards are either readily calculated or are determinable by a long series of trials. They can be established theoretically or empirically.

Problems arise when the developers attempt to assign probabilities to phases of the development process. Objective probabilities determined from coin tosses or picking cards no longer apply. There is no history of observations or mathematical construct from which one can state, on the average, that such and such an event will happen with a specific probability value. The development team must work with situations in which the probability cannot be calculated. Rather, the developer is forced to assign probabilities that "are arrived at by considering such objective evidence as is available and, in addition, incorporating the subjective feelings of the individual" (Parsons, 1978). In short, one can only assign knowledgeable guesses.

Subjective probabilities assigned by developers to the various phases of development must be realistically based on the best available information. They must not be unrealistic probabilities based on an enthusiastic overassessment of the technological skills of the development team. There can be no gut feel.

The development process for a hypothetical product has been broken down into a simple sequence (Figure 5.2). To proceed from a starting raw material, A, to the final desired product, P, requires three intermediate stages, B, C, and D, and four intermediate steps, a, b, c, and d. The steps could be key processing steps to provide a desired characteristic in a product; they could be the likelihood of getting a change in legislation for a permitted additive; they could be steps to undertake the necessary change in some product's standard of identity; or they could be the possibility of penetrating a particular market. They can be represented as logical steps on the way to products or decisions or events for which probabilities have been assigned.

Each step can be assessed a cost figure for its realization. The sum of the costs, $\$(w + x + y + z)$, for each recognized phase in Figure 5.2 represents the total *developmental* costs to go from A to the final phase P. These costs refer only to the costs of the processes involved. The corollary impact of development on other activities in the company processing, warehousing, distribution, etc. cannot readily be factored in.

The time to accomplish this sequence is estimated to be $(T_1 + T_2 + T_3 + T_4)$, the sum of the subjectively assessed time requirements for each step. The probability, the expected cost, and the time expected to go from A to P can then be assessed. It must be remembered that they are all *subjective* estimates in development process.

What is the probability of success? The phases (Figure 5.2) range from a more than moderately difficult one, reaction step "c" with a probability of success of .3, to the very easy last reaction step "d" estimated at .9. The overall

	a	b	c	d
A → B → C → D → P				
Probability	.5	.8	.3	.9
Cost	$W	$X	$Y	$Z
Time	T_1	T_2	T_3	T_4

FIGURE 5.2
Probability, costs, and time as factors for consideration during development stages for a hypothetical food product.

success of the entire sequence, the probability of reaching *P*, is a disappointing .1 arrived at in the following fashion:

$$.5 \times .8 \times .3 \times .9 = .1$$

This poor probability of success in association with the cost and estimated time to success may suggest that abandonment of the project is the wisest move. Much depends on the company's objectives and its strategy to get to these goals.

If the product, *P*, is highly desired, the technology team may be tempted to tackle the difficult "c" process first. This may be the most economical approach to the problem for ingredient developers; it avoids the input of time and money in solving the initial phases if it should be determined that the project is not feasible within the timeframe of the company at the C to D stage (Holmes, 1968). Probability analysis does serve a useful purpose.

Again, if these were percentage yields in the manufacture of some new food ingredient and not probabilities of success for processes, one would anticipate only a 10% yield for the entire process. Such a low yield can only be acceptable if the product is highly desirable for which customers will accept the high cost.

New and improved products are almost a dead certainty to be successfully developed. For example, a breakfast cereal can be improved in several different ways. The probability of better flavor is .5, better crispiness is .7, higher fiber content is .8, and longer shelf life is .6.

To improve this breakfast cereal with respect to one of the above quality characteristics—but without specifying which one—the chance of success is 1 (complete success) less the product of all the probabilities of failure or:

$$1 - (1 - .5)(1 - .7)(1 - .8)(1 - .6) = 1 - (.5)(.3)(.2)(.4) = 1 - (.01) = .99$$

The product is almost certain to be a new and improved one.

5.5 Summary

Only in the very large companies is there a separation of duties as presented here. In small companies, members of the development team wear several hats. I found the development team in one company to consist of one person, the quality control manager, and in another, two people were involved, head of quality control and the president. (Managers of quality control are often the only technically trained personnel in small companies.)

In the small company then the duties respecting product development as related here are not distinct but are amorphous. This gives greater flexibility (there are rarely conflicting opinions on one-person teams), but the team of one responsible person has fewer resources respecting marketing, technical, and engineering skills. New products are confined to those that can be made with existing materials and processing facilities.

Because of the intimacy of smaller companies, personalities, usually that of the company president, can dominate, and strategic growth decisions and, therefore, tactical development decisions may or may not be wisely taken (Rothfeder, 2007).

6

The Legal Department: Protecting the Company—Its Name, Goodwill, and Image

6.1 Introduction

Two very different departments protect the safety of the company's customers and consumers and the company itself and its real and intellectual properties. These are the legal and the quality control departments (or whatever name the latter function goes under). These are the support groups, as they play supportive roles for the other departments but have very active roles in the development process. Areas in which they are required are as follows:

- Protecting the consuming public from any injury or intoxication resulting from the preparation of, use of, and eating the company's food products
- Assisting in setting standards and controls to ensure the stability of the high quality attributes of the new product and its wholesomeness throughout its shelf life
- Advising on and monitoring for adherence to customer protection legislation
- Reviewing label nomenclature, advertising statements, and promotional claims and seeing to their adherence or regulations to norms established by government
- Protecting the company's rights in contracts, leases, and partnership transactions
- Protecting and registering trademarks, copyrights, or filing patent applications and protecting these from illegal use
- Investigating fraudulent damage claims against the company or its products

The size of the company determines how these tasks are divided between the two support groups. Some are more obviously one than the other, but there is a murky middle ground. In many smaller companies, the quality control manager often advises on label statements and claims (particularly nutri-ent claims), and many smaller companies do not have legal departments in-house but resort to legal firms specializing in food legislation. Advertising is often handled by outside agencies with (usually) good knowledge of the legalities of food advertising and claims.

The topics, legal matters and quality control, will be treated in separate chapters.

6.2 The Law and Product Development

A Canadian company has developed a better tasting ice cream with less fat, sugar free, with high-fiber content, and having all the creamy mouthfeel of the real thing. Unfortunately, there is a standard, a regulated standard for ice cream. This ersatz ice cream contravenes that standard and cannot be called ice cream. It will not have the cachet of ice cream. A University of Guelph professor developed such a product (Bascaramurty, 2009). Another (hypothetical) Canadian company has developed a market in organic products, organically grown turkeys in particular. However, the Turkey Farmers of Ontario (the industry marketing board) decided in 2008 that all turkeys must be raised indoors to reduce the transmission of avian influ-enza from wild birds, but the Canadian Food Inspection Agency (a branch of Health Canada) had mandated that turkeys (along with other provi-sions) had to be raised outdoors to be organically raised (Webb, 2009). (This issue is under review.)

Such contretemps abound in food legislation and are not unique to Canada.

Legislation, government regulation, and food have had a long history of association. The Code of Hammurabi (Hammurabi: 2123–2081 BC) con-tained legislation regulating food standards and trade; parts of his Code are believed to predate Hammurabi's life. The Emperor Shun in China is reported to have controlled the production and distribution of grain about 2000 BC (Spitz, 1979). Food and government regulation have been intimately entwined: "The role of the state in building up and controlling grain reserves is nowhere better illustrated than in ancient Morocco, where the same word—mahkzen—was used for both granary and government" (Spitz 1979).

There were, historically, sound political reasons for government interven-tion in food matters. For example, governments used their public treasuries to buy up surplus grain when grain was abundant and cheap; this action maintained a stable fair price. Farmers did not suffer; customers did not grumble. When grain was scarce and expensive, the government inter-vened to prevent starving by doling out grain from its stores, and the people did not suffer. Farmers were happy, and the urban population was kept fed and happy also. Most importantly, for politicians, nobody became restive and rebellious. (Today, marketing boards control the supply of many agricultural products and nobody is happy.)

Government, through its laws and regulations, influences the food microcosm in many ways (see, e.g., Table 1.4); many of these ways have a direct bearing on new foods, new food product development, and the expansion of markets into new geographic areas. In Table 6.1, attention is drawn more closely to legislation that affects foodstuff, its harvesting and processing at the harvest site, its subsequent manufacture into food including the safety of the processes used, and its subsequent sale.

To understand this intervention into and regulation of product and ingredient development more fully, a quick overview of how food legislation is developed and is influenced is useful. Awareness is useful as a tool for developers to anticipate possible regulatory developments with an impact on the progress of development (cf., repercussions from the ban on saccharin). A simple overview of a generalized food legislative system is shown in Figure 6.1.

Groups that influence the policy-making process and hence indirectly influence the development process are in the upper half of this figure. They exert their influence, which can be formidable, by lobbying elected representatives, by presenting briefs at hearings called by policy makers, or by overwhelming attendance as opponents at such hearings, with organized write-in campaigns to elected representatives, or by overt disruptive demonstrations. The ultimate result is that any decisions of the legislative body bear some imprint of these groups. The departments that are influenced occupy the bottom half of the figure.

6.2.1 Nongovernmental Organizations

Marketing boards (which shall be considered nongovernmental organizations [NGOs] but are really quasi-governmental agencies) regulate the supply, price, and other rulings pertaining to food commodities (see Section 6.1); they limit the number of suppliers and regulate the volume of raw material they can produce (i.e., supply management). They set the commodity prices to further processors; through supply licensing arrangements, they control who is allowed to further process the commodities. They control, for example, the supply of industrial milk to boutique cheese makers in Quebec; unfortunately, the latter cannot afford the licenses, and, therefore, there is deep resentment to this board. Boards influence governments to impose quotas on imports of foreign commodities or on products containing the commodities that the boards control. Product development with such controlled raw materials results in more costly added value products. Controls even prevent the development of some new products.

Consumers' associations and trade associations representing food processors invariably oppose the higher prices that are forced upon their members by

TABLE 6.1

An Overview of the Extensive Reach of Legislation and Regulations Pertaining to Foodstuff

Area of Impact	Elements of Foodstuff Regulated and Impact on Food Development
Agriculture and fisheries	Siting of farms and fish corrals: restricted land use for industrial, agricultural, or residential use.
	Restricting the use of pesticides, herbicides, and fertilizers, as well as regulating the use of antibiotics, pharmaceuticals, and feed nutrient supplements: inspection costs.
	Establishing quotas for raw materials: controlling availability and cost of materials through import controls or subsidies.
	Requiring odor abatement at farms in or near municipal areas: demanding environmental controls at primary production and manufacturing sites.
	Restricting or preventing waste water or runoff into lakes, rivers and streams: restricting agricultural practices respecting factory farming and waste treatment from these: indirect impact on raw material costs.
Processing	Establishing zoning requirements, thus restricting site location for plants.
	Requiring waste water recycling at plant sites and noise and odor abatement programs at food plants in municipal areas: costs for environmental control affect plant overheads.
	Specifying design of, and requiring the use of certain construction materials for, plants: certain product types require specially designed and constructed facilities.
	Requiring adherence to processing codes of good manufacturing practice.
	Imposing import and export permit control that affects both sales and availability of goods: government-imposed restrictions on imports limit material availability and marketing's plans for sales abroad.
	Imposing traffic regulations of plants situated in or near residential areas. Local regulations may limit manufacturing hours for plants.
	Challenging the safety of many new foods and innovative processing techniques. Onus on manufacturer to establish safety.
Product	Establishing commodity grades for raw produce. Grades influence pricing schedule of produce grade.
	Establishing standards of identity for produce and some products. Standards of identity may restrict the use of certain ingredients and additives. Formulation changes for cost reductions are restricted.
	Establishing lists of approved additives or lists of restricted additives.

TABLE 6.1 (continued)

An Overview of the Extensive Reach of Legislation and Regulations Pertaining to Foodstuff

Area of Impact	Elements of Foodstuff Regulated and Impact on Food Development
Package	Regulating package sizes. This imposes inflexibility respecting container sizes and could have a possible impact on marketing plans.
	Restricting composition of packaging materials. Some packaging materials are prohibited because of environmental concerns.
	Imposing weights and measures controls.
	Establishing proper nomenclature for products; requiring product information for safe use and nutrition information. Product nomenclature may have impact on marketing plans.
	Requiring label declaration for ingredients. These lists could bring possible consumer backlash concerning ingredients, additives, serving sizes, and nutrient content (or lack of).
Marketing	Establishing guidelines for advertising claims respecting nutritional and health benefits.
	Establishing guidelines to prevent misleading promotional tactics respecting product and claims for it. Severe consumer backlash (bad publicity) and penalties for misleading claims.
	Restricting advertising targeted for children. This restriction applies in many countries.
	Restricting the display of some forms of advertising, particularly the siting of advertising.
International trade	Establishing trade alliances and treaties with foreign nations, e.g., NAFTA.
	Applying tariffs and nontariff trade barriers on imported products.
	Proclaiming antidumping regulations that are attempts aimed at supply management.

Source: Adapted from Fuller, G.W., *Food, Consumers, and the Food Industry: Catastrophe or Opportunity?*, CRC Press, Boca Raton, FL, 2001.

the supply management policies and other restrictive activities of marketing boards. Trade and consumer associations pursue their own agendas respecting food legislation in their advisory reports to government.

6.2.2 Advocacy Groups

Advocacy groups are often more militant than the above NGOs and manage to polarize groups of people; animal rights activists are pitted against farmers using factory farming techniques; those against food irradiation challenge, often violently, those for food irradiation. There are groups that demand food laws that restrict the movement and importation of food products manufactured by socially, ethically, and environmentally irresponsible companies or countries; or that restrict the importation or sale of products that do not meet certain religious laws. The strongest weapon of these

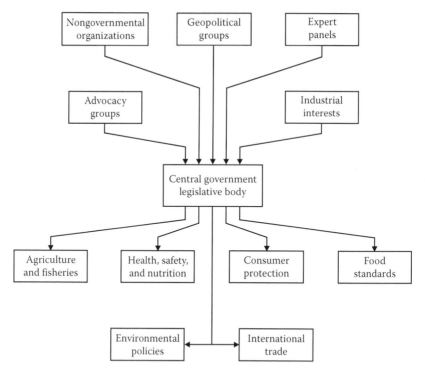

FIGURE 6.1
Generalized overview of a food legislative system with the bodies influencing it.

groups is their ability to organize and mobilize vociferous segments of the population to sometimes violent demonstrative—and widely reported in the media—action or to organize their supporters for letter-writing campaigns to their representatives.

6.2.3 Geopolitical Groups

These groups are vested interest parties, united because they have in common closely defined economies based on their geography or natural resources in their regions. For example, maritime regions have a fisheries industry as a key economy and the Prairie Provinces in Canada or the midwestern states in the United States have cereal crops and livestock production as major common interests; a lively debate regarding imports and exports results as each defends its turf. These groups have specific regional interests, which are defended fiercely. Such partisan activities influence food legislation.

6.2.4 Expert Panels

Expert food panels are made up of prominent food scientists, nutritionists, biochemists, agronomists, marine researchers, processors, indeed all those associated with food, its production, its manufacture, its retailing, and its consumption. They provide "informed opinion" on matters of health, safety, nutrition, and agriculture and fisheries. In this manner, governments expect to receive a rational basis for any legislation dependent on scientific or other pertinent issues. There are at least two inherent weaknesses in this appeal to scientists posing as experts; science cannot prove the absence of harm; therefore, absolute safety cannot be guaranteed and they should not say it can. They can and have been wrong, cf., lead in gasoline is safe, safety of silicone implants, or derision by his peers of the work of Stanley Prusiner, who eventually won the Nobel Prize, on prions. Scientists are guided by risk analysis (an intensive investigation of the risks as they see and interpret them) and the so-called precautionary principle (a euphemistic term for Murphy's law—i.e., if it can go wrong, it will—and is also the principle underlying the hazard analysis critical control point program). Second, each expert brings their own biases and prejudices that are predicated on the financial support of their sponsors or their own commercial territory. "Science is not neutral" was a slogan that shocked the public back in the 1970s when it was published by the British Society for Social Responsibility in Science.

Where there are conflicting scientific opinions or disagreement over the interpretation of scientific data, governments have only the following options:

- They delay enactment of legislation until a clear opinion can be had. This is an option that satisfies no one and certainly intensifies feelings between the parties concerned. The consuming public is simply left confused by the opinions of the scientific authorities unleashed by both sides.
- They pass legislation based on the best available information even if it proves later to be wrong. This option, too, satisfies no one. Those whose expert opinion it went against are not happy, and laws, once passed, are very difficult to have overturned (cf., the Delaney Clause).
- Akin to the first option, governments can simply quash a report until a more politically expedient opportunity to release it.

Interpretation of what is the "best available information" evokes fierce debate in lobbyists and other vested interest groups.

> No lesson seems to be so deeply inculcated by experience of life as that you never should trust experts.
>
> **Lord Salisbury**

6.2.5 Industrial Sector

The food processors have their trade associations, their commodity-based associations (quite apart from their marketing boards), and, to an extent, marketing boards to protect their interests and make pleas to governments. Meat interests in the United States were disturbed at the government's recommendation to curb meat consumption. Sugar interests are upset with recommendations to cut refined sugar consumption, and the soft drink association has raised concerns about efforts to curb their availability in schools. Even the broccoli interests became upset when President Bush proclaimed his aversion to the product.

Changes are made to laws or recommendations in accordance with the political influence groups have with their local representatives.

6.2.6 Summary

These, then, are the major influences on the legislation and regulation of the food product development process. The anti-this and anti-thats, the pro-this and pro-thats, the vested interest groups, all attempt to see their causes espoused and all are voters. Governments hear and enact on the basis of their own self-interest; they wish to stay in power (therefore the quashing of reports). These groups should be recognized for their influence on food and their potential impact on novel foods and food processes in development. Development takes time. Developers should recognize that any turmoil within the legislative arena involving safety or health effects of products or processes under development will be bound to raise the ire of some group or invoke some restrictive legislation.

The bottom half of Figure 6.1 depicts the governmental department that is affected by the influences in the top half. It is from these that blue papers and white papers flow to get reactions that eventually find their way into legislation. Legal departments must be cognizant of all food laws pertinent to their clients. This becomes complex when export of products demands adherence to foreign legislation.

6.3 Food Regulation and the Development Process

6.3.1 Legislation, Regulations, and Safety: A Dilemma

In Chapter 1, three objectives of government legislation were expounded (Wood, 1985). At best, however, this legislation with its resultant regulations and the interpretation of these by government bureaucracy supply little absolute assurance of a safe and wholesome food supply but rather one that meets economic standards. This is perhaps a somewhat harsh statement but exploring Wood's three objectives further, one finds that a government publishes regulations that

- Establish limits for the presence of toxic or proscribed chemicals or agricultural residues in a food. This requires inspection and analytical facilities that is an economic burden for governments. However, only known hazards can be regulated and tested for. New and hitherto unknown hazards are impossible to predetermine. This is a major concern for new products developed using novel technology (new processes), materials (the products of biotechnology), and ingredients (phytochemicals and nutraceuticals).

- Set limits for the presence of known microbiological hazards in foods. Again, marketplace inspection and subsequent analysis are required, and bureaucratic expansion and costs grow.
- Set limits for the presence of extraneous matter in foods. The result is more growth in bureaucracy for inspection and analysis.
- Define composition of traditional, standardized foods, that is, establish standards of identity for foods. Standards of compositional identity thus define some common foods, but these may limit or prevent any improvements in these foods, for example, a more nutritious ice cream would be outside the standard and couldn't be called ice cream.
- Establish grade standards for commodities and many other processed foods. These are minimum standards. Products exceeding these minimum standards would be too costly; therefore, all processors target the minimum.
- Combat economic fraud such as underweights, short counts, and deceptive packaging. They regulate weights, measures, and package sizes to eliminate deception, and regulate profusion of container sizes that make the calculation of unit costs difficult and eliminate false impressions of contents through overpackaging.
- Provide product information with standard nomenclature of foods, itemized lists of ingredients in descending order of magnitude, and nutritional data.
- Publish advertising and promotional guidelines to prevent misrepresentation of products and their properties to customers.

These regulations describe finished-product standards. They have little to do with either safety, quality, or wholesomeness. Foods that contravened any of the above standards should not have been produced in the first place. If defects are discovered by food inspectors in the marketplace, these faults represent gross incompetence, negligence, or mismanagement on the part of the manufacturer. Yet these errors do occur.

To overcome the shortcomings of the above respecting safety, industry and government moved to the regulation of the manufacturing process by creating codes of practice or programs (e.g., HACCP program) for the safe handling, manufacture, storage, and sale of food. These establish that foods must be processed in a manner and with control points such that they are not contaminated during processing or ensuing distribution, storage, and retailing. Legislation extends to the premises where food is prepared or stored. Building codes, processing regulations, or codes of practice are attempts at ensuring that the environment in which food is prepared is such that that food is not contaminated during production or that the preparation or the processing was designed to assure a safe and wholesome product. They assure that the purveyors of foods (or their distributors) in the many different marketplaces received a safe product. Customers and ultimately consumers should receive safe food products if the regulations and in-plant procedures pertaining to safety were followed. Such rules were followed in the plant in Canada where a listeriosis outbreak occurred in 2008.

By enforcement of these practices through inspection and analysis or by self-regulation (processors regulate themselves), food is considered to be safe. Or is it? Countries with strict food legislation or with a long history of such legislation or employing strict food regulatory and enforcement agencies do not necessarily have food supplies that are free from problems of public health significance. Food legislation, no matter how large a body of it there is, nor how strongly it is written, nor how vigorously it is enforced, cannot guarantee either the maintenance of quality or the safety of all food manufactured and consumed within any country's borders, to wit, the listeria outbreak in Canada and the meat recalls in the United States (*vide infra*).

For example, in the United Kingdom, in 1997, more Britons suffered food poisoning than had been recorded since records were kept (Coghlan, 1998). Britain has a long history of and an excellent library of food legislation. The Agriculture Department of the United States in April 1999 ordered a recall of meat products including hot dogs, luncheon meats, and various sausages made by an Arkansas company; the products were declared unfit for human consumption. The United States has probably the largest government organization in the world dedicated to food safety and inspection as well as one of the most comprehensive bodies of food regulations and legislation in the world. In Canada in 2008, there was a major recall of delicatessen meats involving several brands processed by Maple Leaf Foods despite this plant following strict guidelines as laid down by the government—indeed, even surpassing those guidelines.

Food legislation can only be developed to answer to known hazards. Knowledge of food hazards is based solely on processes with a known history of maintaining food safety. Nevertheless, the process of making salami, a food with a long history, is being questioned regarding whether it is a safe process against new variants of an old microbiological hazard. Existing regulations cannot safeguard the public against

- New hazards that may be associated with novel or innovative foods
- Food processes with an unknown history of safety or that lack an established theoretical basis for safety
- Microorganisms that have mutated into more virulent and more resistant forms

The lack of history respecting the long-term safety of genetically modified foods is one argument that those opposed to genetically modified foods have used.

Given that legislation alone cannot ensure safety, it must be complemented by adherence to codes of good manufacturing practice that are accepted by all and the need for developers to design safety into their products. There are, unfortunately, some food manufacturers willing to shortcut procedures for profits. This is the dilemma faced by those trying to safeguard food supply and ensure safety in food products. It is doubly difficult in new food products where new hazards are unknown.

6.3.2 Role of Lawyers

The legal implications in new product development, and there are many, cannot be overlooked by the product development team. Advice from a lawyer knowledgeable in food law or from a resource specializing in food legislative matters early in the process will prevent costly surprises later. The developers need to know the implications that current food legislation or pending legislation might have on the development process. This resource advises the team against wasting time and money pursuing developmental goals that contravene the regulations: for example, development of a product imitative of a standardized product but is compositionally different and contravenes the standard. The implication for marketing is enormous; if the standard is not followed respecting the permitted ingredients, the product may have to be labeled "imitation product" or have some quirky computer-generated name found for it. An imitation label frightens customers off, while the latter option requires education of customers about the meaning of the new name.

The legal department serves in two essential areas in product development:

- Interpreting the law and its regulations as these pertain to label and advertising copy for misleading statements (Anscombe, 2003): They review ingredient statements and investigate the appropriateness of product nomenclature and the registration of patents and copyright matters. They advise on and otherwise protect the intellectual property of the company (Newiss, 1998; Garetto, 2003).
- Overseeing contractual arrangements: Any contracts with co-packers, for consulting services, negotiating partnership arrangements, licensing arrangements for new products, or sales of technology are best left to lawyers. They also prepare employment contracts for executives brought in to manage new divisions, etc.

Another important service provided by lawyers that is not directly related to the new product development process is the defense of the company against any litigation.

6.3.3 Legislating Quality and Safety

The quality of food cannot be regulated by fiat. Grade standards for foods such as meats, canned fruits, and vegetables and standards of identity for composite foods are merely statements that the finished product meets the stated standards for that product or may be called by the common name for the standard of identity. This is not quality; it is avoidance of fraud. Since most composite new products (multicomponent foods) do not have a grading system or a standard of identity, these foods do not meet a standard for quality, and their quality resides with the brand name and its concept of quality that exists in the minds of consumers for that brand. Neither developers nor consumers are protected if novel, innovative food products that adhere to *existing* regulations are introduced (EU, 1997; Huggett and Conzelmann, 1997). There are not any regulations in most countries for *new* foods; it will be up to developers through their lawyers to determine what constitutes a new food.

Quality is associated, in both the customers' and consumers' minds, with respect to some characteristic, for example, quality with respect to color, flavor, chunkiness, creaminess, or with respect to nutritive content, but, more often, quality is associated with a brand. Many companies have their brand of pride and, less widely known, their "off-brand," still very good but not up to the brand of pride. It is the manufacturer's brand that identifies quality in the eyes of customers and consumers.

Labeling regulations permit only statements that the product within the container adheres to certain minimum characteristics specified for that product. In essence, they are standards of commerce. Few manufacturers would exceed these minimal standards in view of the fact that their costs would increase if they did so and soar over the costs of their competitors as price-conscious shoppers buy.

Regulations for ingredient and additive usage vary from country to country; therefore, a product developed for export markets may have as many formulations as there are foreign destinations. These requirements can have a protectionist flavor to them: export to Italy of a pepper sauce whose base was vinegar, for example, had to be made with wine vinegar. Meal replacements as a new product category must meet nutritional standards of the meal they are replacing in the country they are destined for. Formulations of products for export must be screened against all legal restrictions that pertain at their destinations. Legal firms specializing in food law have resources and databases available to them that provide such information.

There are legal implications involving patent or copyright infringement as new equipment is designed or old equipment adapted for new processes during development. Conversely, the team must also recognize when equipment or processes they develop are patentable and protect their intellectual property by filing for patent protection through their lawyers. Thermal processes must be verified for safety before being filed with the proper governmental agencies and lawyers must see to this.

Packaging regulations require the use of only environmentally friendly materials to recyclable packaging. They may restrict the use of certain laminated films and regulate materials that transfer packaging materials to the food it is supposed to protect. Governments, very sensitive to the concerns of environmentalists, have decreed that all packaging be recyclable or require a pickup program for empty containers.

Marketing departments work closely with their legal departments on label statements, packaging design, and trademarks; on guidelines for promotions and advertising; on the legality of the claims they wish to make; and even on the sizes of containers they wish to use. There are unwritten laws too. Not only can advertising not be misleading, but it must also not be sexist or racist—this invites protests. Legal advice is necessary to walk the thin line and avoid bad publicity during product introductions.

Names chosen for new food products cannot be misleading. Goldenfield (1977) presents, in a now-dated article, an excellent example of how the impact of food regulations plus company-based restraints influences product development. Goldenfield uses as an example a refrigerated whipped fruit-flavored puree that the marketing department insisted contain only fruit pulp and fruit juice to take advantage of declaring all natural ingredients. Naturalness was the key to the product concept. It was to be marketed under the product name "Fruit Fluff." Problems began when the technical team attempted to keep within cost parameters established by the strategic development team. The "FRUIT FLUFF Orange Dessert Whip" underwent a metamorphosis from "FRUIT FLUFF (Natural) Orange Flavored Dessert Whip" to "FRUIT FLUFF (Natural) Orange Flavored Dessert Whip With Other Natural Flavor" to finally end up as "FRUIT FLUFF Artificially Flavored Orange Dessert Whip" as technical and budgetary considerations played havoc with the original market concept.

Not all names of brands can move globally. In short, advertising and promotions in one country cannot travel to another; some things do not translate well. For example, General Motors went with the slogan "Body by Fisher" that was translated to "Corpse by Fisher" in the Netherlands; Imperial Margarine's "Magical Crowns[TM]" drew antagonism in countries that had overthrown their monarchies; Chevrolet's Nova[TM] drew derision in Spain where "no va" means "doesn't go"; this latter has been disputed. Advertising is not global. This aspect of nomenclature and promotion must be researched carefully lest it could derail an otherwise well thought out introduction in a foreign country.

6.4 Environmental Standards

Food packaging is not the only environmental hazard of food manufacture (Akre, 1991). Food plants are heavy polluters, largely because of the trim waste and water used in cleaning and preparation of fresh produce, the considerable amounts of water and detergents required in plant sanitation, and the odor and noise produced. To process seasonal crops, they must operate night and day, which can be an annoyance to local residents. Plant and animal waste must be sent off-site to either rendering plants or landfill sites, and this transport causes both congestion and noise. Garbage disposal is expensive because of transportation costs, a license fee, and usually a load fee for dumping at landfill sites. Ideally new products should not contribute unnecessarily to the already heavy burden plants must pay to remove or treat waste in an environmentally friendly manner nor lead to the company being condemned as a bad corporate citizen. This is an issue to be considered in the development process.

Legal departments advise companies when they are not complying with local regulations and provide advice on alternatives that are available for them respecting environmental compliance programs. These alternatives are themselves expensive and often shift the onus of environmental compliance elsewhere, for example,

- Sending the waste away, which involves transportation, handling, and tip charges; this does not solve the disposal problem as suitable dump sites must be found.
- Recycling the waste to recover any valuable by-products or converting the waste to a marketable product (e.g., mulch, fertilizer, fuel). This too involves an expensive, parallel development process and leads the company away from its core business.
- Where possible and permitted, using the waste as fuel for heating and power generation.
- Where possible, preprocessing of produce at the farm level to minimize processing waste at the plant. This simply moves the disposal problem to someone else's backyard.

Boudouropoulos and Arvanitoyannis (1998) discuss at length the impact on food industries of environmental standards published by the International Standards Organization (ISO 14000), which have been adopted by many chemical and automobile industries.

6.5 Summary

New products present an interesting legal dilemma for developers and their legal departments. New processes for which there has been no historical establishment of safety will always be challenged in the law courts by some groups who have found some real or imagined grief with products made with the new process. New ingredients (e.g., for nutrified foods, genetically modified foods and organisms) will meet a challenge. Environmentalists will challenge, indeed are challenging, biotechnological applications as these are incorporated into traditional food production systems. Companies urgently require a legal presence within their development teams to be aware of the legislative climate worldwide as they plan to introduce innovative, never-before-seen products derived from unconventional sources. One international soft drink company is so sensitive to the harm that legislative changes could do to their image, product, and promotional programs that it employed a gentleman whose sole responsibility was to attend any and all meetings with food legislation on the agenda.

> Edible, adj., good to eat, and wholesome to digest, as a worm to a toad, a toad to a snake, a snake to a pig, a pig to a man, and a man to a worm.
>
> **Ambrose Bierce**

Edibility—and, one can assume, safety and wholesomeness—is apparently something for the law to interpret. Somewhere in Bierce's chain, someone is sure to raise a legal objection.

7

Quality Control: Protecting the Consumer, the Product, and the Company

Look beneath the surface; let not the several quality of a thing or its worth escape thee.

Marcus Aurelius Antoninus

7.1 Introduction

There has been much discussion about, and hence confusion over, the nomenclature regarding this function within food manufacturing. It was referred to first as quality control; this was then followed by quality assurance and then it became quality management and, finally, total quality management also referred to as TQM. Somewhere in between these terms, there was product integrity, a term that, frankly, I preferred. However, by quality control, I refer to that function in a company responsible for assuring that all processing, product, environmental, and worker safety standards are adhered to and that all reasonable and practicable precautions to protect the product from hazards of public health significance have been taken. I include sanitation, worker training, worker hygiene, pest control programs, observance of GMP (good manufacturing practices), and establishment and observance of hazard analysis critical control point (HACCP) programs. Some make a distinction between quality control and process control. For simplicity, that distinction will not be made.

7.2 The Ever-Present Watchdog

The quality control department, an internal policing unit, satisfies itself that all the safety controls designed into the product and the process to protect the product's wholesomeness, integrity, and quality are adhered to within the specified written limits and to remain alert to possible emerging hazards to company products. It must, as the above quote says, look beneath the surface. They must, on the basis of experience, present the development team with all their concerns regarding the sensitivity of the process and the product to hazards and provide guidance in putting in place practical and practicable solutions. Its role should not be adversarial either in its day-to-day functions or in its contribution to product development.

Engineers and the research technologists prepare flow diagrams of the process and describe operational standards at critical points; they recognize limitations of plant equipment and identify modifications that will be required in the plant. With the cooperation of the production staff, developers must establish specifications for raw materials and ingredients. It is now up to the quality control department to

- Select reliable and rapid analytical procedures to monitor the maintenance of the product's desirable quality attributes and to monitor that practices for the safety the product have been followed. Procedures and instrumentation usable in-line and in real time are preferred.

- Verify the safety of the sterilization processes with process specialists and determine the shelf life of the product under anticipated storage conditions.

- Develop a practicable HACCP program with control features based on data and process line layouts supplied by the engineering and research departments that provides the desired level of product integrity.

- With the production department train staff in new analytical procedures and provide training of new skills to line personnel if the new products require processing technologies and ingredient handling for which there is no prior history in the plant.

- Conduct sensory evaluation tests and microbiological analyses throughout development to monitor organoleptic values and effectiveness of stabilization processes.

7.2.1 Sensory Analysis in Product Development

A major requirement of food products is that they are flavorful, tasting typical of the product or as advertised. All the sensory characteristics to be associated with a new food must be met; therefore, its sensory appeal must be measured. Herein begin problems in measuring something as subjective as sensory appeal.

7.2.1.1 Sensory Techniques

Peryam (1990) described the historical development of the understanding and application of sensory evaluation to product development. Sensory appeal is difficult to measure. An excellent review discussing sensory analysis is presented by Piggott et al. (1998). Sensory testing must be carried out rigorously with proper experimental design and correctly selected and trained sensory panelists to minimize the errors that can arise in sensory analyses; *this is why technologists knowledgeable in the techniques should carry these out*. Data from organoleptic tests require special statistical skills for their analysis and interpretation; *this is why sensory measurements should not be left to those who have had no training in the field of sensory analysis*. Some populations of tasters, such as children, present unique problems in assessing sensory appeal; *this is why special skills are required in sensory analysis*. It should be apparent that sensory evaluation measurements and their analyses should be carried out by technologists trained to conduct them properly. It is often necessary to farm these tests out to specialist firms. At the very least, a company needs some member of the research and development staff familiar enough with the technology to be aware of their own limitations in conducting the tests and the limitations of the tests themselves.

There are four superb references, any one or all of which sensory technologists or any person contemplating undertaking sensory evaluation measurements should have as a vade mecum for easy reference:

- *Laboratory Methods for Sensory Analysis of Food* written by L.M. Poste, D.A. Mackie, G. Butler, and Elizabeth Larmond (Research Branch, Agriculture Canada, Publication 1864/E, 1991) is a compact handbook written in an easy to assimilate style (Poste et al., 1991).

- *Sensory Evaluation Guide for Testing Food and Beverage Products* prepared by the Sensory Evaluation Division of the Institute of Food Technologists (1981) is a concisely written guide describing the tests that can be used and providing references where one can find further details.

- *Tasting Tests Carried Out at the Leatherhead Food Research Association* (Williamson, 1981) prepared by Marion Williamson is a very readable document giving many of the descriptive terms used in sensory analysis and providing, in an appendix, concerns for the safety aspects of tasting panels (expert panels) and the need to have a clearly defined statement of policy on the use of tasting panels.

- The Bureau de normalisation du Quebec (Ministere de l'Industrie), du Commerce et du Tourisme, Gouvernement du Quebec (Quebec Bureau of Standards, Ministry of Industry, Trade and Tourism, Government of Quebec) has published an excellent series of documents (in French) covering

    Vocabulary (BNQ 8000-500; 84-03-06)

    General methodology (BNQ 8000-510; 84-03-06)

    Sample preparation (BNQ 8000-512; 84-03-05)

> Scaling techniques (BNQ 8000-515; 1982-10-08)
> Triangular tests (BNQ 8000-517; 84-03-06)
> Paired comparisons (BNQ 8000-519; 84-03-06)
> Designing a sensory analysis location (BNQ 8000-525; 1982-08-27)
> Determination of taste acuity (BNQ 8000-560; 1982-10-08)
> Methods for determining flavor profile (BNQ 8000-570; 84-03-05)

Those seeking guidance should look to see if any of these publications have been updated.

There are two types of sensory evaluation:

1. Objective sensory evaluations. These are all generally difference methods, that is, can the tester pick out the odd sample or the difference between samples?

2. Subjective sensory evaluations. These are descriptive tests (i.e., "tell me what you taste") usually used with trained panelists. Piggott et al. (1998) identify the flavor profile method, texture profile method, quantitative descriptive analysis, and spectrum method.

Each serves quite different purposes; they cannot be interchanged. Far too often, the purposes for which the sensory tests are meant are either misunderstood or are ignored. Companies will use the results of an objective test as indicating a sensory preference for one product over another. This is wrong and can lead to incorrect and disastrous decisions in product development.

7.2.1.2 Objective Sensory Testing

Objective sensory evaluation tests are used for just that, to get an objective evaluation of some sensory appeal. Other names used for objective tests are more descriptive of their purpose: analytical sensory tests; expert panel tests; difference tests. The questions asked of panelists are as follows:

- Does a difference exist between the samples?
- How would you rank the samples with respect to the strength of some sensory characteristic?
- How would you describe and rank the sensory characteristics you can identify in a sample (profiling)?

Objective tests are used to determine if a difference exists between products with respect to some sensory quality or between a reference sample and a test sample. For example, Jeremiah et al. (1992), using a procedure called consensus profiling, followed changes of chilled pork loin packaged under carbon dioxide and with vacuum and stored frozen for up to 24 weeks.

They used highly trained flavor/texture panelists and in discussions developed a consensus profile of the pork changes during storage. Details of their procedure of analysis are in the reference.

Objective testing requires trained panelists. Powers (1988), Rutledge and Hudson (1990), Poste et al. (1991), and Rutledge (1992) provide advice on the selection and care to be taken in the training of panelists. Their training includes how to recognize different sensations, to be discriminative, and to quantify these sensations against recognized standards using a common vocabulary that they can describe sensations.

Many food companies select their panelists from staff; this practice can have dangers. Sensory analysis is obviously not the only job of those selected. Indeed, sensory analysis may intrude upon the panelist's day-to-day company responsibilities. Sensory specialists must be careful both with selecting panelists, keeping them in top tasting form and not intruding on their workloads so that they are not irritated nor feel threatened by their supervisors if normal workloads are tardy. Panelists, on their part, must like what they are doing, be in good health, have no genetic or psychological sensory biases, and not be harried into taking part in panels. Williamson (1981) reviews at length the cautions to be observed in selecting panelists.

A large pool of panelists of sensory technologists is necessary to compensate for the usual absenteeism and nonavailability that occurs. Another reason for having a large pool is that not all panelists are suitable for testing all sensory characteristics under study; there are genetic factors influencing taste acuity. It profits sensory specialists to keep profiles of each panelist, listing availability, threshold levels of discrimination and sensory record in the various testing sessions the panelist has attended (see Powers, 1988).

7.2.1.3 Subjective or Preference Testing

In preference testing, also called subjective testing or affective testing, panelists are presented with a choice of samples and must state which sample is preferred. The word "preferred" is understood to mean *most acceptable, tastes best, looks best, would buy*, or any other expression indicating greater satisfaction.

There are three main variants in the way preference testing can be carried out:

1. A variant of the focus group
2. A central location test
3. An in-home test

Focus groups were discussed in Chapter 4. Panelists, representative of the targeted consumer, are asked to taste the product and fill in a questionnaire concerning the product. This process will be repeated several times and always with selected consumers.

For a central location test, the new product development team will take the product under test to some group meeting location where the team will be assured of a broad cross section of potential testers. Many companies select church groups, veterans associations, shopping malls (if the surrounding district is appropriate with respect to the target), ethnic clubs (if such are the target consumers), or social organizations for their testing. These groups can be selected from locations all over the country. A carefully prepared questionnaire is used by the new product team to evaluate consumer preferences.

Focus type tests and central location tests provide developers with control over the preparation and serving of the samples. Qualified personnel carry out the interviews or are at least available for interpretation of questions on self-administered questionnaires.

In the third variant, the in-home test, preselected consumers (there is some control of the consumer) are sent coded samples of a test product to be prepared at home. Developers have no control of the preparation of the samples. Nor is there control of the environment in which the test is carried out. How well were the instructions followed, with all the distractions that are possible in the home? Were the time, place, and other circumstances in the home conducive to a good test? These are unanswerable questions for the survey takers. After testing the product, consumers then fill out and return the accompanying questionnaire or they are interviewed by telephone.

Preference testing techniques are compared in Table 7.1.

7.2.1.4 Panelists

Choosing who should be panelists is influenced by what kind of data are wanted and what the data are to be used for. Trained panelists (including company personnel) should not be used for preference (subjective) testing; they are simply not at all representative of consumers the company wishes to target.

An example of how an expert panel missed the mark with respect to consumer tastes has already been described in Chapter 5 where the trained, expert panel of a tortilla chip manufacturer had determined that an incipient taste of oxidative rancidity was offensive in an established tortilla chip product. The expert panel was proven wrong: consumers preferred that flavor.

Unfortunately, many companies often use their expert panelists gathered from amongst their staff to get data on the acceptability of a product. Company employees, whether from the plant floor or from the offices, have too intimate a knowledge of their products, of what they expect their products to taste like and may be expected to be prejudiced. The preferences of knowledgeable but untrained panelists have no marketing validity with respect to the population. These people are not trained tasters but they do have an expert knowledge of what taste image their company's products have or should have; this may bias their opinions. That is, they are tasting a brand, an image, that they know and are familiar with. They are not typical consumers. They have a bias related to a pride in their work. They are,

TABLE 7.1

Preference Testing: Three Main Variants of Each with Characteristics

Test Variant	Characteristics
Focus groups	They require 8–12 carefully selected, usually untrained, participants. Tests are repeated two or three times.
	They can be easily repeated but over repetition of qualitative data can be unrewarding.
	They require a professional leader to conduct tests.
	Control over display and preparation of product is good.
	Results are obtained quickly.
	Only qualitative data is obtained and this data is non-projectable.
Central location tests	These tests involve larger numbers of people; i.e., social clubs, church groups, etc.
	There is poor selectivity of respondents but the test can be easily repeated and comparatively large numbers of respondents can be reached.
	They are somewhat more expensive to conduct; they require a well-prepared questionnaire but testers have ability to explain questionnaire.
	Control over product preparation is excellent.
	The results are quickly obtained and quantitative and qualitative data are obtained but data is projectable with caution.
In-home tests	These can involve several hundred respondents and there can be good selectivity of respondents.
	These are usually a one-time test (a mini-test market) and they are more expensive to conduct than either of the preceding tests.
	These require intensive follow-up with well-prepared questionnaire; there is poor control over how product is prepared and over the test circumstances under which product was used.
	Results are slowly obtained (hence intensive follow-up needed); both quantitative and qualitative data are obtained, which is projectable with caution.

in all respects, experts on that brand image. I have found that comments such as "This isn't our company's flavor" or "I wouldn't want our company to put out a product like this" will prevail and confound whatever results are obtained.

These testers know nothing of the company's objectives in this test. The company may be working on developing another distinctive taste image in a very different marketing niche. Where, however, companies have a strong brand image these same panelists may have a better idea of what the brand can carry than does management. The point must be made strongly that company workers, particularly the workers in the plant handling the product, can bring a very definite brand (image) bias to preference sensory testing.

MacFie (1990) and Gutteridge (1990) discuss sensory techniques affecting consumers. Macfie describes characteristics affecting consumers' choices and explained "free choice profiling" and preference mapping. Gutteridge

(1990) discusses the technique, repertory elicitation with statistical treatment (REST) (Mathematical Market Research Ltd., Oxford, United Kingdom; see also Thomson, 1989) as a method for finding the most appropriate market niche for products.

7.2.1.5 Other Considerations in Sensory Analysis

What is the team conducting a taste testing for? There must be a clear notion of what is required from the test. Does the development team want to determine whether they have produced the best formulation of a product as compared to other test samples or compared to some absolute criterion? What is that criterion? Or do they want to determine whether this particular product is as good as, or better than, the competition's product? That is, are the results of technical product development under investigation or are marketing personnel trying to understand the product in its full marketplace context, where branding may be a factor? Answers to these and similar questions have a decided impact on how the tests will be conducted and who will do the testing.

Martin (1990) broke product characteristics into three components:

1. Physical characteristics of a product, which consumers readily recognize as poor, acceptable, or high quality; the technologist is trying to formulate only the high quality (ideal) attributes into the product.
2. Image characteristics: Martin recognizes these as most dominant in perfumes, fragrances, and tobacco products. They are not unknown in food products, for example, liqueurs, liqueur-flavored instant coffees, and exotic gourmet sauces.
3. A combination or interaction of physical and image characteristics. This comes very close to defining the essence of a brand.

This interaction is interesting since it introduces a problem. Martin (1990) describes it as follows: when testing similar products in "blind" sensory tests (e.g., various brands of beer), consumers would place the product ideal (the "best" product) some distance away from the test product and with other unbranded competitive products. When, however, *branded products* are evaluated (i.e., the test is no longer conducted "blind"), all products are ranked closer to the consumer's ideal product. Branding is an important factor in sensory testing.

7.2.1.6 To Test Blind or Not?

Should a product be tested branded or unbranded, based on the foregoing? After all, the product will have to face other similar branded products in the marketplace. No-name products are sort of branded and many retail chains now sport their own brands; these are often simply no-names with the chain's label. Brands have uniqueness: they communicate images of quality as perceived by that customer, a product persona. Therein lies a brand's value. A brand is comfort, known values, security to consumers.

There is overwhelming evidence that branding does influence tasters in ranking similar products and in picking a preferred sample. Martin (1990) provides evidence of vastly different assessments of various characteristics of ciders, beers, and chocolate confectionery when the products were tested branded or blind. When one beer was tested blind, Martin found that it was rated higher than when it had been tested identified.

Moskowitz et al. (1981) describe an interesting analysis of consumer perception, magnitude estimation scaling, carried out on chocolate bars. They found that branding encouraged a product's acceptability; furthermore, for some products, it would be branding and not a product's quality that lifted a product's acceptability. (One must be aware of a difference here: branding influences acceptability; branding is image, a "persona," a product has. It does not influence the evaluation of objective characteristics posed by questions such as "Is product A smoother than product B? Sweeter? Sourer?")

Schutz (1988), Scriven et al. (1989), and Gains and Thomson (1990) describe sensory techniques to evaluate consumer attitudes to foods and determine the circumstances under which consumers would use or serve particular foods (contextual analysis). Contextual analysis is a useful guide to marketing personnel in determining market niches for products.

The questions must be repeated: What is the purpose of the sensory test? Does the company want to know whether it has the best formulated product, or does the company want to know whether it has a product that is preferred over the competition's product? This represents a technological capability question vs. a marketing capability question: technology vs. psychology. As Martin (1990) aptly put it, "It may not be necessary to develop a clearly superior product in sensory terms, if reputation can sufficiently enhance one which is the equal of the competition." Hardy (1991) is more adamant and maintains that superior tasting food and beverage products are no surety of a loyal consumer base.

When faced with unbranded popular products, most tasters cannot distinguish between the unbranded products nor even successfully choose their favorite brand. Hardy (1991) found that most tasters not only could not distinguish consistently between competitive products but did not improve with experience without formal feedback (i.e., training). Sensory testers must decide what is the purpose of the test and consider whether the biases that branding may introduce will adversely affect what they are trying to find out.

Another problem arises with blind testing that is associated with wines and perhaps other foods and beverages. Mantonakis et al. (2009) studied preference testing on wines to determine whether the order of presentation of the wines influenced tasters. They found a large primacy factor (first up is best) over the last in the sequence (sequences were of two, three, four, and five wines). There was another interesting result. After all the tasters (students at Brock U and from the local community) had been tested for their wine knowledge, the novices showed a large primacy effect when their data had been separated out. On the other hand, the more knowledgeable tasters showed a tendency to recency as the number of samples increased. The tasters all had the same wines.

7.2.1.7 Can All Tasters Discriminate?

Through the application of their magnitude estimation scaling technique to the survey of chocolate candy preference, Moskowitz et al. (1981) note—at least for chocolate candies—that there could be age differences and sex differences in the response to branding. Age and sex of the taster influence how a product is perceived. Clearly, this suggests that great care is required in the selection of panelists. Whom are the developers targeting?

There are other taste factors besides age and sex to be accounted for in tasters. Reaction to the genetic taste marker, 6-n-propylthiouracil, marks tasters. Some cannot taste it; some have a slight ability to taste it (medium tasters); and there are those with violent reactions to the taste (strong tasters). Drewnowski et al. (2000) found a strong statistically significant relation between strong and medium tasters who disliked the cruciferous and other raw and green vegetables and non-tasters who liked them. Duffy and Bartoshuk (2000) found women's liking for sweet and high fat foods declined as their ability to taste the genetic marker increased. Men did not have this taste preference association with the marker.

There can be problems in the taste perception of people with total or partial anosmia. Odor is a prominent factor in taste and flavor perceptions.

Ideally, sensory panelists should be representative of those for whom the new product is targeted. Ideally too, all geographic areas where the target consumer can be found should be represented in consumer research tests; taste preferences vary geographically. However, the ideal panel tasters can seldom be gathered without assistance from specialists.

Consequently, food companies enlist the aid of product testing/consumer research companies. These groups keep extensive files of potential panelists categorized by age, sex, ethnic background, economic status, and other characteristics that may be important to product developers. McDermott (1990) discusses recruiting for sensory testing, problems encountered with setting specifications on recruitment, and how these involve cost considerations.

Money limitations may provide constraints on both the number of subjective tests that can be carried out and the size of the panels used in them. When to test, how to test, what to test and how big a test to conduct are decisions companies undertaking product development must be prepared to make.

7.2.1.8 Using Children

Products designed for young children present unique problems for sensory analysts. Young children are difficult to work with. They present problems in communicating with them and in determining how to measure their preferences. Skilled panel leaders are required and it is virtually a necessity that the relationship be one leader to one child. Obviously, such testing is expensive for the extra skills that are required to elicit responses unbiased by the leader. Then there are decisions concerning the test itself. How big should the test be with children? How often can the test be conducted?

Kroll (1990) describes another difficulty with testing children: what scaling system to use? She tested children using one-on-one interviewing, a self-administered questionnaire and three different types of rating scales. She found that while all three scales discriminated at the 10% level, a simple in-house scale used by Peryam and Kroll (Peryam & Kroll, Chicago, Illinois) was better than a traditional hedonic scale and (surprisingly) better than a face scale.

7.2.2 Using Electronics: The Perfect Nose?

The advent of gas chromatography to analyze volatile or volatizable materials must inevitably lead to the use of the "electronic nose" to analyze the many components that make them up flavors. Warburton (1996) described one commercial unit and its use in identifying nut varieties and in distinguishing good from bad walnuts. Several types of units are described in detail in a review by Schaller et al. (1998). They describe their use with meats, grains, coffee, beer and other alcoholic beverages, fish, fruit juices, and soft drinks.

Neugebauer (1998) discusses the electronic nose in detail, comparing it with the human olfactory system, neuronal networks and sample recognition. Advantages and disadvantages of the noses in quality control and sample recognition are discussed with comparison to classical methods.

The difficulty with electronic noses is that they do not provide information on how or what the consumer's reaction to the stimulus is. They separate and allow the measurement of quantities of volatile material and permit their identification but do not give any idea of how the consumer might integrate all these volatiles to describe the consumer's sensation and assessment of that flavor. An example might help here: gas chromatography was used to measure the capsaicinoids in hot pepper sauce. Uniformity of the heat principle from batch to batch was wanted. Unfortunately, the capsaicinoids are a group of compounds not all of which have the same heat-stimulating property. Thus total capsaicinoids was no measure of heat and capsaicin, the hot one, had to be identified and measured separately for control purposes.

Electronic noses have been used to identify sources of spices and even adulteration of flavorants.

7.2.3 Shelf Life Testing

Developers use terms such as *high-quality shelf life* or *acceptable-quality shelf life* or *useful storage life*. These terms have no meaning to either customers or consumers.

A product's shelf life is a verification of the stabilizing systems designed into the product. Companies certainly hope their product does not stay on shelves beyond its shelf life; that indicates an unsuccessful product. Food companies cannot release new products into a market, especially a test market, without knowing how stable that product will be. This is especially true for short and medium shelf life products. The seemingly simple requirement to provide a shelf life estimate is fraught with many difficulties, which Curiale (1991) discusses with particular reference to microbiological shelf life. His and other points to consider in shelf life determinations follow.

7.2.3.1 Selecting Criteria to Assess Shelf Life

First, some criterion that changes as quality or safety degrades can be measured which is appropriate for the product must be selected. A characteristic that changes suddenly or abruptly without some measurable antecedent or precursor is not satisfactory as a criterion of change; for example, by the time the peroxide value of fats rises perceptibly, the fats have pronounced odors. Criteria that may be chosen are

- Microbiological changes (Curiale, 1991): where appropriate, total plate counts, psychrophilic counts, or counts of specific microorganisms of public health or economic significance may be monitored for the estimation of shelf life.
- Nutritional change: the loss of a nutrient such as vitamin C might be chosen. This nutrient should be one for which the food product is a significant source.
- The loss or change of color or the production of breakdown color compounds can be followed. Other changes might be exudation or drip loss, moisture transfer; shrinkage; malodor production.
- Change in some functional property: the loss of some functional property for which the product is noted for can be used to follow shelf life, for example, its ability to whip, to color, to flavor, to foam, to leaven, or to set.

- Undesirable textural change: hardening, softening, staling; loss of crispness; development of graininess, viscosity, etc., might all be suitable criteria to follow the course of shelf life.

Selection of a criterion for stability presents different degrees of complexity for developers. It is comparatively simple to follow the destruction of a nutrient like vitamin C in a food if the loss of vitamin C parallels the loss of a major but difficult-to-measure quality characteristic of the new food product, in which case vitamin C is a good standard. If vitamin C is a simple congener unrelated to deterioration and the food is not a major source of the vitamin, it is not a likely measure.

Rarely is only one quality characteristic of complex systems as foods are the sole determinant of a product's shelf life. Several characteristics will breakdown concomitantly, for example, color, texture, and flavor will all degrade and frustratingly, they will degrade at different rates. Indeed, the breakdown of one characteristic may accelerate the breakdown of another characteristic, for example, we found an off-flavor in a jalapeño sauce caused by the breakdown of a yellow artificial color, part of a blue-yellow dye mixture. Therefore, choosing the correct criterion or criteria to follow during the determination of shelf life stability becomes very complicated. Figure 7.1 depicts three quality attributes that deteriorate at different rates with respect to increasing temperature. The problem is to choose to follow the one most characteristic of the product and the attribute most valued by consumers.

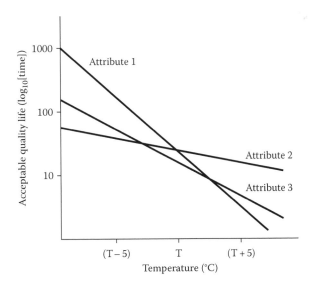

FIGURE 7.1
Comparison of different rates of deterioration of quality parameters for a hypothetical food product with respect to increasing temperature.

Second, there must be some judgmental decision respecting how much loss of quality characteristics can be accepted (and by whom?). An off-flavor in tortilla chips was bad for the technical staff but apparently acceptable to consumers. How much loss of a quality characteristic can be accepted before spoilage is declared or customer acceptance stops? The loss of some nutrient cannot be seen or tasted but if a label declaration has been made for that nutrient, unacceptability acquires a new meaning, that is, a label violation. Losses of color, flavor, or texture are assessed by consumers. The result is dissatisfied consumers; a label violation is an offense punishable with a fine and public notoriety.

What constitutes an acceptable loss of quality? Is it the loss of 60%, 50%, or whatever percentage of a quality characteristic of a product? Is it when the total plate counts of microorganisms reach a particular level? At 10^5–10^8 microorganisms per gram of product, there will be a distinct malodor for most products and obvious slime production as well. Obviously, there can be no toleration of a loss of safety. We are concerned only with economic losses.

Most jurisdictions now require a best before date or some similar statement indicating the product's shelf life. A misstated shelf life has serious economic implications. A conservatively stated shelf life will cause retailers to return wholesome product that is past its stated expiry date believing the product to be either deteriorated or unhealthy. This constitutes a loss to the manufacturer. If, on the other hand, an exaggeratedly long shelf life is stated, a large number of consumer complaints may arise from failed product being on the shelves.

A rough rule of thumb that is frequently used, although many may deny that they do, is the 2/3rd or two-third rule. The rule works as follows: shelf life tests show that a particular packaged product has a good quality shelf life of 90 days at refrigerator temperature. According to the 2/3rd or two-third rule, the shelf life is stated as 60 days. I have never found any practical or scientific justification for this but I have found, nevertheless, that it is used. Since the publication of the first edition of this book in 1994, several people have commented that they too have used this rule. No one has suggested any scientific basis for it. I do not recommend its use.

7.2.3.2 Selecting Conditions for the Test

A final decision: under what conditions will the shelf life test be carried out? If ideal temperature conditions of frozen or chilled storage are used for either a frozen or a chilled new food product, the resultant shelf life will bear no likeness to what will happen in the real world of handling, distribution, retailing, and in the hands of consumers. For frozen or chilled foods, this real world may include any or all of the following:

- Temperature changes encountered in factory warehousing that abuse the product.
- Temperature changes during transport and transfer to a wholesaler's or a retail chain's warehouse, storage in this warehouse, subsequent

transport, and transfer to a retail outlet (Slight, 1980) may all contribute abuse to products.

- Inadequate temperatures in display cases in retail stores that abuse the product. This display temperature is often set at one temperature that may not be optimal for the great variety of products held in the cases. There is often placement of product stocked well above their limit line in display cases and this coupled with improperly maintained temperatures causes further abuse.
- Storage in poorly maintained vending machines can be very damaging for frozen or refrigerated goods (Light et al., 1987).
- Further temperature abuse results from transport and in-house handling by customers and consumers.

Is the product to be tested under nonabusive conditions, that is, following recommended conditions of storage throughout or under abusive conditions such as encountered in the distribution chain? This, then, is the dilemma to be faced in determining the shelf life of a product.

Three factors help developers to determine which test to use for the quality shelf life of a food product:

- The preservative systems designed into the food product, which are protected by the packaging material selected for this ability
- The physical abuse that the food will encounter from factory floor to consumer's table
- Environmental abuse (very different from the above) that the product and its package can be anticipated to encounter from manufacturing and packaging until it is opened and consumed

A fourth possible factor has been omitted—the microbiological load on the food after processing and prior to packaging. Good manufacturing practices and a sound HACCP program that were in place during manufacturing are critical to chilled and minimally processed foods. The initial microbiological load is an important factor in an acceptable shelf life.

Several factors affect a product's shelf life adversely from factory door to consumer's table have been discussed earlier (Chapter 5; Mossel and Ingram, 1955). Temperature, for example, influences the rate of chemical reactions as every secondary school student knows (the familiar Q_{10}) that for every 10°C increase in temperature, one can anticipate a two- to fourfold increase in the rate of a chemical reaction (see Labuza and Riboh, 1982). Temperature changes will cause breakdowns in food structure and lead to loss of quality characteristics (Slade and Levine, 1991; Goff, 1992). They affect biological reactions and alter growth rates and growth patterns of microorganisms and the activity of enzymes (Thorne, 1978; Williams, 1978).

Seasonal temperature changes occur in warehouses from doors being opened and closed during receiving and shipping. Heaters (for the comfort of personnel and to avoid condensation and sweating of containers) in warehouses can warm product stacked near them. Changes occur during transportation in refrigerated vehicles with improperly set or maintained thermostats. Retail cabinets with their defrost cycles impose further temperature changes. Humidity changes closely associated with temperature changes can cause sweating of packages resulting in rusting of metal containers or label damage on the package. Incident light striking exposed containers on ship decks can cause large temperature changes in foods. All are factors that developers must anticipate and have control programs for.

7.2.3.3 Types of Tests

Three general approaches are used to determine shelf life:

1. Static tests, in which the product is stored for a given period of time under a given set of environmental conditions selected as most representative of the conditions to which the product will be subjected
2. Accelerated tests, in which the product is stored under a range of some environmental variables (for example, temperature)
3. Use/abuse tests, in which the product is cycled through some environmental variables

At intervals through these tests, samples are taken and subjected to sensory, chemical, and microbiological assessment.

To these there should now be added another. Good estimates can be made for the shelf life of foods by resorting to the technical literature describing the shelf life dynamics of similar products. An excellent review of shelf life and techniques to predict it can be found in Robertson (2000).

7.2.3.3.1 Static Tests

Static tests have obvious shortcomings. First, they are prohibitively long to produce noticeable changes in many foods. Second, because of this fault, the tests are too costly to undertake given the paucity of information they provide. A static test can be likened to a one-point viscosity measurement or a one-point moisture sorption curve: it tells nothing of the behavior of the product under other stresses. It provides no kinetic data.

The shortcomings are amply illustrated by Okoli and Ezenweke (1990). They used a static test to determine shelf life for a new product, pawpaw juice, for consumption in Nigeria. Samples were subjected to sensory, physical, and chemical analyses at intervals over 80 weeks. This is much too long a test period for practical research and development and would certainly frustrate the marketing program of most companies.

7.2.3.3.2 Accelerated Tests

Accelerated tests are preferred by most researchers since such tests provide information about the kinetics of a product's deterioration. A range of conditions of some environmental variable (such as temperature) are carefully chosen to cover the range that could be anticipated in distribution. Samples of the packaged product are stored under each of these temperature conditions and analyzed at intervals for the loss of some quality characteristic. If the variable chosen is temperature, a simple application of the Arrhenius equation allows researchers to demonstrate graphically the relationship between temperature and time in days until an undesirable loss of quality occurs. Then researchers can calculate the number of days of good shelf life to be expected if one assumes similar storage conditions to prevail.

Factors to be considered in accelerated tests are well documented by Labuza and Schmidl (1985). If temperature is the only environmental variable, one must remember that temperature changes cause freezing and thawing in foods; they can melt or solidify fats; and alter food structure. These changes can seriously skew kinetic modeling data. Water, for example, can still be liquid in a food system well below its freezing point as pure water.

Labuza and Schmidl (1985) provide a table of recommended storage temperatures for frozen foods (−5°C to −40°C for the control), for dry and intermediate moisture foods (0°C for the control to 45°C), and for thermally processed foods (5°C for the control to 40°C).

Some quality characteristics of foods vary widely in reactivity with respect to temperature. That is, at temperature T°C (Figure 6.2), attribute 2 of a hypothetical product is the limiting quality characteristic. At a higher temperature $(T + 5)$°C, quality attribute 1 is the limiting quality characteristic to a good quality shelf life. At a colder temperature $(T − 5)$°C attribute 3 limits acceptability. Again, reference to Labuza and Schmidl (1985) will provide guidance in determining both storage times and intervals between samplings.

Temperature is not the only accelerating factor that can be used but it is the usual one. Chuzel and Zakhia (1991) used adsorption isotherms at different temperatures to derive an equation describing the shelf life of gari. (Gari is a semolina prepared from cassava, which has been fermented, cooked, and dried, and is popular in West Africa and in Brazil called "farinha de mandioca.")

Hardas et al. (2000) describe accelerated stability studies on encapsulated milk fat in which they studied peroxide value, hexanal production, fatty acids, and emulsion droplet size distribution.

7.2.3.3.3 Use/Abuse Tests

Use/abuse tests are as varied as an imaginative mind can make them as a tool in assessing the shelf life of the food and its package as a unit.

Frozen food developers commonly cycle new frozen food through the temperature range of −10°F to +20°F that approximates the temperature range of freeze-thaw cycles of frost-free frozen food cabinets. Cycles are set to correspond to those that could be anticipated under store conditions. Some developers go so far as to purchase freezer display cabinets, stack them in the manner seen in most supermarkets (with a good portion of the product above the recommended fill level) and thus simulate real supermarket conditions.

In one instance I am aware of, a company's frozen product was stored in freezer lockers set to cycle at the temperatures that their experience had taught them would be encountered from factory warehouse to frozen wholesale warehouse through to the retailer. In between these stages, product was removed to simulate the transfer from storage to ambient temperature to frozen transport to dock transfer at the next stage and so on.

In another example of a use/abuse test, a pallet of cased product was dispatched on a journey around the country by truck and by train. The rigors of transportation and of temperature and humidity changes due to weather on both the condition of the product and the effectiveness of the package could be studied on the pallet's return to the plant. Application of this data to shelf life calculations can be limited but information on what one can expect of the product and its package in handling and distribution can be useful in protecting the validity of one's shelf life statement.

Cardoso and Labuza (1983) studied the moisture gain and loss of egg noodles packaged in typical packaging materials for this product as they cycled the packages of noodles through varying conditions of temperature and relative humidity. From their data they developed a kinetic model to predict moisture transfer, an important factor in product stability.

Porter (1981) discusses the unique problems the military has in shelf life prediction under conditions from stateside controlled temperature facilities to Arctic bases or to arid desert conditions. One also has the problem of packaging material to survive air drops. However, food manufacturers face many of these same problems in exporting their products to foreign countries with vastly different environmental conditions as well as inadequate warehousing and shipping conditions. Temperatures in container shipments on ship decks can reach levels that can be very stressful to a good quality shelf life.

Static tests and accelerated tests challenge the first and simulate partially the last, of three factors stated earlier that determine the shelf life of a food product, that is,

- Its preservative system and the package protecting this
- Abusive handling treatment within the factory, from factory to consumer, and within the possession of the customer until use
- The environment that the food and its package encounters throughout its shelf life

Use/abuse tests answer to all three factors.

No use/abuse test can simulate all the stress that may be heaped on the food product and its protective packaging from manufacturer to consumer. How can one duplicate the military mind that pinholed film-wrapped food packages so that they would fit better into the ration cartons? This actually happened while I worked at the Canadian Food and Drug Directorate. The officer in charge informed us the soldiers while out on manoeuvres suffered no ill effects. Or how simulate the damage done to the package and ultimately to the product itself by retail shelf stock clerks who slash the outer cartons and puncture unit packages inside or practice basketball shots with packaged frozen chickens?

The above abuses, which I have personally encountered or seen, can be eliminated only by education of the food handlers in the safe handling of food products at all levels from the plant to the table. Food manufacturers have relied on retailers to do this but unfortunately many retailers, with part-time, temporary help, find no profitability doing so and rarely undertake to train staff.

Transport of product can result in abusive treatment. Transport of product in containers where the container is exposed to the hot sun during a trans-Atlantic crossing have been already mentioned. Temperatures inside such containers can go well over 120°F.

Vibration caused by transport results in product changes. Positions in the hold of a ship in relation to the engines can produce a gentle vibration that can destabilize some food suspensions. Surface transport with its gentle rocking action has been known to produce subtle, and in one case unexpected but desirable, changes in a chocolate product.

A chocolate couverture supplier shipped bulk chocolate by tank car from its factory in the East to a customer on the West coast. Demand was so good it was felt advantageous to all parties to produce this product closer to the customer in a new facility. This step turned into a disaster when the customer complained over the loss in quality of the product although the satellite plant rigorously followed the parent plant's procedures for making the couverture. Investigations revealed that the extra conching action caused by rocking of the tank cars developed the better flavor that the customer preferred. Problem solved with longer conching times.

7.2.3.4 Guidelines to Determining Shelf Life

Shelf life is required by many governments or is required by contract between a co-packer and a buyer or insisted upon by a retailer. Consumers, at the end of the distribution chain, certainly expect food to have a good quality life until it is consumed, no matter how long and under what potentially abusive conditions they may have stored it. At the start of the chain, manufacturers of sensitive food products are at the mercy of

- Contracted distributors and warehousing companies
- Retailers and their storage and handling practices in the rotation of stock

It is also unfortunate for manufacturers of sensitive products that many retail food chains misinterpret the product statement that reads that a particular product has, for example, a 40 day refrigerated shelf life. They assume that a 40 day refrigerated shelf life starts when they decide to stock their shelves or to plan a promotion.

My client, a packer of a fresh pack salsa, had shipped product to a retailer for a special promotion. The retailer delayed the promotion until the product was beyond its expiry date. Product was returned. My client was devastated.

Few manufacturers, especially small manufacturers, would dare to refuse to take back product beyond its expiry date for fear of losing future orders. These are the facts of life that developers determining shelf life must live with.

Estimating shelf life is a guessing game. Many scientists who model kinetics of spoilage reactions may take umbrage at this description of estimating shelf life: it is, nevertheless, true. The kindest that can be said is that the stated shelf life stamped on any product has a strong element of guess in it along with the element of hard data that food scientists can add. Neither consumers nor, for that matter, retailers know what previous treatment a product has had when they read the "best before" date. As data on the deterioration of foods are collected, and knowledge of growth mechanisms of microorganisms in foods progresses, this guesstimate can be refined.

Let it be clearly understood: no one can say what the shelf life of a particular product will be. The best that can be done is to predict what the shelf life should be if.... That "if" encompasses all the precautions that should be taken in manufacturing, distribution, warehousing, retailing, and home storage and preparation. This becomes a big if.

The following points should be considered. First, there are no tests that can absolutely be relied upon to allow one to predict the shelf life of a given food product. *All that such tests can do is provide an approximation.* Experience with similar products in the same product category (e.g., chilled foods or dried foods) can help provide some initial estimates. Data in the scientific and technical literature can help refine these initial estimates. Audits of similar competitor's products drawn from the retail showcases will provide more data to complement one's own findings. Here too, the complaint records of a company can provide information on a product's stability that may be applied to another similar product in development. This marks another reason for documenting the complaint files. The development team should not waste their time looking for the perfect test.

Second, shelf life tests should ideally be carried out only on finished product manufactured on the production line (and equipment) to be used in regular production and packaged in the container that will be placed on the shelves for consumers. Product prepared in test kitchens or pilot plants does not simulate the product prepared in a plant at the height of the packing season with adjacent packing lines running products capable of being a source of cross contamination. Bailey (1988) discusses these problems of scale-up from the pilot plant to the production floor and describes attempts using predictive techniques to minimize the discrepancy between the pilot plant and the manufacturing plant.

The following examples illustrate the fallacy of relying heavily on test kitchen samples for reliable shelf life data.

A potato processor wanted to develop a line of chilled prepackaged, prepeeled potatoes. A shelf life of the product was determined to be an excellent chilled shelf life exceeding 30 days. I had determined this on pilot-plant-prepared samples made from potatoes purchased on the retail market as part of my graduate research studies. In full plant production trials, a shelf life of 10 to 15 days was the norm. The microbiological load between the samples prepared in the well-kept, sanitary test pilot plant and those prepared in the plant factory under full production with field grown produce was very significantly different.

In another case,

The unavailability of the desired cannister with metal ends from a supplier and the pressure exerted by marketing "to get on with it", led researchers in one company to substitute a cannister with plastic ends for a spiced and herbed bread crumb mix for shelf life trials. The supplier suggested that the plastic substitute would provide a more rigorous test "since plastic breathed." Tests were successfully conducted and shelf life determined. Complaints began to pour in when the final product used the cannisters with the metal ends. Off-flavors were noted; rusting was observed. Constituents (principally citral) in the spice and herb blend reacted with the metal ends to cause a breakdown of the flavor and initiate detinning.

Scale-up from the test kitchen to the pilot plant to the factory floor has always produced changes, some subtle and others not so subtle, in a product for reasons discussed earlier. Two products, one factory-produced and one pilot plant- or kitchen-produced, should not be expected to have the same storage properties because they are different.

Third, once the shelf life of a product has been determined, any change in the recipe, in the suppliers of the ingredients, in the water treatment system in the plant or in the water used in batch preparation occasioned by plant relocation, or any other change can have a major impact on the acceptable quality shelf life of a food product. For example, the mineral content of plant waters can have a profound effect on the flavor of a product, not only immediately but over a period of time.

It was found that the flavor of our marinated, fried peppers was adversely affected when the supplier of the frying oil, a fractionated peanut oil, changed from one antioxidant for their oil to another one. We and the oil supplier were surprised at the flavor change. Problem solved with a return to the original antioxidant.

Fourth, interpretation of data obtained from accelerated storage tests for chilled foods can be misleading if due care is not given to the types of microorganisms whose growth, and hence their contribution to spoilage, is affected by temperature: spoilage can be overt with obvious indications of slime, off-odors, loss of color, etc., or it can be covert with no obvious outward signs of spoilage but with undesirable increases of microorganisms of public health significance or even of toxins.

Fifth, the customer, the consumer, or even the retailer is the ultimate determinant of a product's shelf life; they reject it at point of purchase or display. Nevertheless, in basing shelf life on sensory analysis by these judges—perhaps on odor or taste—their ability to recognize spoilage by their senses varies considerably. Dethmers (1979) corroborates this and discusses the evaluation by sensory panels for failure criteria for open dating. Curiale (1991) and Beauchamp (1990) both point out some of the shortcomings that can arise in the use of sensory panels for shelf life determinations; Beauchamp, in particular, cites intraindividual and interindividual variations that can confound the use of sensory panels as an evaluation tool. The use of taste panelists in determining shelf life not only poses health hazards for the panelists, it does not produce reliable estimates of shelf life. A clear understanding of the criteria used to assess the end of acceptable (or even high quality) shelf life must be established.

Finally, determining the shelf life of products involves measuring the differences between control samples and test samples that are subjected to some stress over time. Wolfe (1979) discusses the advantages and disadvantages of different reference standards especially with respect to sensory studies but which are equally valid for shelf life studies. See also Labuza and Schmidl (1985) for recommended storage temperatures for control samples.

7.2.3.5 Advances in Shelf Life Determination

Added value provided to traditional foods (especially the customer's desire for naturalness) alters traditional paths of food spoilage and intoxication (Williams et al., 1992). Predictive techniques in microbiology would greatly assist food manufacturers who provide added value to their food products to predict the course of microbial spoilage. Predictive methods allow an assessment of the duration of a product's quality and safety.

Instead of observing a food spoil over time under static conditions, could one predict with accuracy its expected shelf life from a knowledge of spoilage mechanics and a product's composition? That is, a knowledge of the kinetics of the spoilage mechanics ought to make it possible to estimate shelf life.

The use of predictive models is not new. For example, the botulinum cook established for the safe thermal processing of low acid, high pH foods is an application of such a model, an inactivation model, of the destruction of spores of *Clostridium botulinum* by heat (Gould, 1989). The value of predictive models of microbial spoilage is that they provide a data gathering tool to apply to many types of food stabilized by various techniques. Based on this data, rapid predictions of duration of quality and safety are possible with a greatly reduced need for testing and therefore with reduced costs (Roberts, 1989).

Roberts describes two types of predictive models, the probabilistic model and the kinetic model. Each serves a different purpose. Probabilistic models, as their name suggests, predict the probability of an event, for example, probability of toxin development in a food; however, they provide no information about how quickly the toxin develops or the amount produced. Roberts (1989, 1990) describes a mathematical model able to predict botulinal toxin production in a food as a function of the presence or absence of preservatives, thermal treatment, and storage temperature.

Kinetic models predict the rate of growth of microorganisms and therefore are useful for anticipating time to microbial spoilage or time to growth of critical numbers of food-intoxicating microorganisms. Roberts (1989) discusses several examples of kinetic modeling in his paper. Figure 7.2 is

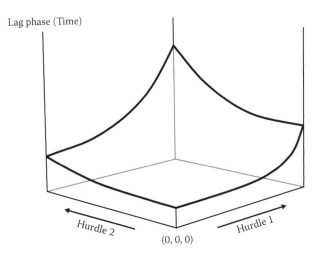

FIGURE 7.2
Three-dimensional representation of the influence of concentration of two hypothetical hurdles on the lag phase of a hypothetical microorganism.

a three-dimensional hypothetical depiction that shows how the lag phase of microbial growth behaves under the stress of increasing concentrations of two different hurdles. Choosing the conditions that provide the longest lag phase of microbial growth gives the greater security. A visual presentation assists technologists in designing stabilizing systems for new products. It would obviously be advantageous to select concentrations of "hurdle 1" and "hurdle 2" that extend the length of the lag period of microbial growth (vertical axis), if these are consistent with other quality factors such as taste. Likewise, one could determine which stabilizing system causes the slowest rate of microbial growth and this would again contribute to the product's stability. Predictive models enable technologists to formulate the product or design its process and be able to predict the length of microbial lag phase or rate of growth.

The Arrhenius equation has been used for both assessing nutrient losses with temperature (Labuza and Riboh, 1982) and microbiological growth with temperature (Gibbs and Williams, 1990). The Ratkowski square root equation has also been used:

$$r^{1/2} = b(T - T_0)$$

where
  r is the growth rate constant
  b is the regression coefficient
  T is the storage temperature (°K)
  T_0 is the temperature at which the growth rate is zero

Gibbs and Williams (1990) describe the use of this equation for plotting the growth of *Yersinia enterocolitica*.

The government's and the public's demand for minimally processed foods has meant foods with low salt, low acid, nitrite reduced, sulfite-free, or sugar-free variants of food products; these demands have multiplied the number of deteriorative routes of conventional products (Williams et al., 1992). These ingredients had contributed to their stability. As multiparameter stabilizing systems are reformulated, new risks are introduced. The need to understand the mechanisms of these systems and to be able to predict both quality changes and microbial activity in a complex matrix of variables is imperative (see, e.g., Chuzel and Zakhia, 1991).

In his review of predictive food microbiology, Buchanan (1993) attributes the growing interest in mathematical modeling to three major factors:

1. Easy access to powerful, number-crunching software programs and computers to process the data.
2. A growing desire of consumers for minimally processed, just-like-fresh foods. This desire usually favors the chilled foods category.

3. The need to organize quantitatively and systematically the wealth of microbiological data on the vast array of foods potentially at risk as either economic or public health hazards.

Buchanan discusses and describes various classifications of models. Examples of the use of Arrhenius models, Ratkowski, and Gompertzian equations to model growth are available (Buchanan, 1993; Gibbs and Williams, 1990). Andrieu et al. (1985) used an Oswin type relation ($X = f \{a_w, T\}$) to model pasta drying.

Walker and Jones (1992) explain predictive microbiology in general and describe the coordinated research program then current in the United Kingdom. This program, originally under the direction of the Ministry of Agriculture, Fisheries and Food, had a host of participating scientists at various laboratories. Private food companies were invited to contribute data. Campden Food and Drink Research Association would house this U.K. Predictive Food Microbiology Database. Since the writing of the first edition of this book, this program has had a rather checkered history. Commercialization was undertaken by Leatherhead Food International; then the intellectual property rights were sold to the U.K. Food Standards Agency. In a press release dated June 16, 2003, this agency describes an international collaboration between itself, the Institute of Food Research and the U.S. Department of Agriculture to provide access to the database free of charge: "The new common database, called ComBase, already contains around 20,000 growth and survival curves and 8000 records containing growth rates."

In the United States, it is available at http://wyndmoor.arserrc.gov/combase/ and in the United Kingdom at http://www.ifr.ac.uk/combase. Such a database resource is useful to all in-product development for predicting food safety and quality.

Lund (1983) cited three major reasons for modeling food processes:

1. Models allow developers to optimize processes with the minimum amount of costly trials.
2. Models provide better understanding of processes, which leads to better processes and safer new products.
3. Models permit better prediction of shelf life and quality changes in foods.

Kinetic studies and predictive modeling techniques have been applied to measuring the deteriorative rates for food quality, specific food components or optimization of processes (Labuza, 1980; Norback, 1980; Rockland and Nishi, 1980). Description of modeling techniques and applications to food quality deterioration can be found in Lenz and Lund (1980), Hill and Grieger-Block (1980) and Saguy and Karel (1980). Heldman and Newsome (2003) review papers on microbial inactivation kinetic models presented at

IFT's second Research Summit held in Orlando, Florida in January of 2003. These papers center on microbial survival during processing. Erkmen (2000) used high pressure carbon dioxide to develop a predictive model for inactivation of *Listeria monocytogenes*.

7.3 Designing for Product Integrity

The elimination of public health hazards as well as hazards of economic significance from food products is best done by designing quality and safety into products at the start of development, a concept comparatively new to food processing but well established in instrument manufacturing. Mayo at AT&T (1986) argues that quality by design applies equally to products and to services. Four elements are essential to the program:

1. Design to the correct requirements of the customer. If the customer's needs have not been clearly identified, product design will be faulty. Market research should have identified the customers' needs.
2. Design using the right technology. This influences the cost of the product and the customer's satisfaction with the product.
3. Design for manufacturability. As Mayo states it, the product and its process should be designed "to be insensitive to 'noise' such as conditions of customer use, drift in components, or variations in the manufacturing environment."
4. Design for reliability.

The latter two elements are less important in food products but, arguably, noise could be interpreted as abusive treatment a food product can be expected to receive. Coincident with development's progress, feedback mechanisms should continually monitor and analyze the product with respect to quality and safety. This is Mayo's design for reliability.

Huizenga et al. (1987) at the PerkinElmer Corporation, stress the need for co-operation and communication in safe product design described in six steps:

1. The needs of both internal and external customers must be identified.
2. Quality characteristics require identification.
3. Means to measure these quality characteristics must be obtained.
4. Quality goals that satisfy customers and suppliers at reasonable cost need to be established.
5. A process to attain the stated goals must be set in place.
6. Processing capability requires verification.

Neither Mayo's nor Huizenga et al.'s steps pertain specifically to food products but they have some elements that apply. Pearce (1987) encompasses both sources into six principles for a design assurance policy for food products:

1. Design work for quality must conform to marketing concepts and regulatory needs.
2. Design work for quality should conform to properly established procedures and standards.
3. All design work for quality should be properly documented with changes recorded and regulated.
4. Challenges to the design should be carried out at each stage of scale-up including production trials. (This reflects the need to be constantly in touch with marketplace and market changes during development.)
5. Third party review of design work for quality is required at critical stages before advancing to subsequent stages.
6. A feedback system must be prepared to collate all activities and planning that support manufacture of new products.

Pearce (1987) further broke these principles into 12 subsystem requirements and developed a responsibility matrix for product design assurance with primary responsibility delineated.

Wilhelmi (1988) considers the following requirements to be important in product design:

- Product composition
- Safety considerations for the product
- Regulatory compliance
- Knowledge of product stability
- Packaging considerations
- Considerations in the marketplace

Wilhelmi's and Pearce's points have obvious similarities.

Indeed, all the above authors have essentially stressed the same points concerning designing for quality. That is, they emphasize formulation for quality, stability, conformance, and safety based on known spoilage mechanisms and anticipated abuse and protected by suitable packaging. Such design is a built-in safety net: it is error prevention in food product development. Technologists must incorporate safety and quality design into their products through the judicious use of ingredients, processing, and packaging technology.

7.3.1 Safety Concerns

Equally important systems to monitoring quality characteristics are control systems to monitor safety with respect to hazards of public health significance, and to maintain the integrity of these systems throughout processing, storage and distribution, and retailing. Monitoring systems (plant's good manufacturing practices, quality control/quality assurance systems, and HACCP programs) should already exist; they should be in place for all the company's other products. Each product needs a HACCP program unique to itself.

All these existing systems for maintaining integrity of existing products must be reevaluated every time a new product is introduced into a food plant. New products bring in potential hazards through the introduction of new raw materials with new and unknown microbiological hazards and ingredients. The result is that the safety of all the company's products and processes may be jeopardized.

7.3.2 Concerns in Designing for Food Safety

Two major questions require answering when prototype recipes or products have been obtained:

- What hazards of public health significance are, or could be, associated with these products during manufacture; during warehousing, storage, and distribution; and with their home use? In short, a complete HACCP program must be developed based on whatever experiential evidence is available.
- What desired quality attributes are to be built into these products that require stabilizing or maximizing? (Attributes could be organoleptic, specific dietary or nutritional, functional, or convenience of preparation.)

These beg other ancillary questions:

- What are the major (probable) spoilage routes of products of this particular composition and what health hazards are associated with them?
- What duration of acceptable quality shelf life is desired for these products?

Elements of quality and safety (HACCP programs) are designed into products at their inception. Concern over the safety of any food product centers around all of the following:

1. Possible presence of food intoxicating microorganisms exceeding recognized norms in the ingredients and raw materials forming part of the product or in the product itself that, when ingested by susceptible consumers, may cause illness or death.

2. Presence of preformed biological toxicants (enterotoxins, mycotoxins, or naturally occurring toxic compounds such as domoic acid of algal blooms) that may have entered with the ingredients or be formed during processing.

3. Presence of chemical hazards (pesticides, herbicides, growth stimulants, or even fertilizer uptake) resulting from carryover in the food product or in any of the ingredients.

4. Development of chemical hazards arising from processing through inter-compound reactions within the food.

5. Presence of miscellaneous extraneous matter (stones, glass fragments, metal pieces or wood in the food product) that can cause serious injury if ingested.

6. Presence of insects and insect parts, an esthetic hazard, that often has a shock effect on consumers resulting in illness.

Items 3, 5, and 6 should be prevented from entering the food or removed from the food by a total quality management program (Shapton and Shapton, 1991). Item 4 may be circumvented by a more judicious choice of ingredients and processing conditions, or both.

Many processing steps (trimming, cleaning and blanching) plus plant support systems (good manufacturing practices, HACCP programs, cleaning and sanitation) reduce the numbers of microorganisms jeopardizing the safety and stability of products (see, e.g., Shapton and Shapton, 1991). For minimally processed products, extreme attention to HACCP program is essential.

7.3.3 New Concepts of Safety

The concept of safety is evolving; its interpretation is broadening and coming under closer scrutiny from regulatory agencies in all countries. Part of this keener interest is due to some well-publicized food intoxications (*Listeria* in deli meats in Canada in 2008, pet food contamination with salmonella infected peanut products in the United States in 2009, contamination of wheat gluten and vegetable proteins contaminated with melamine fed to pets arising in China). The globalization of food and food supplies has increased the need for closer controls.

Safety concerns have traditionally centered on susceptible consumers, that is, the very young and the elderly. But the concept of the susceptible consumer must be expanded now to include immuno- or health-compromised individuals. Brackett (1992) describes these as

> people with underlying chronic health problems such as cancer, diabetes, or heart disease; individuals taking certain immunocompromising drugs, such as corticosteroids; and individuals with immune deficiency diseases such as acquired immunodeficiency syndrome (AIDS).

The latter group is a growing proportion of the world's consuming public. They are more susceptible to much smaller numbers of infective microorganisms than would harm other uninfected consumers and their reactions more severe. AIDS-infected male patients are 300 times more susceptible to listeriosis than are AIDS-negative males (Archer, 1988).

This new susceptible consumer is a cautionary consideration to shape technologists' thinking in designing safe stabilizing systems for new products.

Food microbiologists now recognize the ability of both some exotic and some well-known microorganisms to grow and thus become health hazards in stored chilled foods. One such exotic microorganism is *Mycobacterium avium* spp. *paratuberculosis*, thought to have a role in Crohn's disease (Williams, 2003). Indeed, as growth and viability of food-associated microorganisms are studied in multi-parameter stabilizing systems (e.g., hurdle technology), some anomalies to accepted knowledge of the limits of growth and viability are appearing. Pathogenic psychrotrophs capable of growth around 5°C—considered a safe refrigerator temperature—are *L. monocytogenes*, *Y. enterocolitica*, type E *C. botulinum*, *Vibrio parahemolyticus*, *Vibrio cholerae*, *Bacillus cereus*, *Aeromonas hydrophila*, *Staphylococcus aureus*, enterotoxigenic *Escherichia coli*, and some *Salmonella* species (Farber, 1989). Any temperature abuse of susceptible products in the chain from processor to consumer is potentially dangerous.

Where the contamination comes from is important. *B. cereus* was the culprit in spoiled Japanese meat products and the most likely sources of contamination were the additives used in their production not the meats (Konuma et al., 1988). Beckers (1988), reviewing the incidence of foodborne diseases in the Netherlands (1979–1982), found *B. cereus*, *Salmonella* sp., *Campylobacter jejuni*, *S. aureus*, *Clostridium perfringens*, and *Y. enterocolitica* to be major causative microorganisms. Meat and meat products, fish and shellfish, snacks, and foods prepared for immediate consumption as for food service outlets catering to hospitals, old peoples' homes, cafeterias, and restaurants were the primary carriers. Hemorrhagic colitis associated with *E. coli* has been frequently reported in nursing homes (Stavric and Speirs, 1989). Processed packaged meat products were found by Tiwari and Aldenrath (1990) to be contaminated by *L. monocytogenes*. Slade (1992) considered the presence of the *Listeria* microorganism to be ubiquitous as a result of an extensive review of food processing environments. The outbreak of listeriosis in processed deli meats in Canada in 2008 affected nursing homes for the elderly.

Keeping the stabilizing systems designed for food products safe and free from compromise is not as simple as one might think. Interactive packaging, modified atmosphere packaging and controlled atmosphere packaging (MA/CAP) control the growth of some spoilage microorganisms but certainly not all microorganisms. For instance, Brackett (1992) noted that spoilage was retarded in MAP produce but *A. hydrophila* and *L. monocytogenes* were unaffected, visual spoilage was stopped but toxic microorganisms were able grow.

Many gas mixtures used in MA/CAP chilled products are also the same composition as those used to culture anaerobic pathogens (Idziak, personal communication, 1993); thus anaerobes (if present) could grow with no visual indication of microbial spoilage being apparent. Park et al. (1988) confirmed this possibility in a study of MA-packaged commercially processed wet pasta stored at good and poor refrigerated temperatures; the poor temperature they used is frequently encountered in open refrigerated shelves in retail outlets. At the end of the recommended shelf life, sufficient staphylococcal enterotoxin had formed in a significant portion of those samples held at the poor refrigerated conditions to have caused illness in sensitive individuals.

Many products, for example, anchovies, unpasteurized acidified foods, beef jerkies, are "semi-conserved." They are safe, stable, wholesome products in their uncompromised stabilizing system—for anchovies this is their very high salt-on-water content. By themselves, anchovies can be eaten safely; it is a preserved product. When, however, they are used as ingredients in another product, for example, an all-dressed pizza, a quiche, or a flan, the stabilizing system of the semi-conserve is compromised; the potential for a food hazard to develop is very real. If the pizza, quiche, or flan is held for a long period of time in a warming oven or in a poorly maintained refrigerator, the microorganisms contributed by the anchovies, some of which are spore formers held in bacteriostasis by the high salt content of the anchovies, are no longer in such inimical conditions. Growth commences and spoilage or intoxication of the multicomponent product can occur.

The unusual behavior of sublethally injured bacteria during processing has been reported by Archer (1988) and Rowley (1984) as presenting problems with the newer preservative technologies. Sublethal injury to microorganisms, regardless of the nature of the injury, evokes an adaptive response. In some instances injured microorganisms become more resistant (Archer, 1988). The very action of pretreatment, that is, processing, of foods may alter the characteristics (sensitivity) of microorganisms to stresses.

This desire on the part of customers and consumers for minimally processed foods opens up a Pandora's box of new considerations concerning safety and stability that technologists must deal with. These new safety considerations are paramount concerns for development technologists, with the assistance of quality control and processing departments, to address.

7.3.4 Costs of Quality and Safety Design

Quality and the preservation of quality characteristics in new food products have costs. These costs can be crudely distinguished (Table 7.2) as intangible and tangible costs of quality (previous editions referred to these as indirect and direct costs). Intangible costs are hidden in the development work to design new products to minimize hazards and to design or seek suppliers of instrumentation and equipment to manufacture products with the desired features, which market research was needed to find. Modifications to plant

TABLE 7.2

The Costs of Quality in New Food Product Development

Intangible Costs	Tangible Costs
Market research	Increased inspection; more bodies or more equipment are required
What product?	
What target?	Additional analyses are required for newly introduced product
Where?	
Quality and safety design of product	Increased costs for:
Product standards	Maintenance
Ingredient standards	Sanitation
Processing standards	Hygiene
Process design	Warehousing
HACCP programs	Control systems
SPC programs	In-line instrumentation
Operator training for	On-line instrumentation
Maintenance staff	Offline equipment
Sanitation personnel	Rework costs
Hygiene personnel	
Equipment operators	
QC line inspectors	
Laboratory analysts	

equipment for the new product involves much costly work. Retraining of operators and technicians in new procedures (sanitation, preventive maintenance, process and quality control, storage) that novel processes and products require contributes to indirect costs. Staff must be trained in new inspection routines to recognize and report hazards associated with something novel to them. In some development processes, consultants may have been required. These are usually one-time costs.

Tangible costs include (a) salaries of additional inspectors and analysts for the increased need for grading, collection, and interpretation of data for incoming materials; (b) development and installation of in-line, on-line, or off-line process controls; and (c) the inevitable costs of rework of failed (not meeting standards) material, returned goods, or lost customers. The more time, effort, and money spent on prevention in the design phase of new product development, the less the losses will be for failed products.

Not to be forgotten is the possible indirect cost impact that newly introduced products may have on existing safety-related systems. Part of a plant's quality control systems and complementing its HACCP program are several programs that add to the company's quest for safety and quality. These are

- Preventive maintenance
- Pest control

- Plant sanitation
- Product complaint handling policy and analysis
- Worker-related programs
  a. Worker safety and health (ergonomics)
  b. Hygiene training for food handlers
  c. Training programs for food handlers and operators
- Grounds maintenance
- Good manufacturing practices
- Environmental concerns
  a. Waste management
  b. Water reclamation and effluent control
  c. Odor reduction

Food companies have some quality control policy from which are derived procedures for attaining quality in purchasing, processing, warehousing, indeed, for every aspect of the company's business. The end result is a manual of operations documenting all the product and process procedures that affect quality. Implementation of these procedures is the responsibility of the quality control department. All new products must conform to the company's policy. These procedural manuals reflect management's interpretation of its company's ethos.

Plant personnel unfamiliar with new ingredients, with strange raw materials, with unusual products, or even new plant routines can themselves introduce hazards that bring new microflora into a plant. Purchasing departments may not have the expertise to purchase wisely on the commodity market, or the plant may not have the facilities to properly store ingredients they are unfamiliar with. Each of the above programs needs to be reviewed for its adequacy respecting these hazards.

My client, a developer of breaded coatings, had a moth infestation that had spread into other parts of the plant and, more damagingly, into other finished products. The cause was simply explained. The purchasing department had purchased a quantity of difficult-to-obtain crumb for a new product; they had "bought long." They bought several months' supply; they stored these in an area of their warehouse made of porous cinder block. The crumb was infested; moths had ideal breeding spots in the cracks and crevices of the cinder blocks; they spread rapidly. A cleanup and disposal of the crumb had preceded my visit but the problem had persisted. The plant manager was aghast during my inspection tour when I insisted an electrical control panel in the main factory be opened. It was alive with moths. A more thorough clean-up followed.

That which is new or unusual introduced into the controlled environment of a food plant introduces a new hazard when the nature of the introduced hazard is not understood or anticipated. The hazard cannot be controlled adequately.

7.3.5 Hazard Analysis Critical Control Point Programs

The early history of the development of the concept of hazard analysis critical control point system is described by Bauman (1990).

An HACCP program is required for every food product that a plant produces and before any new product is introduced into a plant for production. One blanket HACCP program for all established products is not adequate even for similar line extensions. Each product must be treated as having unique HACCP program requirements or should be so treated. HACCP programs must be able to evolve continually to meet constantly evolving concepts of hazards of economic and public health significance. Archer (1990) cites two emerging microbial hazards, *C. jejuni* and *L. monocytogenes*, that can challenge an inflexible HACCP concept.

Members of the development team prepare a flow chart of the process and equipment to be used for the new products. They assess the effects that plant processes might have on the product's sensitive characteristics and identifying points in the process where, if control were to be lost, a hazard might arise that jeopardizes the new product's safety or characteristics. Once identified, hazards to either safety or quality are either eliminated or minimized to safe levels at that critical point in the process (or in the plant environment) or the product is protected from the hazard at that point with remedial action. All responsible personnel now elaborate how this control for elimination or reduction of identified hazards is to be carried out.

The final step, the determination of process parameters and control limits for the guidance of production can be established with confidence to monitor product integrity. A sound statistical process control program will then warn of deviations from accepted norms. Stevenson (1990) elucidates the steps and considerations that should be applied when introducing HACCP principles to foods.

7.3.6 Standards Necessary for Safety

Development teams are responsible for setting standards and specifications for raw materials and ingredients; for packaging materials; and for the truthfulness of labels, label statements, and promotional materials. The team is required to produce a detailed description of the entire process with the various pieces of equipment used and the support systems for the new products including everything from the analytical procedures to be used to the detergents and sanitizers necessary to clean equipment. In my experience this is rare; the result has been unacceptable changes in a product when, out of necessity, a different pump was used than that specified.

First, specifications for the ingredients and raw materials are clearly defined with precise descriptions of ingredient characteristics to obtain exactly the style, grade, cut, color, flavor, heat level, or whatever other criteria that are required at the lowest cost or at a cost consistent with the original cost estimate for the product. For this reason, all the important, essential characteristics of ingredients used in new products must defined as closely as possible.

Only the essential characteristics of an ingredient in a new product formulation should be identified. If color is important, it should be so specified and the color identified; if particle size is important, the range of acceptable particle size is identified. After the essential and important attributes of each ingredient are identified and specified, then a list of suppliers whose products meet these specifications can be prepared for the purchasing department.

A second requirement of specifications is that the quality control laboratory have analytical or functional tests that permit assessment that ingredients and raw materials meet these standards. Included with this methodology there should be sampling procedures suitable for the nature and form of needed ingredients. If attributes are specified for any ingredient or raw material, they must be checked frequently enough for safety's sake and frequently enough to keep suppliers honest. Good and trusting customer/supplier relationship need still to be verified; suppliers have been known to make mistakes or to take liberties if vigilance is lax or to make substitutions in formulations. It is the wise customer who subjects suppliers to the same degree of quality inspection rigor that they apply to their own products.

Any special storage requirement for ingredients and more particularly for raw perishable materials need to be specified.

There should, of course, be finished product specifications. These are not the same as quality standards. Quality was designed into the product. These finished product specifications identify whether the product has met the specifications required in the trade of that product.

7.3.7 International Standards

Ever-expanding markets are a goal for every company introducing a new product. Ultimately, this will mean the exportation of their product. There are no specific international standards for foods per se but two bodies, Codex Alimentarius and the International Standards Organization (ISO), provide guidelines for contracting companies, that is, between manufacturers of products and buyers.

Codex Alimentarius publishes such documents as *General Principles of Food Hygiene* and *Recommended International Code of Hygienic Practices for Canned*

Fruit and Vegetable Products. It would be advisable for any company introducing new products to be guided by the general principles outlined in these documents. Walston (1992) discusses with reference to Codex Alimentarius many problems in international food trade. The suggestion that Codex be the standard for food safety has sparked controversy because of perceived shortcomings in Codex. There are three criteria for rejection of food in Codex: quality, safety, and efficacy. However, many governments, especially those of Muslim countries, and many nongovernment organizations believe that there should be cause for rejection on religious or ethical grounds. Such non-quality- or non-defect-related rejection of product could have serious implications for exporting companies. The product developer with an eye to exploring export markets for new product expansion would be well advised to be aware of international standards and regulations.

The ISO documents, ISO 9000 group, and the 14000 group describe quality management systems. The 9000 series are more pertinent to food companies than the 14000 series that apply to environmental management practices. ISO 9001 describes requirements for a quality system in which a contract between two companies requires proof of the capability of the supplier to produce to the required level of quality. ISO 9004 is a guideline for establishing total quality management (TQM) in a company. It provides the bases for establishing and maintaining a quality management system. TQM is described by Shapton and Shapton (1991) and by Taylor and Leith (1991). The latter reference is accompanied by descriptions of the application of TQM at several food plants (pp. 21–26).

Direct application of either the Codex Alimentarius or the ISO series of documents may not be pertinent in the self-manufacture of newly developed added-value products for domestic consumption. Where exporting is a major objective or where companies wish to contract out the manufacture of their products to co-packers domestically or in other countries, the ISO documents, in particular, may be very useful in negotiations. Many companies now require that their suppliers and co-packers be ISO certified. Possession of such certification may provide a powerful marketing edge although there seems to be less of a cachet to the adherence statement on labels presently. There was a great hoopla several years ago with the publication of the ISO documents and companies raced to erect signs claiming their ISO certification; I have seen fewer of such signs lately. Ingredient manufacturers should be aware, particularly, of the ISO documents and their possible impact on export sales.

7.4 Summary

The foregoing description of the support roles of members of the development team shows that there are overlapping areas respecting the maintenance and confirmation of product integrity. Lawyers (Chapter 6) ascertain that the product is "legally safe" and that the company has the best advice and guidance in any contractual arrangements associated with the new product and arrangements for its manufacture, distribution, and sale. Quality control serves to assure all the necessary actions in procurement, manufacture, warehousing, and distribution have been taken that ensure the maintenance of safety and quality at the desired levels throughout its shelf life.

There are gray areas of responsibilities between the roles of engineers, food technologists, quality control personnel, and production personnel. It is the support groups' roles to cover all contingencies in maintaining the integrity of the product and the safety of the customers and consumers.

The point that is important in the management of the development process is that no opportunity should be missed to bring all the skills of the team to the process of screening and producing the best product that meets the needs of customers and consumers at a price the customer is willing to pay.

8

Going to Market: Success or Failure?

> ...with few exceptions, marketers generally stub their toes with new product introductions...
>
> **Gershman (1990)**

8.1 Final Screening

Binkerd (1975) reviewed the variety of tests that typically would have been applied to the product so far: focus group interviews, concept screenings, blind product tests; concept tests, mini-market tests, and finally test markets. Table 1.5 also provides an overview of activities during development.

Development has progressed through several stages of screening in which the product may have changed from what was initially conceived for many reasons: continuing market research saw a need for change, limitations of technology or production facilities, restraints imposed by regulatory bodies, and requirements for the assurance of safety and product integrity. Test kitchen samples and the final factory run product are very different, and consumer tests on them are expected to be different. Data from all these screening tests result in a changed product but one that is as close to meeting the needs of consumers as possible.

8.1.1 Test Market: What It Is

Test markets are introductions of new products into regions carefully selected for a variety of geographical, marketing, and company reasons. The product is introduced, and after a predetermined sell-in period of time for *a*wareness, *t*rial by consumers, and *r*epeat sales (the ATR period), results of the test are analyzed. These results dictate whether marketing is either continued for further data collection, or marketing is extended into other regions, or the product launch is dropped and the whole project reevaluated.

Test markets take many forms; indeed, it is often difficult to distinguish between extended use tests or mini-market tests. Lord (2000) describes three classes of market tests:

1. Simulated test markets: essentially a concept testing technique similar in many respects to a focus group
2. Controlled testing: similar to the traditional test market but the entire test is farmed out to a market research company, which manages the entire test from distribution through to promotions
3. Traditional sell-in test marketing

However, as Lord does point out, there are many variations of each and they are not mutually exclusive testing techniques.

8.1.1.1 Examples

Two examples illustrate the many differences in test markets that exist. Mazza (1979) test marketed native fruit jellies (a seasonal product) in a gourmet gift pack through one retailer's stores in two cities in the first year of introduction. In the second year, three retail outlets were chosen and two additional cities included. Marketing and consumer evaluations were carried out through a questionnaire accompanying each gift pack sold in both introductions. Customers completed and returned the questionnaires. Response averaged approximately 20% of sales; data obtained included why the product was purchased, what attracted the purchaser to the product, whether the purchaser would repeat purchase of the product, and how the purchaser would rate the product. This was not a highly competitive marketing environment and demonstrates a very simple test market for the cottage trade. The two year introduction period is too long for most companies.

Clausi (1974) describes how the General Foods Corporation moved to a test market. After making modifications arising from the results of in-home testing, the product was put into a test market in one or two cities. Then, based on market data, a move to a larger section of the country was made to evaluate both consumer reaction to the product and awareness of the advertising and promotional campaigns.

Test market situations for small and large companies are very different. Development teams of large companies are more conservative, cautious, and concerned with all the ramifications that mistakes in the introduction might have. Big mistakes have consequences and reflect badly on company image. The consequences of error permeate collectively and individually throughout the development team in large companies; more is at stake should a blooper have been made. People have something to lose—their jobs.

Small companies have more flexibility in their introductions; they can literally deliver and sell off the back of their company station wagons to small independent grocers. Their introductions are also more hands-on operations with all members of the team participating. They use sales at country fairs or local sporting events to introduce their products or have tasting sessions at church socials. My first introduction to retort pouch entrees was at a local country fair where a senior member of the research and development team introducing the product served me. McWatters et al. (1990) used a mobile kitchen travelling about to local events to evaluate consumer response to akara, a snack product.

Small companies get immediate feedback of consumer reaction to their products' characteristics. They have the patience, the time, the intimacy with, and the proximity to, their customers to develop a market; this is something the big company, driven by short-term gains, does not have. These smaller markets also do not require introductory fees (slotting fees) to be paid to retailers.

8.1.2 Test Market: Its Goals

A test market is the first, large scale, controlled experiment to evaluate how customers, consumers, retailers, and the competition will react to a new product. It is the final phase in the development process prior to a more formal, perhaps national, introduction. Many companies bypass a test market and introduce a new product directly into their intended markets. If so, the financial and marketing departments now assess the results of the work of the development team on the basis of sales and market penetration.

Advertising and promotional strategies based on the targeted consumer are in hand. The production department has filled the distribution channels, and the timing is right for a market launch into a test market. The ball is in the marketing department's court. The next and final stage in development is theirs.

Test markets are a significant part of the screening process. They are a unique opportunity for verifying the stability of the product in the marketplace, for verifying the ability of the package to protect its contents, and for determining the effectiveness of promotional and advertising materials on the responses of customers, consumers, and retailers. Costs can amount from hundreds of thousands to millions of dollars, all spent to find out everything pertaining to the introduction of the new product:

- What are the reactions of the targeted customers and consumers to the product and those of the retailers to the product? Equally important, what retaliatory action has come from the competition?
- How well has the product stood up in handling, distribution, and retailing? Have the quality features been stable and the shelf life justified? How accepting are the public to its valued characteristics?

- The production department's ability to consistently manufacture a uniformly high quality product and in a timely fashion is being tested. They are tested to keep distribution pipelines filled with both regular product and the new product.
- The package designer's skill in creating a package that sells and protects the contents is challenged.
- The warehousing and shipping department's ability to store and distribute the product in top quality and on time is put to the test.
- The test market assesses the marketing department's skills with its advertising and promotional campaigns designed for the targeted customer.
- The skills of the sales department to use the materials and to sell the product to retailers are tested.
- Management's strategic and tactical skills at countering competitive action, with the support of all the other members of the new product team, are tested.

A test market introduction is a very complex experiment involving emotional, intellectual, political, and people issues. It offers developers, with the support and analysis of market data, opportunities to evaluate their efforts, to explore ideas for the next generation of products, to plan comprehensive product maintenance programs, that is, looking forward to how the product can be improved, to what variations (line extensions) can be added to support it as its growth falls off (see life cycle curves, Chapter 1), and so on.

The people element of new product introductions cannot be ignored; the personal careers of all members of the team, but especially the technical members, are under scrutiny and are often forfeit. It has been my experience that new products seldom fail technically in test market, but as Gershman (1990) notes at the start of this chapter, marketing often "stubs its toe." Nevertheless, the technologists face more directly the stigma that somehow it was their fault, that they could not duplicate the concept, and that the concept was good but the research and development group could not match it. The other members of the team glide silently off to other positions within the company; the technologists are stuck in their laboratories and pilot plants.

Consumer research is as active during a test market as it was during earlier stages of development but is focused to answer many questions: How good was the original research? How and when does the consumer use the product? Does the consumer misuse or misunderstand the product? Are preparation instructions clear? Is the product's message being misinterpreted? How is the customer reacting to the message? What is the competition's reaction? How are retailers reacting? Consumers' reactions to, and their usage of, the product may suggest new opportunities for repositioning a product or indicate possible line extensions. Test markets provide excellent learning opportunities for companies.

8.1.2.1 Some Cautions

The nature of test markets varies widely depending on the product being tested and the goals of companies doing the testing. (Types of new products and their characteristics are discussed in Chapter 1.)

Improved (reformulated) or repackaged, established products for which new market niches are being explored or for which new marketing strategies are being tested present unique test marketing situations to marketing personnel. These products are already established. Marketing departments are using the test markets to seek answers to such questions as follows: Will changes incorporated in established products be accepted by established consumers? Will they attract new consumers? Will new market niches be opened? These rarely rate a market test.

The introduction of line extensions into test markets may backfire with the overall marketing strategy. If already established products are valuable cash cows to companies, then new line extensions (new flavors) may cannibalize the company's existing bell-ringer products. For example, a company with an established pungent hot sauce introduced a milder version of its hot sauce. Rather than opening up a new market, the newly introduced product cut into the sales of the established product. In such a situation, marketing personnel must interpret consumer reaction carefully for the market value of new product but also for economic impact of the new introduction on other members of their product line. The data from test markets must clarify what activity is going on in the marketplace.

In addition, in any introduction of a new product where heavy advertising and promotion accompany the introduction, this advertising also gives impetus to other competitive me-too products wanting to capitalize on the promotion. The promotion carries all similar products including those of the competition.

8.1.2.2 Costs: A Deterrent

Many companies question the value of a test market although in truth, much of the challenge may center around how companies define test markets. A successful test market does not guarantee a successful national launch or even a wider regional one; therefore, what does a test market accomplish? A test market is costly; consequently, many companies get the same information that allows the company to refine its product and its packaging and refine its marketing introduction strategies with alternative mini-market tests at a much cheaper outlay of money than a traditional market test (Lord, 2000).

Besides the expense, there is possible loss of face, an embarrassment, in the event of a launch of a product by a major company that is a distinct and loudly publicized flop. Social network sites rapidly make the failure a widely known fiasco. Management must weigh the marketing risks of foregoing

TABLE 8.1

Advantages and Disadvantages of Test Markets

Advantages	Disadvantages
Information about the effectiveness of product, pricing, packaging, and marketing strategies is obtained.	They can be very costly ventures.
	They are time consuming.
Information about retail reaction is obtained.	Sales force is diverted to new product launch possibly to the detriment of regular, bell-ringer products.
Information about competitive counter action is seen, and protocols can be developed to thwart the competition.	Test markets warn competition of company activity.
Development protocols are justified.	A successful test market does not foretell a successful full-scale launch.
	Loss of face occurs if test fails. This could result in possible poor trade reaction for other products.

a traditional test market vs. alternatives such as mini-market tests or omitting it altogether. If the risk of omitting the test is small and if there are existing financial constraints, it very well may be worth risking dispensing with the test (Kraushar, 1969). If getting into the market early without alerting the competition is important, dispensing with this expensive and time-consuming exercise is a wise tactical move. The advantages are that six or more months of lead time to build a dominant market share are gained and expenses in excess of several hundreds of thousands, indeed, millions, of dollars are saved. Advantages and disadvantages of test markets are presented in Table 8.1.

8.1.3 Considerations for a Successful Traditional Test Market

The where, when, and how to introduce are all closely interwoven. Understanding the principal elements in what Lord (2000) refers to as the traditional test market enables a better understanding of the characteristics, concerns, shortcomings, advantages, and pitfalls of simulated market tests and controlled test markets. Separating these in this manner should be recognized for what it is, an explanatory device. The nature of the product being introduced will greatly influence the answers to these questions.

8.1.3.1 Where to Introduce

The location of the test market should not introduce a bias in the data obtained. There is no area that represents a cross section of the population with all its ethnic, religious, cultural, and economic diversity. Therefore, any area chosen for the launch introduces a bias that market researchers must be aware of.

It is, therefore, necessary to be aware of what biases are introduced by the area selected. Some issues that require consideration are as follows:

- Is the area chosen for the test market peculiar to the company? A launch into areas where the company and its products are well known (company town and environs) can lead to equivocal sales and marketing data. In areas where the company has not been a good corporate citizen or is in a labor dispute, an introduction may be influenced by the company's reputation.

- Is the area for the introduction peculiar to a competitor company and its products? Introducing a new product into marketing areas that are heavily saturated by a major competitor's products is foolish unless, of course, that head-to-head confrontation with the competitor is deliberate. This is not usual practice. A test market is not the time or the place to challenge competitors. A market dominated by a competitor will require heavy advertising, promotions, and trade allowances to get any shelf space or significant market penetration. Indeed, one can expect the competition to disrupt the test market with their marketing tactics.

- The area chosen for introduction should be one where there is a competent sales force in position and a competent distribution system already established. The sales and distribution team should be representative of the company's skills. It is the strength of the product that is being evaluated not a particularly skilled sales force in the chosen test market area.

- If the area chosen for the launch is dominated by large retailers or by a single large retail chain, it may not be possible to evaluate advertising, promotions, and sales efforts for the product. Dominance by large retailers or by a single retailer in the test area restricts the activities of the marketing personnel in planning promotions and advertising campaigns. Campaigns may not be conducted as a company wishes but as retailers want.

- Is the targeted consumer in the chosen test area? Introduction of products with a strong ethnic appeal in areas devoid of that ethnic group is remarkably stupid. Likewise, introducing products aimed at an elderly population into a growing suburban area dominated by young families does not make much sense either.

- Does the product have a style, flavor, form, etc., foreign to food peculiarities in the test area chosen? Some geographic areas have distinct biases for flavors, colors, forms, or styles of products. For example, there are many different styles of "authentic" chili: meatless; *con carne*, with ground meat or with chunk meat; and heat levels from mild to very hot, with each popular in different areas of the country.

Likewise, pizza has many variations of styles of crust and shapes, for which there are regional preferences. Each area with its unique style considers their style to be the authentic one. Strong regional preferences based on the local variant will weigh against new products that do not conform.

- Test market areas should not be dominated by mono-economies; that is, the area should have a mixed economy. The economic health of an area influences the buying habits of consumers. In economically depressed areas, consumers may not be willing to try new products that are perceived as being luxury items or may not purchase higher priced products in which the added value is not appreciated. However, consumers can be very perverse; an old marketing adage is when the going gets tough, people eat chocolate. Nevertheless, marketing departments should take cognizance of the economic mix of communities where new products are planned when interpreting market data.

Test markets are carefully designed experiments based on a marketing plan to obtain as much marketing, sales, and consumer research data about the product as possible. They are not tightly controlled science experiments: a test market launch will alert competitors who can be expected to retaliate in some manner to seriously bias the test market results.

Small companies rarely have any geographic areas that are peculiar to them. They usually test market locally within range of their factories to keep costs low and to avoid stretching their distribution resources to the limit. Usually too, they use their own sales forces. They are rarely poor corporate citizens, and, often, their tactics in test markets, even those dominated by a single competitor, go unchallenged or even unnoticed. This is not so with large companies.

Small companies are at a distinct disadvantage in trying to introduce products where a large retailer dominates. The manner in which large retailers conduct their purchasing does not favor new introductions by unknown or one-product companies without heavy promotional and advertising support. For small companies, introductions are usually done with little fanfare in small independent grocery stores.

8.1.3.2 When to Introduce

The seasonality of products dictates when test markets are carried out. Promoting a soup to be consumed hot in the summertime or promoting ice cream or frozen yogurt when the snow is flying outdoors are inappropriate times for introduction of those products. It is not weather alone that determines seasonality. Products associated with national, ethnic, or religious festive occasions should be introduced at their appropriate times. Promoting seafood products during, for example, the American Thanksgiving period is inappropriate. The Christmas period presents some interesting anomalies, and the development team had better be aware of these should their product be a Christmas-oriented one. There are geographic, ethnic, and traditional variations (perhaps these could all be classed as traditional). For some communities, fish is traditional fare at Christmas; my daughter informed me of the scarcity of turkeys at Christmas time in the Hamilton area around the end of Lake Ontario, where ham is the traditional fare. In other areas, either goose or turkey is the festive fare. It would be unwise introducing ham-based products (ham rolls, smoked ham, etc.) in a fish fare area or, alternately, attempting to introduce turkey rolls, smoked or otherwise, in a ham preferring area.

The return to school and the need to pack school lunches are ideal times to introduce nutritional snack products. Leisure summer activities are associated with foods like salads, prepared meats, dips and the like, marinades, and barbecue items designed for outdoor activities or patio living. Wintertime outdoor activities bring in an entirely different range of products. The timing for the test market, the food itself, and the activity associated with it must fit as appropriately, for example, as hot and hearty soups, cheese fondues, and liqueur-flavored coffees or hot chocolate drink mixes for the après-ski crowd.

8.1.3.3 *Length of the Test Market Period*

When to market is closely related to another question concerning time: how long should test markets be continued before an evaluation is made? Mazza (1979) tested a gourmet pack of wild fruit jellies over a 2-year period. The simple answer is that the test market should continue until reliable data have been obtained to evaluate sales volumes, the effectiveness of advertising and promotional strategies, and customer and consumer and retailer response. This period includes sell-in to the trade, promotion, first purchase, and repeat purchases. Time is necessary to establish a pattern of purchasing both by the trade and customers. The nature of the product determines its usage rate and hence determines the frequency for repeat purchases to be made. Further time is required to analyze data and get information back to marketing personnel and to the other members of the development team for any refinement of marketing strategy.

Too long in a limited test market without capitalizing on the advantages of early introduction serves no useful purpose. Lead time is lost. Copycat products are introduced in other market areas by the competition; they get market share in these new areas and make further market penetration and expansion difficult. Timing is very important, and lengthy test markets are expensive without adding greatly to information.

8.1.3.4 *Disruptive and Unexpected Elements in Test Markets*

The best laid plans, however, can be sabotaged by having a monkey wrench tossed in the works to mix metaphors badly. Events in international trade, world agricultural pricing structures, or other events in the food industry produce a short-term alteration in political events, in consumption patterns, or in the economics of an industry. Shortages occur for many reasons, and raw material prices increase. Company management need to have their antennae out for unusual developments in the supply and prices of commodities, government activities respecting legislation, and any untoward activities that could thwart markets in which they are introducing products. For example,

---

The introductory test market launch of a pre-cooked (micro-waved) bacon product was seriously disrupted for a subsidiary of Imasco Foods Ltd. when the availability of pork bellies declined and prices rose. Estimated costs for the finished product went above a price which marketing felt would discourage purchases.

   In a similar occurrence, pricing for a newly introduced blended salad and cooking oil (sunflower seed oil with olive oil) was sent tumbling when Russia flooded the market with sunflower seed oil that before had been high priced; we were left with expensive sunflower seed oil in stock when the price dropped. Countries can place trade embargoes on goods or buy up stocks causing temporary shortages; consequently, prices rise. Then they flood the markets at the higher prices with the shorted commodities and reap the benefits of the higher prices; then prices are driven down.

Nature can play a role, but how to be prescient about nature is unknown:

Natural disasters or, in less sensational fashion, weather events play a short term economic role in raw material shortages. A highly successful hot sauce was seriously disrupted in its second year of production when rainy weather destroyed the pepper growing areas in California and the specific variety of very hot peppers was unavailable.

Many such short-term events greatly influence the timing of test markets and can be disruptive of tests already underway. Shortages of raw materials cause extensive reformulation and pricing is thrown off.

8.1.3.5 *How to Introduce*

How to get products introduced can be a problem for both small and large companies. Small or new and comparatively unknown companies find it difficult to get shelf space, sometimes difficult even to get an appointment to meet with the purchasing agent of large retail food chains.

   Many chain stores are becoming much more hard nosed regarding new product introductions because these take space away from products with proven sales records. They are reluctant to give space to new products unless they are assured of good margins, rapid inventory turnover, and advertising and promotional support of the new products. Stores want to eliminate slow-moving items with poor margins. A Catch-22 situation can result: products that stores want can only be developed by market testing, and these new products need space in stores for market testing that stores are reluctant to give.

New product introductions by large companies are accompanied by extensive advertising and promotions with in-store demonstrations; couponing in magazines, newspapers ads, and door-to-door fliers; or piggybacking offers and special pricing offers. Small companies cannot afford these.

Advertising and promotional activities must be measured for their impact on the introductions of products in any marketing area. Introductory promotions to consumers (and to the trade) are one-time events; they are to get the introduction known and pique the targeted consumer's interest. Such heavy introductory promotions cannot be carried on beyond the introductory phase and certainly not throughout the product's life cycle.

The marketing department has many tasks during the introduction and test market. It must understand the effectiveness of its own adverting and promotion on sales and determine what level of promotional maintenance is needed to encourage repeat sales; it must analyze the actions of its competitors during this period and how this influenced sales and decide on defensive tactics. It is also tasked with determining consumer reaction to the product and what future maintenance may be required.

8.1.3.6 What Product to Market

First impressions can be very difficult to change these; this is equally true of first impressions of people, places, or products in the marketplace.

All too often, a company will introduce a product into the market in a specially designed package or even as a specially manufactured product. It is not the normal product that consumers will see in repeat sales. The use of specially packaged product during an introduction can be a disaster. Consumers have been introduced to, have become accustomed to, or have come to expect something specific; customers have been educated to a particular price, package, and product. Retailers have come to expect certain price deals and promotional support. A new package or modified product constitutes a new product. Data obtained on the sales of the product originally introduced may not be valid for this changed final product.

The test market must be carried out with the same product that all the testing and research have been done; the same product as the factory will run. Any retooling of the package or product during introduction and test marketing will certainly skew results and perhaps compromise the test market.

8.1.4 Evaluating the Results

At some stage during the test market period, inexorably, there comes a time of reckoning, of evaluation while marketing of the product continues. Kraushar (1969) discusses problems with using test markets as predictive tools and makes apparent why many companies choose to skip them as too expensive for the information they provide; these companies prefer to use other means to get predictive data.

8.1.4.1 The Market: Misinterpreted and Misunderstood

Data from test markets can be very easily misinterpreted despite all the precautions and care in conducting the test. There is a natural desire of all involved in development to see the product succeed; this desire must not color judgment of the results. Clausi (1971), detailing errors in interpreting marketing data during the introduction of a dry cereal with freeze dried fruit, commented that "the strong initial purchase pattern coupled with overwhelming consumer acceptance of the concept tended to obscure the significance of the negative evidence." They erred in their interpretation of initial success. Repeat sales were flat. Negative signs were underestimated and misunderstood.

> …wisdom comes to us when it can no longer do any good.
>
> **Gabriel García Márquez,** *Love in the Time of Cholera*

Enthusiasm, perhaps, for the project carries the team away.

8.1.4.1.1 Dynamism and Interrelationships in Marketplaces

Why are consumer research and test market data so easily misinterpreted? First, all the forces at play in the marketplace are difficult to research practically, to measure quantitatively, and to understand intellectually. The data are often subjective and therefore subject to misinterpretation. There is the behavior of both the customer and the consumer, the cooperation of the seller, the activities of the manufacturer, the activity of the competition during the test market to determine, and a changing technology that affect all to understand and to counter. All data obtained must be read against the backdrop of this complex behavior. In Figure 8.1, the major protagonists, the food manufacturer, the seller, the customer, the consumer, and the all-surrounding competition, interacting within any marketplace and adding to the complexity of the marketplace (see also Fuller, 2001), are depicted.

In Figure 8.1, the complexity surrounding test markets that must be understood begins to emerge. Very distinct marketing arenas become evident. These are as follows:

(A) The seller–customer interface: There is direct interaction between the seller and customer. There can be several types of interactions here depending on who the seller and customer are and in what marketplace they are interacting. The interaction between customer and seller in a fast food outlet is vastly different from those in a white table cloth restaurant from those in a chain store from those in a mom-and-pop convenience store.

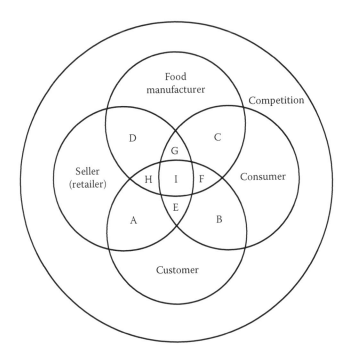

FIGURE 8.1
The major protagonists and their interactions within the various marketplaces. (Adapted from Fuller, G.W., *Food, Consumers, and the Food Industry: Catastrophe or Opportunity?*, CRC Press, Boca Raton, FL, 2001.)

(B) Customer–consumer interrelations: The gatekeeper and user interactions that define usage, desire, cost, and other product characteristics. Again, there are multiple interactions, all very different: mother and child, wife and husband, dating couple, purchasing agent and sales person, or institutional buyer and captive user.

(C) The consumer–manufacturer arena: This is the dominant interaction in new product development. Opinions expressed from the consumer to the customer determine whether there will be repeat purchases. Here, there must be gratification of needs and wants.

(D) The manufacturer–seller interaction: It is here that selling policies are dictated, pricing and distribution are defined, and, in general, trade relations are established. These can be very complex according to who the manufacturer and the seller are. Sellers, especially the large chain stores, may have supply chain management or order management policies in place with manufacturers. Ingredient supplier and baker, for example, relationships differ from farmer producer and manufacturer.

(E) The customer–consumer–seller interface: Most selling takes place here as typified by the mother with children in tow shopping in a supermarket. However, it can be much more complex, for example, the interrelationships between the engineer, his vice president of finance, and the technical sales representative of a food equipment manufacturer at an exhibition booth at a trade fair with their goal to negotiate the purchase of equipment for product development.

(F) Customer–consumer–manufacturer interrelations: Most marketing research takes place here; promotional campaigns are developed here to attract the customer (gatekeeper) or attract the consumer or do both. This area is often regulated especially in the promotion of products to children.

(G) Consumer–manufacturer–seller arena: Media and promotional campaigns are tested and used. Again, this is an area where restrictions may apply.

(H) Manufacturer–seller–customer arena: Here, there are mutual efforts by the manufacturer and the seller to attract the customer.

(I) The main selling arena where all interactions are present.

Surrounding all these interrelationships, there is the ever-present competition, watching, researching, and, perhaps, interfering. Within this diagram, all the interfaces can be seen that influence and hence must be reckoned with in any new product introduction whether into a test market or into a major launch.

How active or inactive (a rare event) is the competitor?

A colleague informed me of a test market conducted by one company when that company's competition was suffering through a strike. No competitive product was on the market; of course, sales of the newly introduced product were excellent and repeat sales good throughout the lengthy shutdown of the competitor. This strike at the competitor's plant was ignored in the assessment of the new product launch – a disastrous miscalculation. When the strike ended the competitor entered the market vigorously and drove the new product off the shelves. If the competition is inactive during another company's launch of a new product, then why?

Competitors will be expected to buy up product either for chemical analysis, for their own test purposes (to get their own test information of consumer reaction through taste tests or mini-market tests) or simply to foul up the sales volume figures and so cause misinterpretation of sales data. Knowing what consumers are doing and how they are reacting to a new product is valuable information both to the company doing the test market and to the competition. It cannot be overemphasized: knowing what the competition is doing during a test market is vitally important.

8.1.4.1.2 Personal Opinions, Biases, and Self-Deception

There is a danger that product managers and marketing personnel, in particular, become emotionally attached to new product projects that, perhaps, they, or others, overpromoted to superiors during earlier phases of development. Justification for past actions may be read into their interpretation of the introductory results. Emotions must be kept out of the interpretation of data (see Clausi, 1971; Kraushar, 1969).

The science of consumer research is still not strong enough to prevent misinterpretation of the information obtained from all the data obtained in a test market. The complexity of the test market, no matter how carefully designed the test is, is subject to the multitude of interfaces seen in Figure 8.1. This, combined with the imprecision and ambiguity of applying marketing science to the interpretation of test market data and the strategic goals of senior management, serves to confound many test market results. In short, errors in interpreting test market results arise from a lack of objectivity by the interpreters of the data, from the highly subjective views and feelings for the project held by the interpreters of the data, or from marketplace interactions that confound the data and the tools and their imprecision.

8.1.4.1.3 *Criteria for Evaluation*

Criteria for evaluating whether the launch was a success or failure reflect the objectives of the company performing the test market. Accordingly, different criteria will be used by different companies to judge whether or not a launch merits expansion.

Four measures can be used:

1. Payback: When will there be profits? Is the company strong enough financially to wait for greater returns or does a short-term payback mentality prevail in senior management? An unrealistically short-time frame to achieve a satisfactory rate of return has been the downfall of many new product ventures. The interpretation of "unrealistically short" rests with the company's strategists.

2. Sales volume: Will sales volume goals or targeted percentage share of market or even significant market penetration be achieved? Will these criteria meet the company's objectives? Case movements must mean profit for the company and not disappearance of product in continued ill-advised promotions.

3. Consumer reaction: Did consumers like the product? How can this be capitalized upon (product maintenance) in the future? Can simple strategies be applied to improve consumer reaction?

4. Tactics: Being there in the market is the thing. Did this introduction preempt action by the competition, increase market share, provide market penetration, or otherwise satisfy the strategic marketing goals of the company?

The first two are very similar. One says it in money; the second uses case volume and share of market. The third is technical and asks whether product maintenance can initiate a family of products to capitalize on and support a new product line based on the introduction. The final measure is much harder to assess.

The second measure is perhaps a better measure of trade acceptance than the first. Nevertheless, it does require some caution in interpretation. If projections indicate the volume of units sold are related to enthusiastic consumer acceptance and repeat sales, then appreciable economies of production can be anticipated by scaling up manufacturing. Usually, on the evidence of learning curves, more units of a product can be made more economically than fewer units of the same thing (Malpas, 1977). This, in turn, will influence the rate of return of investment as manufacturing costs go down.

Case volume sales must be examined very carefully to determine precisely what it means. If these sales are consumer sales, this is a very positive factor in interpreting the results of the test market. If they are merely case movements between warehouses or buy-up by the competition, sales volume could be deceptive as a measure in interpreting the results. The following incident describes one such event:

I was on an acquisition study on the West Coast. My object was a company that was test marketing a line of pouch-packed entree items. My company was interested in the product and the company. The product was being test marketed in three large supermarkets in Vancouver. Prior to my meeting with the principals, I purchased two cases each of the four flavors for shipment to our laboratory in Montreal. Much to my surprise, during my meeting the next day with the president of the company, he regaled me with the tale of more than eight cases of product being sold in one store so great was the demand for the product! What chaos would I have made of this company's sales statistics had I been a rival!

The third measure is an assessment of consumer reaction to the product. The test market is a commercial experiment in consumer studies. The reason why the consumer is buying the product and how the consumer is using the product, that is, the context of product use, is valuable. Can the context indicate strengths of the product to be capitalized upon? Are refinements needed or future line extensions justified? Are promotion and advertising directed properly based on the context of use? Should other market niches be explored? Weaknesses in the product need to be ferreted out by consumer research and eliminated before new marketing strategies are undertaken.

The last measure is more in the nature of a business tactic, a strategic tactic forced on the company. A product may be introduced or positioned to counter the activity of a competitor or to establish a position in a particular niche. Success would be measured by whatever small share of market could be gained as a foothold for strategic or tactical purposes at a later date. The company feels that it cannot relinquish a position in the marketplace—that is, the company must figuratively be seen and heard in this particular market niche.

8.1.4.1.4 Applying the Criteria

Small companies apply the criteria to evaluate a test market more flexibly. They are satisfied with a weekly increase in units sold indicating customer acceptance and reports back from the marketplace that say consumers like it. That product is usually considered a success. Small companies are generally less concerned with market share that large companies use to define success. Repeat sales with an indication of growth and customer satisfaction are their criteria for success. Small companies demonstrate more patience.

New product development costs in small companies cannot be determined with accuracy; consequently, the payback cannot be determined with accuracy. Management of small companies are quite content to pay their bills and have an increasing bit left over each week. Financial criteria are not stringently applied because budgets for research and development in small companies are frequently not separated out as in large companies. These costs are bundled together with either quality control (usually the seat of research and development) or as production expenses.

A note about the importance (or lack of importance) of market share must be interjected here. A company wants some market share for its product, but does it need (or want) the biggest market share? Having the major share of a market with some product will only attract competitors trying to evict the major share owner. Attempts for dominance with a major market share are usually the result of ego trips of company presidents. Having some share provides a stable presence in the market from which the company has future options to develop in different directions. A large enough profitable share with a product that is not a drain on resources should content most companies. Unfortunately, it often does not.

8.1.5 Judgment Day: The Evaluation

At the conclusion of any test market, all subjective and objective data related to the product, the marketplace, and customers and consumers are gathered and analyzed. Questions related to the product itself, its protective package, the label, preparation or recipe instructions, pricing, and positioning of the product need to be asked. Salesmen must be interviewed. The need for changes in the product, the process, or the package must be reviewed to determine what changes, if any, should be made based on customer, consumer, and retailer reaction.

Whether the test market was a success or a failure, reasons for its success or failure need to be examined and documented. Where and why a test market went wrong provide guidelines for future development teams to learn from prior mistakes. Similarly, how success was achieved points the way for future development projects.

In the analysis of a product's failure, the questions are, "What went wrong?" "What elements of market research, technical development, and analysis of data led to the failure?" These exercises should not be witch hunts for a scapegoat. Unfortunately, this backward look turns into a witch hunt with everyone trying to obfuscate the facts, pointing fingers all too often at the research and development group with the result that nothing is learned. The search should be for flawed systems or faulty information that led to incorrect decision making.

An apologia is required here. Often, in a product failure, it is overtly or covertly suggested that the product could have fitted the concept better somehow and research and development is blamed for why it did not. The stigma of failure sits with the technologists. Others on the development team, who may equally have contributed to the failure of the product, usually have other avenues to pursue within the extensive framework of the large company; the failure does not blight their future. This is not so for the technologists; to pursue their chosen career paths, they must stay within research and development. The stigma remains until it is washed off with a success; the pressure is on the technologists, but the reader should note Gershman's remarks at the start of this chapter.

8.1.6 Failures in the Marketplace

Most frequently, much of the success or failure of a new product hinges on elements outside research and development. These elements are all in the realm of marketing. Best (1989a), Gershman (1990), Kraushar (1969), Morris (1993), and Wang (1999) have each described these (Table 8.2) with both Best and Gersham facetiously calling them the "P"s of marketing. A closer examination of all the tabulated elements reveals they are all variations on a theme; Best takes the hallowed four "P"s (product, place, price, and promotion) to add "perspective." Gersham splits the traditional four and subdivides them—one can only with difficulty distinguish between promotion, publicity, piggybacking, and premiums.

Kraushar's "lack of objectivity" deserves some further comment. Simply put, it is the inability or unpopularity to say "No." Saying "No" in the face of the development team's enthusiasm is hard especially if some problems appear in the research (see earlier Clausi). Often, as Kraushar comments, these snags are minimized in the spirit of keenness of the moment. One must be objective in product development.

Wang (1999) weighs in vaguely against management as a dominant factor in product failures. He provides a murky distinction between placing the blame at senior management's doorstep and at the product or brand management's level (the vagueness is due to the varied nomenclature of positions within companies).

TABLE 8.2
Elements in the Success or Failure of New Food Product Introductions According to Some Sources

Kraushar (1969)	Best's 4 plus 1 "P"'s (1989a)	Gershman's 12 "P"'s (1990)	Morris (1993)	Wang (1999)
Product not appropriate	Product	Perception	Inadequate market research; no market need or changing needs	Lack of funding for long-term innovative research; short-term goals
Product is faddish	Place	Pitch	Management; no commitment with budget and resources; no clear strategic focus	Time pressures in days or weeks rather than months or years; emphasis on line extensions
Timing is wrong	Price	Packaging	Risk aversion and short-term orientation; me-too products; line extensions	Career risk in pursuing innovation
Pricing is wrong	Promotion	Price	Poor fit with in-house capabilities	Poor recognition of skills for development
Too high or too low	Perspective	Promotion	No formal new product development process	Poor management of development skills
Product is wrong		Promises		Cannibalization of existing products
Does not perform		Piggybacking		Wrong research; established brands cloud new research thinking
Not significantly different from competition		Positioning		
Poor communication of suitable image		Placement		
Lack of objectivity		Premiums		
		Publicity		
		Perseverance		

Hollingsworth (1994) reported on the top four reasons for product failure as reported by Group EFO Ltd.:

1. Strategic direction
2. Product not delivering promise (more "P"s)
3. Positioning (still more "P"s)
4. No competitive point of difference

Nothing has changed since 1969 when Kraushar lashed out at these same reasons for failure. He could not understand how failures could happen in large companies with money and supposedly sophisticated and experienced marketing teams and could only lump the reasons together in his lack of objectivity category above (see Section 8.1.4.1.2; Kraushar, 1969).

Silver (2003a) lists seven deadly sins of product development in a true biblical style with "thou shalt not" phrases. Her "laws," less biblically rephrased, are as follows:

- Management commitment is essential. Proceeding without the full support of management is a signal for trouble. (Compare Snow's "absence of their passive resistance" regarding the need for full support.)
- The formulator's dilemma in attempting to replicate cooking techniques should not be ignored. Scale-up from kitchen top samples to the pilot plant to the plant-scale production has difficulties.
- The sophisticated palates of consumers should not be belittled or ignored. They know quality.
- Cost-cutting methods should not be applied to testing. This is not the time or place to cut costs and take short cuts with necessary testing.
- Packaging should not be skimped on since it must sell the product. In addition, with packaging "form should follow function."
- Marketing should not be skimped on.
- Weak projects should be killed early.

Despite Silver's light-handed approach, her points are not trivial.

Most new food products fail to survive their first year in the marketplace. This is a staggering loss of money, of physical resources, and of the efforts of skilled and professional personnel. If all development work and research data had been done thoroughly and analyzed dispassionately, failure should have been most unlikely. Hindsight, unfortunately, provides better vision than foresight. Hindsight allows one to make generalizations or speculations on what went wrong, and these observations from hindsight need to be used to improve techniques for future development.

8.1.6.1 Causes of Failure

An in-depth study of why a new product succeeded would have been a much more valuable contribution to an understanding of new food product development. Causes of failure merely provide developers with a series of "don't"s when what is needed are a set of "do"s. One would then be equipped with guidelines to follow for future product development. However, predicting the success or failure of any product against the volatility of the consumer in a changing marketplace is still an art. As Clausi (1971) might have said, reading all the evidence correctly, including the negative evidence, will lead to success.

The reasons for a product's failure (Table 8.2) are difficult to classify except generally and harder still to pinpoint a particular product's failure (but see Gershman, 1990): the causes cannot easily and neatly be pigeonholed, as Clausi (1971) found. Clausi, in describing one particular failure, could only suggest that the signals from the marketplace were misinterpreted. Signals are rarely objective and quantifiable. They require subjective interpretation with all the baggage this implies. How is people error in judgment to be pigeonholed amongst those listed in Table 8.2? An examination of failed product introductions provides a very broad overview of probable causes from which only generalizations arise. One cannot apply the generalizations at the start of the development process or at any other point up to and including the test market: they can only be applied in hindsight. One cannot predict failure based on this table and say that this product or that product will fail because....

Simplistically, the causes for failure are broadly classified as those beyond the control of the company and about which they could do very little and those they could have controlled but did not. The former are reasons external to the company; but should not the company have been aware of weaknesses—or were they blind to them?—and have had some contingency plan in place? Blindness to weaknesses is not always manageable for various company reasons, but again is blindness an internal or external reason that caused failure?

Separating reasons into external and internal categories cannot always be done with clarity. For instance, too small a market (an external reason) can be a cause for a product's failure (see next section). Smallness of the market is beyond the company's control but then should not marketing personnel have seen there was too small a market? Were the marketing capabilities and resources within the company either incompetent or inadequate or both (internal reasons)? How else would marketing research have failed to determine the magnitude of the market beforehand?

Thus, if one states baldly that one reason is external, one must, equally, understand that an internal reason may have contributed directly or indirectly to it.

8.1.6.1.1 External Reasons for Product Failure

After products have been introduced, marketing personnel may find that markets for them are too small. Growth potentials would be limited;

possibilities of recovering development costs would be minimal. In certain markets, this knowledge may come unexpectedly. For example, changes in the purchasing policies of governments with respect to institutional buying for the military, for government-run correctional institutions or prisons, or for school meal programs may suddenly and abruptly be altered, and the size of a market may change. Nevertheless, companies servicing such markets should keep themselves informed of pending government changes by close liaison (networking) with their government contacts.

Markets controlled by a dominant competitor are difficult to get footholds in; companies introducing new products find themselves not battling just for customers but battling with competitors. A dominant competitor has power to control, influence, or buy retailers and limit shelf space exposure for rival products. Consumer acceptance is too costly if advertising and promotional dollars have to counteract retaliatory action by a dominant competitor.

Domination of markets by a single customer (e.g., a major retailer or the government, i.e., the military, penal institutions, or large fast food chains) can present severe challenges to companies introducing new products into those markets. The cooperation of customers (retailers) is always essential, but when suggestions from customers become directions, then the situation can be fraught with stumbling blocks. After all, companies, not customers, have spent the development dollars and will risk most in a failure. But the dominant customer has the greatest say in pricing and marketing strategies in general. The tail, that is, the customer, wags the dog, the developer. The food service industry is one where this problem is apt to arise. The immense buying power of some quick food and retail chains has permitted them to dictate to producers what products and what development they want and at what price they want this for their marketing purposes.

There are product-related reasons for failure in the marketplace. With the introduction of a me-too product into a market that is saturated with similar products, consumers will refuse to buy another brand or variation if they cannot see a point of difference between it and already established products. The problem is perhaps beyond the control of the developing company; no one could have foretold the flooding of the market with copycat products, but market research in the marketplace may have indicated saturation. Markets do fragment and may provide a special marketing niche for new me-too type products with a measurable point of difference over competitive products.

Where a novel added value has been designed into a product, this novelty dominates the message to the consumer and they must be educated to this novelty. This is, so to speak, an introductory fee a company must pay for novelty, educating the consumer. Consumers can be forgiven for questioning, "So what?" or "What advantages are there for me?" They must be educated to the novel added value and its advantages. Irradiation raised the specter of a new process not of advantage to consumers but to processors who now would not maintain high levels of sanitation. Novelty in a product, for example, frozen for canned, tablets for powders, aerosols for liquids, and so on, may fail disastrously if the consumer cannot see the advantage being offered or if the advantage (point of difference) over other similar products is insignificant. Flavored ketchups have not been successful for this reason, and all indications that I have received indicate that colored ketchups are not the success they were touted but this may be regional.

Products ahead of their time or for which consumers are not previously prepared have poorer chances of getting market acceptance. They meet consumer resistance because consumers are not educated to their possibilities. Arguably, one might consider this an internal reason for a market failure, one within the control of the company; marketing personnel did not promote the product correctly. On the other hand, educating customers and consumers is costly and why educate customers and consumers only to have a competitor reap the benefits with me-too products?

8.1.6.1.2 *Internal Reasons for Product Failure*

Management is often unwilling or unable to recognize the strengths and weaknesses of their company and of individual staff members. Perhaps management failed to understand what business the company was in. Hence, company objectives are misguided and have misdirected growth respecting new product development. This assumes rather simplistically that a series of bad management decisions (intrinsic reasons) are responsible for failure.

Without a strong, competent senior management with a clear idea of what the company is, marketing and research and development resources are not usefully directed or profitably applied. Management must recognize its internal strengths and weaknesses. Areas of weakness must be complemented with outside resources where practicable or removed and replaced with strong and competent personnel. If a company's marketing department is incapable of conducting reliable market and consumer research—a subjective evaluation a company must make—new products are apt to fail. In such a situation, many independent market research companies are available that could complement internal resources. However, one must question whether incompetent internal marketing resources are competent enough to appreciate, understand, or communicate with the external resource.

More damning are situations in family-run and family-operated businesses. The second and third generations frequently do not have or have not gained the innate skills of the founding family. The company's senior members may be unable or unwilling to recognize their children's or other relative's shortcomings (see Rothfeder, 2007).

If the product is a success and there is a growing demand for the new product, then lack of production capacity can cause product shortages. When retailers cannot restock their shelves quickly enough to satisfy demand, they are not happy; customers and consumers cannot get product, and the impetus of the launch has been lost. Buyers, users, and retailers will lose interest. But again, one must question the lack of foresight in management not to have seen the likelihood of this and backed up production capacity with production contracted to co-packers.

Unnatural adherence to and support for a project (unnatural in the face of negative evidence for its continuation) is, perhaps, more readily understood than the other causes for failure because of the human element entangled in it. People, whether in small or large companies, do become emotionally involved in their projects. It is their pet project—their baby. Their rather illogical reasoning goes something like this: too much money has been spent to date to stop the project now, so spend more monies to rescue the project. It is like the gambler who gambles more money to recover his losses. Costs need to be regularly evaluated to prevent their escalation.

A product can fail if it simply does not perform as promised or does not live up to the standards promised or as the customer and consumer had expected from what the promotions promised; customers and consumers expected something else. This is missed communication. Was the cause for the poor performance inherent in the product itself? That is, it was poorly designed. Or was the cause of the product's failure poor communication and promotion? That is, the product was fine; it lived up to the concept. It did not live up to the false or misleading promotion of its virtues; the wrong message was given to customers and consumers.

It is an oversimplification to suggest that there are only two reasons for the failure of a new product. Nevertheless, there can be a great deal of truth in such a generalization. These two reasons are as follows:

1. Expecting too much too soon
2. Not being lucky

The first, expecting too much too soon, does not happen to companies whose objectives are based on a realistic assessment of their companies' strengths and realistic financial and marketing objectives. And good luck, or whatever one wishes to call it, comes more regularly, rather than randomly, to companies that utilize their resources well in order to research markets, consumers, and their products. At the very least, those companies will reduce their margins of error.

8.1.6.1.3 Product Maintenance: Salvaging Failure

Kraushar (1969) makes an interesting point that should have been considered in product design. The growth curve for any product eventually displays a flat, no-growth phase, which is usually followed by a dying away phase when only a prohibitively costly amount of promotion can liven the sales.

He suggests that before any new product is launched—that is, during development, a product maintenance program should have been set up. Product maintenance programs are designed to make improvements in the product and are used, when sales stagnate, to pep the product up. They suggest new uses for the product, appeal to a new customer niche, or contribute some unique added value. Product maintenance programs are an integral part of the development process and useful in retooling a failed product introduction.

9

Why Farm Out New Product Development?

Learn from the mistakes of others—you can never live long enough to make them all yourself.

Anonymous

Joint ventures are almost always bad. At worst, both parents neglect the stepchild in favour of their own.

R. Townsend, *Up the Organisation*

The sage in his attempt to distract the mind of the empire seeks urgently to muddle it. The people all have something to occupy their eyes and ears, and the sage treats them all like children.

Lao Tzu, *Tao Te Ching*, Book 2.

9.1 Introduction

Each of the above quotes contains advice for, warnings about, and uses of going outside for help in new product development and test marketing. Anonymous suggests that going outside can be a learning process. Goodness knows I have seen enough mistakes in this field, heard of more from colleagues, and, indeed, made my own, to be able to impart some learning to others. Townsend gives good advice about ventures with two companies joining forces in the false belief there is strength in numbers. Usually, both bring some skills or resources to the venture, for example, a supplier and a manufacturer (see, e.g., Houston and Johnson, 2000). I interpret the third aphorism as follows: I have yet to see a consultant's (the sage's) report that came to a firm conclusion without technobabble and did not have some disclaimer that further work needed to be done to clarify some findings. Of course, consultants want to stay employed.

9.1.1 A Rose Is a Rose Is a Rose

There are terms that need to be clarified. This chapter is about the following:

- Outsourcing
- Joint ventures
- Partnerships
- Consultants

All have similar features in practice. They all describe an outside resource with which or whom a working relationship may be entered into at any time during development—some are without contractual obligations, and some have lengthy documentation of terms of duties. Customer–supplier relationships (a form of joint venture) or co-packer relationships have contractual arrangements and frequently play a role in new product development when the supplier or the co-packer provides their skill and expertise to the design and development of the new product. As a generic term, outsourcing covers all the bulleted items above.

9.1.1.1 Outsourcing

Outsourcing, as a business term, is only 25–30 years old, but it is a very old business practice that food companies have practiced for many years. For example, it has been normal practice to use trucking firms and customs brokers for distribution. A simple extension of this thinking was to use a third-party warehouse (i.e., distribution center). This distribution center supplies much more than simple storage; it provides inventory control, invoicing of goods, and electronic data tracking of goods that allow improved distribution efficiencies, reduced capital costs for warehousing facilities (since these are shared with other companies), and reduced staffing costs and thereby lower overhead costs. Information technology and, with it, intelligence gathering are expanding so rapidly that, nowadays, they are commonly outsourced to companies more skilled in the operation of these technologies. By outsourcing, it is reasoned that the client has more time and resources to concentrate on its core business while experts take care of the outsourced work.

Thus, the nub of the argument for outsourcing is revealed. For distribution, it means that the task of logistics is left to the logistics professionals; information technology and telecommunications are farmed out to experts in these fields. Other services that are often outsourced are legal matters, accounting, public relation and advertising for radio, television, and newspapers, and more mundane activities such as general cleaning (grounds maintenance, landscaping, and office cleaning), plant security, and laundry services. The general area of outsourcing, then, might be interpreted as the use of expert services (acting similarly to consultancies) for specific tasks.

A company's core activities are defined as servicing its customers. All services that are peripheral to the company's core activities are contracted out to others who do the service more adroitly and efficiently; the practices unique to, or essential for, the core business are kept in-house. Green (1996) suggests outsourcing can optimize return on investment capital. He does caution, however, that this will occur "in the right circumstances." Patterson and Haas (1999) warn that only those functions that have clearly defined boundaries (e.g., performing nutrient analyses is a task with defined borders) that need little cooperative participation between the principals should be outsourced. However, this caution from Patterson and Haas hinges on the interpretation of "little cooperative participation." The buying of produce or ingredients to established standards is no different than buying services to be outsourced; both need some monitoring. Activities such as legal matters (labels, claims, contracts, etc.), warehousing and distribution, and development of promotional material for television or newspaper and magazine advertisements all require close cooperative collaboration and deserve to be in outsourced and expert specialist hands but should be monitored. One does not want promotional materials that management deems offensive or inappropriate to company ethos.

9.1.1.1.1 Cutting to the Core: Advantages and Disadvantages

Advantages and disadvantages of cutting to the core, that is, outsourcing, are presented in Table 9.1. Green (1996) provides a slightly different set of advantages and ignores what many see as the disadvantages. Cutting to the core raises the following questions: What is the core? Where does the core begin? How close does one go to it in cutting? Warehousing and distribution are

TABLE 9.1

Advantages and Disadvantages of Outsourcing

Advantages

Client trims staff by outsourcing specialized activities to service providers with expertise in the outsourced fields.

Resultant staff cuts reduce need to recruit staff and reduce payroll (and all attendant benefits); fixed payment based on one central contracted agreement required.

Liabilities resulting from internal resources have now been transferred to a highly skilled external source. Outside source is expected to try harder to keep business.

With noncore activities outsourced, workforce and its systems are more focused to attend to servicing its clients, i.e., its core business. All efforts are directed to this activity.

Disadvantages

Outsourcing activities with its loss of staff can demoralize remaining staff with each wondering which activity is next.

Company loses out on gaining practical knowledge in fields in which it might see its self-interests unfulfilled.

Activities of service provider still need to be monitored to determine whether they meet standards contracted for.

The service provider provides only the services contracted for. They do not upgrade, modify to fit changing client needs, or necessarily work to client's goals except as contracted.

often outsourced, but the following questions arise: Are customers being serviced with prompt deliveries? Are peculiarities of customers being serviced in friendly and competent fashion? Is it service without the personal touch the client may have provided? These services are not part of the impersonal activities of the service provider. Thus, the company must examine carefully what they outsource. As Green states, outsourcing is advantageous in the right circumstances.

There is an opportunity for greater focus on servicing one's customers with noncore activities farmed out to experts. This begs the question whether the company really understands its core business and hence knows what it should outsource. Service providers do what they are told in the written contract with their client; the client had better know what it is to their advantage to outsource.

Certainly, there are economies as staff are cut, but there also ensues a lowering of morale. How deeply will cutting to the core go? There will still be a need to have personnel *with the necessary skills* to monitor the activities of the service provider. There is a disadvantage in the loss of intellectual property: the lost staff were personally invested in the interests of their company; the service provider is not motivated except to follow its contractual obligations in order to keep the contract. The client gives up managerial responsibility for those outsourced activities to a possibly disinterested second party.

I encountered an example of the latter in a company I consulted for. They had outsourced plant sanitation and clean-up to a company specializing in this activity. No check-up on the efficacy of sanitation was performed; there was no walk-through by senior plant staff prior to start-up in the morning. "It was in good hands." The sanitary cleaning company fell on bad times, cut back on its services hiring less experienced professionally trained help with the result that my client experienced spoilage problems. My client was at fault for not maintaining managerial responsibility over plant sanitation and the activities of its service provider.

The greatest resource a company has is people; if these are eliminated through outsourcing, what does the company possess and where do responsibility, innovation, and creativity come from but from people?

Cutting to the core business does have benefits and should allow a greater focus on servicing customers. A "lean and mean" philosophy should not result in an emasculated structure that is unable to grow and develop new needs but hampered by an inflexible contract with disinterested supplier unable or unwilling to meet the new demands of its client.

9.1.1.2 Outsourcing, Consulting, Partnering, and Joint Venturing

The similarities between outsourcing and consulting are obvious, so much so that one may consider one as a variant of the other. In outsourcing, a client depends on an individual or company to provide a necessary service that meets an agreed-upon standard or level of service, that is, solution of a problem or performance of a service. Outsourced services are usually contracted for periods of a year or more. In some outsourced activities such as telecommunications and information technology, these contracts may amount to many millions of dollars and continue for several years; the contracts themselves can be extremely complex legal documents. For example, it is reported in *The Gazette, Montreal,* for December 3, 2003, that Canadian Pacific Railway, Ltd. has outsourced its computer and technology section to IBM Canada, Ltd. for $200 million in a 7-year contract. The service provider is responsible for delivering a product (service) of high quality and free from error that is both on time and within the standards described in the contract. The buyer of the service rids itself of both responsibility and accountability for performing the function; it is in the hands of experts.

In consulting, a client depends on an individual for help, advice, or resolution of some problem. There may or may not be a contract; I have only been asked to sign a contract once and it was for nondisclosure (for something that was common knowledge). Some consultants are kept on retainers.

Partnering is the joining together of two or more companies in a contractual agreement; the activities of all are complementary and, by working cooperatively, they benefit one another. An example, provided and described by Kuhn (1998b), is a partnering of a flour milling and bakery operation, a processed meat company and a pizza manufacturer with a combined operation centrally located; another example is a cacao bean processing plant entering into a partnership to supply a confectioner with chocolate. Flavor houses often enter into partnerships with their clients in the development of a particular flavor for a new product the client is developing; supplier and customer collaborate and develop a trusting business relationship. None are rivals, and proximity benefits all parties.

Williams (2002) describes partnering between customer and vendor and building a relationship between customer and supplier, in this instance, between a customer (client) and a flavor house (consultant) for flavor development. Williams enumerates in detail the requirements to make such a partnering successful by knowing what both the client and the supplier want from each other for product development. Conditions or situations are elaborated that can undo a working relationship, such as

- Breach of trust between client and consultant supplier
- Poor communication resulting from mixed messages
- Lack of benefit for one party (but see Harvey, 1977)
- Disincentive to add value, that is, one or other of the partners neglecting to arrangement

Poor communication and mixed messages stress partnering relationships. These have consequences in long-term contracts as the client's business goals may change, but the service provider does not deliver services to satisfy the new requirements necessary to the client. There will be problems when the principals do not communicate the changed needs.

Joint ventures are somewhat different. They might very well involve rivals. They are often based on the principle "let's stop fighting one another—there's enough pie for both of us." Such agreements rarely remove rivalries and rarely involve equals. There are always conflicting loyalties that confound joint cooperation. Joint ventures and partnerships do have some similarities; Townsend's comment heading this chapter describes the situation for both succinctly. Many managers feel that if a venture (e.g., construction of a factory for a new product development) is worth doing, it is worth doing alone lest if it were done cooperatively, it might be neglected for vested, self-interest projects either of the parties undertakes individually.

Houston and Johnson (2000) discuss in purely theoretical terms supplier contracts versus joint ventures and when, between a buyer and a supplier, a joint venture may be advantageous. In any buyer–supplier relationship, there are three variants of the relationship: (1) the supplier's value is much higher than the buyer's value (that is, the client); (2) the buyer's value is much higher than the supplier's value (cf., supplier and military purchaser); and (3) the buyer and the supplier are of equal value (horizontal ventures). Their models discuss the conditions where joint ventures versus simple contracts would be mutually advantageous.

Consultants are somewhat more complex entities. Shahin (1995) rather cynically described the consultant as follows.

> When you encounter him, you'll know him on sight. He'll glide into your meeting as radiant as confidence.... If he were a car, he'd be a Lexus. On cruise control.

Scott Adams, through the voice of his Dilbert cartoon character, told a time management consultant that he, the consultant, had become a consultant because he had been fired from every job he had ever had for wasting time. To which the consultant replied, "Welcome to the wonderful world of consulting." Putting cynicism aside, consultants are individuals or companies who are experts in unique areas of knowledge hired for guidance and advisory services in those areas in which the company is ignorant.

In contrast, a consultant or consulting company is seldom kept beyond the duration of the task at hand unless that consultant is on a retainer basis to provide a continuing service or to be available at a moment's notice. A consulting arrangement is often very informal, as simple as a handshake, or requiring signing of a contract detailing the work involved and with a confidentiality clause. A retainer is simply a prearranged agreement to provide an agreed-upon amount of time to a company at a fixed per diem.

9.1.1.3 A Classification of Consultants

I classified consultants as either professional or amateur (Table 9.2) using an analogy to amateurism and professionalism in sports (Fuller, 1999). This classification created a bit of an uproar amongst colleagues. *Amateur* is not meant pejoratively reflecting on the competency or experience of a consultant. Very simply, amateurs are consultants whose consultancy practice is not their main source of income. It is a hobby—a means of keeping busy, of "keeping their hand in," and of supplementing a pension or a regular income by moonlighting—retired school teachers do it filling in for absent regular teachers. It is a means whereby professors at universities obtain monies for graduate students or a stopgap tactic resorted to by many executives

TABLE 9.2

Outside Resources for New Product Development Classified on an Income Criterion

Class	Subclass
Amateurs	Executives and others between jobs: available as the result of outsourcing and mergers of companies. They are job hunting.
	Retirees: early retirees choose to work on aid programs in developing countries or to supplement incomes with consulting. (They often have restrictions on companies for whom they can consult.)
	Academics: academics are allowed to consult with the prospect in mind that this activity will bring industrial projects in.
	Extension departments of universities provide consulting services for local, regional, state, or provincial industries.
	Student training programs: business administration students under guidance of professors consult for small companies as part of their training.
	Research institutes: groups of academics or departments at a university combine to form a research institute. This is a quest for financial support.
Professionals	Individuals: working alone or networking with others to form larger consulting entities.
	Independent companies: more formalized than above with a broad range of consulting services with base of operation and in-house facilities.
	Private research institutes and associations: they are usually contract research groups much like independent companies: companies subscribe by paying a membership fee and are then privy to research activities of the larger group.
	Government agencies: governments have research groups that can be used by local industries or they provide individuals (often retirees) to guide fledgling companies.
	Trade associations: often, they provide consulting services for their members.
	Specialized service providers: these provide unique services, for example, nonroutine laboratory facilities, information retrieval, forensic accounting, decontamination processes, retail sampling, and recall programs.

or research scientists who find themselves between jobs. They use consulting as a tactic to seek a new position.

Nor is *professional* a reflection on a consultant's competency; nor does it suggest that a consultant meets some standard of professionalism. To my knowledge, there are no professional standards in consulting. Professional consultants, pure and simple, make a business of consulting; it is their main source of income.

Both the amateurs and the professional consultants are excellent resources to assist clients in new product development. Nevertheless, the distinction between the two groups must be clearly understood by the client as this distinction may have implications in future client–consultant relationships (see caveats in the next section).

Some university-affiliated consultant organizations (Table 9.3) are described in more detail in Giese (1999; 2000). Universities see a need (and a source for the generation of income) in making their expertise available to food manufacturers. Consequently, new centers of assistance and expertise are being formed regularly; the IFT's *Food Online Newsletters* for August 20, 2003 and November 5, 2003 each describe newly created centers of expertise. Hollingsworth (2001) focuses on U.S. federal research and development programs funded by agencies or administrations and run cooperatively by university centers. In short, resources for advice and research are proliferating.

There are foreign universities and institutes that undertake contract research projects. Any follower of the Institute of Food Science and Technology's (United Kingdom) Information Journal *Food Science and Technology Today* (now defunct) and current *Food Science & Technology* will find any number of government, private, and university research facilities described as the following arbitrary listing shows:

- Food Refrigeration and Process Engineering Research Centre (University of Bristol, United Kingdom); March, 1994, volume 8.
- Swedish Institute for Food Research; June, 1994, volume 8.
- University of Milan, Department of Food Science and Microbiology; March, 1995, volume 9.
- Campden & Chorleywood Food Research Association; September 1995, volume 9.
- British Industrial Biological Research Association; March, 1997, volume 11.
- University of Nottingham, Division of Food Sciences, has four food groups: Flavor Technology; National Centre for Macromolecular Hydrodynamics; Food Microbiology and Safety; and Food Structure Research. The latter group is described in detail in *Food Science and Technology*, June, 2003, volume 17. The *Interactif Club* sponsored by the Division consists of food companies with interests in the research of the Food Division.

TABLE 9.3

A Brief Listing of American University Affiliated Research Centers Available for Assistance in Product and Process Development

Research Center and Location	Professed Specialty
Center for Advanced Food Technology New Brunswick, New Jersey	Problem solving for member companies Cooperative venture between industry, Rutgers University, and government
The New York State Food Venture Center Cornell University New York State Agricultural Experiment Station Geneva, New York	Assistance in all aspects of new product development and introduction
The Food Processing Center University of Nebraska Lincoln, Nebraska	Technical and business assistance to food industry in product development
The Northern Crops Institute North Dakota State University Fargo, North Dakota	Devoted to northern crop development and promotion
The Kansas State University Extrusion Center Kansas State University Manhattan, Kansas	Grain products and extrusion processing
The Spray Systems Technology Center Carnegie Mellon University Pittsburgh, Pennsylvania	Spray systems and atomization
The Southeast Dairy Foods Research Center Center for Aseptic Processing and Packaging Studies Both at North Carolina State University, Raleigh, North Carolina	Provides integrated approach to product and process development from formulation to shelf life studies to scale-up and market research
The Food Innovation Center Oregon State University Corvallis, Oregon	Provides advice and technology for added value products for Pacific Rim markets Lead organization for ValNET (Value Added Network of Export Technologies)
The Food Industry Institute Michigan State University East Lansing, Michigan	Provides research and outreach in food technology; provides workshops; leases equipment
The Food Industries Center Ohio State University Columbus, Ohio	Provides pilot plant facilities, product development, and scale-up for fruits, vegetables, meat, and dairy products
The Center for Food Safety and Quality Enhancement Griffin, Georgia	Provides facilities for sponsored research in food safety and quality studies; conducts consumer attitude and perception of quality studies

(continued)

TABLE 9.3 (continued)

A Brief Listing of American University Affiliated Research Centers Available
for Assistance in Product and Process Development

Research Center and Location	Professed Specialty
Institute for Food Safety and Security Iowa State University Ames, Iowa	Gives assistance to the food microcosm to combat food-borne infections, to prevent contamination of water and food, and to protect animals and plants from catastrophic diseases
The Institute of Food Science and Engineering University of Arkansas Fayetteville, Arkansas	Value-added research for processing of agricultural products
The Food and Agricultural Products Research and technology Center Oklahoma State University Stillwater, Oklahoma	Support for and provision for basic research, training, and advisory services for food and agricultural processing in Oklahoma

Source: Giese, J., *Food Technol.,* 53, 98, 1999, 2000.

To avoid confusion throughout this chapter, the company or any authorized
individual within the company who buys, hires, partners with, or enters into
a joint venture with an outside resource will be referred to as the "client";
all references to the "consultant" will describe any outside resource or any
resource that is outsourced, whether this is an individual, a private research
company, or a government- or university-based research institute, hired by
the client.

9.1.1.3.1 What Do Consultants Do?

Table 9.4 is a tabulation of consulting activities some of which I have been
engaged in. Not all have a bearing on new product development. The political or tactical use of consultants is interesting in that it may have a direct
involvement with new product development. I have on two occasions been
employed to conduct exploratory research in my own name. In one instance,
it was for a client wanting to study the effect of high pressure processing on
their products and the other to contact and evaluate research houses for
work on supercritical carbon dioxide extraction and to undertake some initial studies. My clients did not wish their interest in these fields to become
known. Problem solving—especially breakdowns in newly introduced
products—and advisory services are common tasks for most consultants.
Another more unusual one for me involved a small food manufacturer who
asked me to act as liaison between it and the research institute with which
it had placed a research project; they did not feel comfortable with their
ability to discuss, ask questions, and digest the data and information they
were being fed.

TABLE 9.4

A General Classification of Services Provided by Consultants

Services Provided
Political or tactical use of consultants
The consultant is used by the client to undertake some activity that management or the company prefers not to be seen to perform directly.
Problem solving
The client faces a crisis and requires some crisis management skills, for example, new product failure in the field.
Investigative research
Long-term basic research projects are placed with academic institutions. Projects may be contracted to outside resources with specialized pilot plant facilities, for example, extrusion processing, ultrahigh pressure processing, encapsulation technology, end-over-end can rotation thermal processing facilities, etc.
Specialized services
Training programs for staff, market research, product audits, forensic accounting, etc.
Advisory services
Support in application of novel technologies or strategies.

A cynical use of consultants was depicted in the cartoon strip *Alex* by Beattie and Taylor: the following dialogue occurred:

Geoffrey: Basically, as a result of poor planning and mismanagement, my company is facing bankruptcy. As chairman, I should be able to find some course of action but I can't... I can't cope.

Alex: You've got to get yourself out of this negative mindset, Geoffrey. "Can't" is not a word we recognize in the business vocabulary. Be positive. What **should** you be thinking in terms of?

Geoffrey: Can.

Alex: Can exactly. Who are you going to let carry it? A subordinate? The finance director, maybe?

Geoffrey: Or I could bring in some management consultants and then blame everything on them.

Alex: Now you're thinking!

9.2 Going Outside for Product Development

There are valid reasons for going to a consult for product development work. They can usually be summed up in realizing that the technical, physical, and personnel resources are not present in a company or are fully occupied elsewhere in the company.

9.2.1 The Need

Companies use outside resources to conduct some or all aspects of market research, new product development and research, associated engineering research, and marketing. The reasons for doing so are varied (Table 9.5). Any one of these reasons contributes to a client's need to use outside resources.

A client can order a product designed to its specifications based on its own market research or to specifications based on market research performed by its consultant. If the client wishes, the work may at this time be passed over entirely to an external development company. Then, the client and consultant meet thereafter at scheduled intervals to review progress and to determine whether the product-in-progress still meets the client's business and marketing goals. A regular review of progress is absolutely necessary because, with the passage of time, the client's needs may change or unforeseen difficulties in development arise. Difficulties should be addressed at once as time and money could be wasted on fruitless research. A review process indicates the client's continued interest and allows the client to watch out for any lollygagging by the consultant. Both client and consultant also wish to avoid extensive research trials that lead to dead-ends and, therefore, may jointly decide to change direction with the project.

Alternatively, some consultants work closely with the client's technical staff, visiting regularly to work on the development of a product. The development work is done entirely on the client's premises with the client's staff under the guidance of the consultant.

Often, a consultant will approach a prospective client with either a product or with a product concept with some prototypical products.

TABLE 9.5

General Reasons Why Outside Resources Might Be Consulted for
New Product Development

General Reasons for the Use of Consultants
Company lacks necessary skills or physical resources for customer, consumer, market, or technical research and development.
The skills are available but are already maximally deployed elsewhere in the company for maintenance of existing product lines.
Opportunity in a new applied technology or in a line of novel products has become available. The necessary skills are alien to the company's present strengths. The company wishes to explore these without committing physical resources.
A financial analysis reveals it is cheaper to contract with outside resources rather than to utilize existing manpower and facilities in-house or attempt to develop new skills and facilities in-house.
An extended period of basic research and experimentation is required for the new project. The company does not want to tie up its own resources for long periods of time.
A company may feel more secure if the research and development is conducted elsewhere than on its premises.

Client and consultant come to some agreement respecting ownership, licensing, patents, and any other legal matters.

The similarity to turnkey construction projects is obvious; the finished product, a newly constructed factory, is turned over to the client for an agreed-upon fee. In new product development, the finished product is handed over to the client for a fee or for royalty payments.

9.2.2 Finding and Selecting the Appropriate Consultant

Finding a consultant is usually not a problem: consultants more often find their clients. Selecting the right consultant with the requisite skills that fits comfortably into the company environment is the problem—one does not want or need a Lexus on cruise control, to quote Shahin. Nevertheless, to find them, there are trade directories where consultants of every kind can be found along with brief descriptions of what they describe as their expertise. They also advertise in trade and technical journals. The magazine *Food Technology*, published by the Institute of Food Technologists, has annually published a paid, comprehensive listing of consultants and their services as does The Institute of Food Science and Technology (United Kingdom) along with a brief description of their members' areas of expertise. Many trade associations refer clients to members of their organizations who provide consulting services.

A more successful way to find consultants is through networking. Today, most business people network through the trade, professional, social, and philanthropic associations to which they belong. From their fellow members in these associations, clients looking for assistance hear about and get referrals to consultants. Here too, clients can get word of mouth verification of the caliber of work performed by these consultants.

Consultants are looking for business; they also network. They are constantly alert through business magazines and newspapers to situations where their expertise can be used. They solicit interested parties by telephone, post, or e-mail. They give lectures at symposia, they teach short courses, they provide in-house training seminars for staff, they are prolific writers of articles in trade and technical journals, and they have Web sites advertising their services.

If consultants are found from impersonal sources such as an e-mail contact, a directory, or a referral from a professional or trade association, it is strongly recommended that the client request references and follow these up. Experienced and reputable consulting experts are wanted. Consultants referred through networking come with the personal comments of those who made the referral. If not, further enquiries can be made of others in the network.

An interview with, or a presentation by, the consultant is also recommended to determine the compatibility of client, the client's development team, and the consultant; and to familiarize the consultant with the client's requirements. After this consultation, the consultant should be required to present a proposal carefully detailing what work shall be carried out and within what time frames, it will be performed and clearly outlining all the legal implication of the arrangement (e.g., who owns what intellectual property?). This proposal describing the work objectives should be clearly written without any hyperbole, jargon, bafflegab, or obfuscation of terminology and clearly ought to be reviewed by a lawyer. Clear communication between the client and the consultant should be such that both understand the problem and that both know clearly what will be done and for what both parties have responsibility.

However, in over 30 years consulting in several countries, I have been interviewed only once—and that by a company that was sold less than 3 weeks after the interview! They found me in a directory. All my business contacts have been by word of mouth; this method of making contacts situation I have found to be true for my colleagues. Thus, the real world seems to favor networking to provide referrals for consultants.

The difficulty is usually not in finding a consultant; it is in finding the right consultant who can resolve the client's problems in whatever technical problems the development team has encountered. Clients choose a consultant not on what the consultant can offer or on who the consultant is but on whether the clients' needs can be resolved. That is the basis for selection.

Both parties, the client and the consultant, need to have compatibility, respect one another's capabilities, and, most importantly, be clear on the objectives of the project. The project should be a working relationship based on mutual trust and confidence. This is no different from the relationship with any supplier.

Factors that are important in selecting a consultant are as follows:

- The client must know what area of new product development it requires assistance in. What does the client want from a consultant? These needs should be written out by the client in a clear and concise statement describing what the client wants the consultant to do. Nothing should be vague about the request for the consultant's assistance.

- (A corollary to the above) The client must recognize their limitations for the problem they face. What is the consultant expected to deliver that are not within the client's capabilities? Could the client do it as profitably and as timely alone?

- The consultants' skills and expertise need to be carefully explored for their applicability to the problem. In short, is the client more knowledgeable than the consultant? It is not unknown for consultants with their hyperbole to promise more than they can deliver; they are hucksters selling themselves. Often, they subcontract work without the original client realizing this. Security and confidentiality can be breached in this subcontracting. The circle of people familiar with the product development work grows larger in such instances.

These points can be illustrated by some actual occurrences. As Vice President, Technical Services, of Imasco Foods, Ltd., I observed an instance of having more awareness of the technology than the consultant:

We wanted to measure the water activity of several products prepared in our various manufacturing plants. I had discussed the project, clearly I thought, in advance with a representative of a well-known mid-western research company. They had assured me of their ability to perform water activity measurements. (It was the early days of water activity studies.) Our companies sent in their products from various locations. In due time I received the results: the moisture contents of each of the products! An accompanying letter politely informed me that there was no such thing as 'water activity' and what I obviously meant was water content! A flurry of telephone calls and faxes ensued until finally they admitted I meant water activity; they had erred; they could not measure water activity. Much time, money and goodwill were lost on a simple development project. This research company was never consulted again.

An alert manager of our research and development group at our manufacturing plant caught an example of duplicity on the part of a consultant:

We had hired a New York-based consulting company to prepare some pour-over sauces for pasta. Samples were submitted along with their formulations for our evaluation. Our Manager of Research and Development noted a strong similarity to formulations he had been working on using a supplier's ingredients and the supplier's accompanying book of demonstration recipes. We were expected to pay for what was basically public knowledge.

The next example occurred when I was at the Poultry Science Department at what was then the Ontario Agricultural College. We had quoted for a research project requested by a local turkey grower and processor.

We did not get the contract for this research project; it went to a large commercial research and development company. The owner of the growing and processing operation was a frequent visitor to our Department and during one visit I asked why we lost out. He told me we lost because he felt a bigger and more prestigious research and development company could handle the project better. Little did he know, and I never told him, that his bigger and better development company subcontracted a large portion of the work to us.

Buyer beware, indeed!

Giese (2001) discusses guidelines similar to those above for selecting laboratories for analytical services. He also provides a very brief listing of testing laboratories and a useful checklist for working with outside laboratories.

Of great importance, as Giese points out, is the need to determine the methodology that laboratories will use. It has been my personal experience, and my error, to find that very few outside laboratories use official or standard methods; they often have some in-house method that they invariably find "more accurate" or "less time-consuming" or that "gives just as good results." Unfortunately, I had wanted comparative analyses of similar products from widely distant processing plants in one instance; in the other, we were studying the characteristics of different hot pepper varietals grown in different countries in sauce manufacture. It was necessary to work with different laboratories. Comparison of results between factories, products, or plant varieties was useless despite my injunction to use official methods, different methodologies were used. It is absolutely necessary to spell out what methods are to be used especially if the results are to be used for arbitration in legal work.

9.2.3 Some Caveats in Selecting and Working with Consultants

Patterson and Haas (1999) present a strongly reasoned argument for buying outside consultants to undertake tasks that the client considers to be not their core competencies. Their outsourcing is a broader interpretation of consulting than that which is generally understood. Their reasoning for choosing to outsource certain business activities closely parallels decisions that a client must make in selecting a consultant to undertake new product development (see also Fuller, 1999).

One issue glossed over by Patterson and Haas is the extra cost required to monitor the outsourced activity. No matter how much the consultant and the client trust and respect one another, the consultant's activity needs to be monitored with the same diligence that the client uses to monitor its own in-house activities; that is, the function of process and quality control. Why shouldn't outsourced activities be monitored regardless of how much confidence is placed in the consultant?

As in all things purchased, the cautionary philosophy is *caveat emptor*, buyer beware! Similarly in partnerships (or partnering) and in joint ventures, warnings still apply. Perhaps this is why one outsourcing contract was reported in the *Wall Street Journal* to be over 27,000 pages long. There are several caveats that clients seeking consultants need to be aware of. Patterson and Haas (1999) provide guidelines for the selection of partners in outsourcing.

9.2.3.1 Exposure

Exposure of sensitive business plans to outside parties is a very real danger in using consultants. Clients reveal the nature of the project that, in turn, reveals the direction of their research or business goals. Consultants, particularly those who are executives between jobs, could very well at some later date be hired by a competitor. Consulting is an excellent way to job hunt.

Those consultants, retired from other companies, understandably have strong ties of allegiance to their former employers; they have social and business ties with their old employers. A casual remark or an inadvertent comment can provide a good listener with hints of a competitor's activities. Companies with an intelligence gathering program can glean confidential information from retired employees.

I have been reluctant to place with universities any projects the nature of which I wished to be kept private largely because of the following:

I experienced an egregious breach of security while on a pre-arranged consultation visit with an academic at an American Midwestern university. The professor left to give a lecture and I was introduced to a doctoral student who conducted me on a tour of the laboratories during which work on a competitor's research project was in progress—openly displayed on the work bench. I had clearly identified myself and my company affiliation to both the professor and the student.

Many nonacademic people have access to university laboratories. There are staff and invited outside taste panelists or test subjects, students and their friends, professional staff and technicians, and equipment and instrument sales personnel, as well as representatives of industry. All, at some time, are normal occupants of such facilities. They have eyes and ears.

Borrowing or renting pilot equipment from equipment manufacturers prior to purchase can have its confidentiality hazards:

One of our companies in New Jersey was interested in a thermally stabilized tray pack but required a special sealer. We called in the sales representative of a company manufacturing sealers for a discussion on the possibility of renting a unit for some trial runs. He gleefully told us we were lucky. A sealer was available across town in the premises of our competitor, who had just finished their trials with it.

The use of consultants always presents the possibility of a security breach respecting a client's research and development programs or business plans.

9.2.3.2 Loss of a Collective Learning Opportunity

By turning a product development project over to a consultant, the client loses experience with the product and its ingredients and also the opportunity to develop and grow members of the development team. The client's research and development team does not have the opportunity to work with the ingredients or to note the behavior of the product in different applications. The team loses the "feel" for a project and the knowledge and pride that accompanies the experience of having brought the product to fruition.

The potential client has to weigh the package deal and its costs (often with some tax benefit) presented by the consultant against the intangible benefits accruing to going it alone with minimal outside resources.

9.2.3.3 Employee Growth

Concomitant with the above, the client's technical employees lose an opportunity to learn "on the job" and grow. It is a lost training opportunity for the client's employees. Senior technical staff lose an opportunity to learn new skills or to manage a new product development project. Management, in turn, loses an opportunity to evaluate staff, to reward them, and to select a cadre of future leaders.

9.2.3.4 Dissension

Both the client's technical and nontechnical staff often resent the presence and interference of consultants in the client's routines and their territories. They are indifferent to the project because, having no input into the product development process, they do not see it as "their baby." There is lack of interest in it. A subtle uncooperativeness with the consultant can prevail. An unfortunate example of this occurred to me:

I had arranged several weeks in advance with the company president, the plant manager, and the research manager to run a small trial on plant-scale equipment for a condiment sauce. The research manager who was in charge of this particular plant operation in addition to his research duties made all the arrangements for raw material, staffing and production time. He also helped in designing the research protocol. On the day of the test run, I found that the research manager had taken a day's holiday to attend a bridge tournament. The trial couldn't be postponed, and Murphy's law was fully proven: what could go wrong did.

The client's staff can resent the project in general or the consultant's contribution in particular. They certainly will not be overjoyed if their work schedule is interrupted to assist the consultant. The consultant will earn his fee, but those on the line may lose their production bonuses.

9.2.3.5 Other Obligations: Problems in Academe

The use of academics for consulting has presented me with some difficulties. Academics have obligations that make working to the rigid schedules that prevail in industry impossible. They teach. They advise graduate students in their research programs. They conduct their own research programs. They travel to present papers at conferences and to be guest lecturers at other universities. They have departmental business to attend to. They write up research grants and must network to find sponsors for research dollars.

Many lack any business, marketing, managerial, or practical industrial experience. Two experiences that I have encountered illustrate this point:

I lost contact with an academic hired to work on a project. He simply disappeared on sabbatical in the middle of the project without ever informing me. The project had to be shelved until he returned. He did not think it was important to inform me.

In the second incident:

We wished to improve the shelf life of a processed meat product and thought that lowering the water activity might provide the stability. We sought the help of an academic who was well-published in this field. He advised the addition of glycerol. This, I informed him, was illegal in meat products. His answer was simply that nevertheless it would work and nobody could or would think to detect it.

I hasten to add that I have had excellent assistance and advice from other academics but have found it necessary to spell out requirements respecting the particulars of the work to be done and the timetable for this.

9.2.4 Advantages and Disadvantages

That there are both advantages and disadvantages to the use of consultants in new food product development should be apparent from the foregoing. What is an advantage or a disadvantage depends on what and whose standards are to be used to measure the results and who within the company stands to gain tangibly and intangibly from the results of the services offered.

9.2.4.1 Utilization of Resources

Some claim that a client's research and development dollars go further with the use of consultants (see, e.g., Patterson and Haas, 1999). Costs for consultants are a bottom line expense, a cost of doing business, and, therefore, they are deductible expenses. Where the research monies are placed with a university for long-term research, there are taxation benefits and, coincidentally, goodwill is generated (see Figure 4.1). By farming out research, the client is not saddled with the costs of hiring staff, investing in expensive research and processing equipment, and devoting company time and personnel to explore risky ventures. The disagreeable task of ridding itself of the extra staff that were hired for the project is gone; no sophisticated equipment needs to be rid of. Again, however, the question must be asked: Is the company interested in short-term or long-term, intangible benefits such as growth and development of staff?

9.2.4.1.1 Flies in the Ointment

Companies must understand that a certain amount of their money, when given to universities, is "lost"—that is, it goes to overhead costs, up to 80% of the research dollars. My research showed me that 40%–80% of monies given to universities may go to overheads (see Chapter 4 in Fuller, 1999 for an in-depth discussion of consultant fees). Thus, only 20–60 cents of each dollar given for research actually goes for research. However, one department head of food science informed me that this overhead fee is highly negotiable; another quietly told me to let him handle it. Our company's money went for supplies used and the support of a graduate student.

Harvey (1977) found that joint ventures and partnerships can have flaws. Broadening Harvey's concern that these can be flawed, one can say, in general, outsourcing can have flaws—outsourcing as a buzz word had yet to be placed in the business lexicon in Harvey's time. Outsourcing should be approached with caution. Harvey is particularly scathing of joint ventures to reduce competition, to share a risk ("risk is hardly divisible..."), or to share costs one company alone cannot afford. He discusses good and bad reasons for joint ventures.

Cost savings may be realized by using outside resources; it is true. But as the client's needs change, the services provided must grow and develop as the demands alter. There will naturally be an escalation of fees to the client, for example, where long-term contracts for services have been signed. The outside resource providers may not advance their services as the client's needs grow. They are not pushed to deliver; they have a contract. Technologies change and a client's shifting business plans change respecting product development, brand expansion, and their outside resource needs.

Consultants are considered by clients to be both objective and unbiased. This is not necessarily the case. It must be realized that whether they are a consultant, an outside resource, a supplier/client joint venture, or a partnership, there is a self-interest on the part of one in the relationship to keep themselves employed. In over 8 years, as the client hiring and working with consultants, I have yet to see a report that didn't have a conclusion

- That suggested more work should be done for a more definitive answer
- That suggested a follow-up review should be done to confirm that all was well
- That suggested the client would be advised to follow up certain "promising" avenues of research

All are tactics employed for soliciting further work, and it will be noted that all are veiled attempts to put doubt in the client's mind.

9.2.4.2 The Need to Monitor

Patterson and Haas (1999) gloss over the client's need to monitor the outside resource's activities to assure themselves that their contracted work is progressing toward the goals defined by the client. I must reemphasize that the same care, diligence, and monitoring the client gives to its own products and services must be applied to all its consulting activities. Clients want a consultant who can work successfully, cooperatively, and without upsetting their internal operations too much; who can advise on improvements; and who can help them interpret their ideas for development into new products. It is the wise client that maintains a tight rein on consultants.

Consultants with established reputations and experience in esoteric and emerging fields of technology (see Chapter 5) bring distinct advantages to companies wishing to explore these newer technologies. Clients are not hobbled with struggling within their own companies to establish a foothold in these areas. Wisely undertaken, they educate their staff in the new technologies and, thereby, get a head start for product development.

9.2.4.3 Does the Client Understand Consultantspeak? Communication

Obscurity is the refuge of incompetence.

Robert A. Heinlein

A major problem that presents itself in all consulting activities is communication. The client is engaging an outside resource with an entirely different skill set and language, and that language may be spoken in consultantspeak. Do both the consultant and the client speak and understand the same language? Do they use and understand the same symbols?

9.2.4.3.1 "Speak Clearly, Dammit"

Consultantspeak, bafflegab, doublespeak, gobbledygook, and even governmentspeak are all terms used to describe language that, when heard or read, sounds wonderfully pregnant with erudition at the moment but later when one tries to remember what was said, one is left with a lot of buzz words "full of sound and fury signifying nothing." Margaret Thatcher put it succinctly: "You don't tell deliberate lies, but sometimes, you have to be evasive."

Words can be used for evasiveness, and words may also be chosen to say much and mean little. I have told the following story before (Fuller, 1999):

> I worked with the Canadian Food and Drug Directorate and my first assignment upon leaving university was to build a gas liquid chromatograph (there were no commercial units). This done I proceeded to research food flavors and citrus oils in particular. A research director of a major food company asked my opinion of the method and its potential in food flavor work. I wrote favorably about the technique quoting my own public domain work as well as published work. All mail

out of our department had to go through the department head for approval; he was very much the civil servant. We were after all THE GOVERNMENT. The department head severely edited my letter and filled it with government jargon. I was informed that should I be wrong in my opinions regarding gas liquid chromatography that this would reflect on him, how his department was run and ultimately on the government. I signed his letter but snitched a copy and took it home to my wife. Neither of us could tell whether I was for or against the method as a tool in food analysis.

Doublespeak is not rare. Words begin to have wonderful meanings: for example, restructure, downsize, destaff, right-size, unassign, all mean someone is going to be canned, fired, on the street. I have found all these in business and trade articles. Unfortunately, they seem to proliferate in articles that were written by consultants. I am not alone in this observation; some articles against this loose talk are as follows: "Professor singles out double talk" headlining a newspaper article and "Speak English, Dammit," a title for a business magazine article.

Anyone wanting examples of evasiveness and obfuscation need only watch the BBC television series, *Yes, Minister,* to get excellent lessons as the permanent secretary, Sir Humphrey Appleby, bafflegabs his way around his minister.

The following list of words, individually meaningful, have found use in bafflegab and into much business language where they have become meaningless. This list was prepared from one found on the Internet. It was used to play a game called Bafflegab Bingo. The player took this list into their meeting, seminar, or conference call, and with a bingo card containing these words in each square checked off squares whenever the speaker mentioned a word in the list. When, as in Bingo, five squares horizontally, vertically, or diagonally had been filled in, they had "Bingo" or "Bafflegab."

Synergy	Benchmark	Knowledge base
Strategic fit	Value-added	At the end of the day
Core competencies	Proactive	Touch base
Out of the box	Win-win	Mindset
Bottom line	Think outside the box	Client focus(ed)
Revisit	Fast track	Ballpark
Take that off-line	Results-driven	Game plan
24/7	Empower (or	Leverage
Out of the loop	empowerment)	

The game or ones similar to it are not new. Hammer (1994), a Food and Agriculture Organization consultant who analyzes project proposals and recommendations received by this body, produced a lexicon organized into four columns. By selecting a word from each column, one could make up wonderfully forceful phrases meaning nothing. Prophetically, her paper is entitled "Why Projects Fail...." A few gems from her article are "centrally motivated organizational approach," "radically delegated technical initiative," and "strategically balanced development scheme."

These make perfectly meaningless phrases. My personal favorite for evasive doublespeak was reported in *New Scientist*, 1996. It described the Ariane space rocket, which blew up within 40 seconds of launch. The European Space Agency described it thus: "The first Ariane 5 flight did not result in validation of Europe's new launcher."

So much for the debasement of the language. Any client who spots any examples of bafflegab or evasive language in the reports, language, or discussions with their consultant should quickly show the consultant the door. Communication in a language that both the client and the consultant speak and understand is of utmost importance.

> Beware the Jabberwock, my son!
>
> The jaws that bite, the claws that catch!
>
> Beware the Jubjub bird, and shun
>
> The frumious Bandersnatch!
>
> **Lewis Carroll, "Jabberwocky" in *Through the Looking Glass***

The offbeat approach used here in this section should serve to underline the importance of clarity and understanding in all communications.

9.3 Summary

Working with consultants can be very profitable for a company involved in new product development. Where the consultant brings a new skill or knowledge, the client can be put into a more advanced stage of development and the client's staff can get an opportunity to learn from the advice and guidance provided. There are, however, some pitfalls and the client needs to be fully aware of these to ensure a good working relationship. The consultant is there to assist the client not to supplant management's prerogative to provide sound management decisions. The client knows their business objectives and goals better than the consultant. All the consultant contributes are the tools, some arcane knowledge, and some ideas—all of which for a price can be purchased.

10

New Food Product Development in the Food Service Industry

> Man does not ingest nutrients; he eats food.
>
> **Attributed to R. L. M. Synge**

10.1 Understanding the Food Service Industry

Two areas, the food service and the food ingredient industries, are major components of the food microcosm. They are sufficiently different from the mainstream consumer food product retail marketplaces to deserve separate treatments with respect to new product development.

People are changing their eating habits for many reasons not the least of which are changing social behaviors. Today, fewer meals are prepared from scratch and eaten at home; rarely are meals eaten as a family unit except on festive occasions. Two corollaries arise from this observation. First, meals are more likely to be put together from prepared menu components—it is said we have become assemblers of food—and eaten by the assembler who is "on the run." The second obviously follows: there is an increase in meals away from home with a consequent increase in the number and variety of food service outlets.

10.1.1 Food Service Marketplaces

There is a truly overwhelming variety of food service establishments for which new products are required (Table 10.1). A more extensive listing is provided by Bolaffi and Lulay (1989). Each category in Table 10.1 is subdivisible. Fast-food restaurants (now euphemistically referred to as quick serve restaurants) could be further separated into: independent family-owned businesses; large multiunit chains either all or partly franchisee owned and operated, outlets with sit-down service, and those with only counter or take-out service. Likewise, industrial work-site feeding could be an urban or suburban factory-run cafeteria offering a hot lunch; a cafeteria serving a 24-hour-a-day workforce as in hospitals or a food service organized for an isolated oil rig in the North Sea. Each category of food service establishment

TABLE 10.1

Variety of Outlet Types in the Food Service Marketplaces

Sector	Subclassification
Restaurants	Fast-food (chain) restaurants: chain owned and franchisee-owned
	Fast-food family-owned restaurants (diners, cafes, bistros that may also be chain-owned)
	White-tablecloth restaurants (stand alone)
	Hotel (again white tablecloth restaurants) and motel restaurants
	Tourist travel: train, plane, cruise ship, and barge boat meals
	Isolated work camps and industrial sites: floating oil rigs, mining camps
Food stores	Food malls in shopping malls
	In-store delicatessens with eat-in and take-out facilities
	Sandwich bars in food stores
Catering services	Hospital, convalescent home, and retirement home catering
	Visitors and staff
	Special diets for patients
	Bars, pubs, and nightclubs—foods for noshing, e.g., happy hours, light buffets
	Transportation related: airports terminals, bus and railway stations)
	Charity
	Meals-on-wheels programs; soup kitchens
Institutional feeding	Penal institutions
	Sanatoriums and rehabilitation facilities
	School meal programs
Miscellaneous	Vending machines: ubiquitously sited
	Hot and cold food vending, beverages
	Street vendors: finger foods
	Mobile canteens servicing urban work sites
Military feeding	Officers' messes
	Servicemen's messes
	Combat rations
	Field kitchens
Leisure feeding	Vacation camps
	Roadside eateries

has its own peculiar needs and menus to suit their diners' needs and the economic class of its diners.

Food service establishments are further partitioned into two sectors with different characteristics adding a further challenge to product development:

- The commercial sector: this sector runs the entire gamut of restaurants to the ubiquitous vending machine dispensers. There is a free choice of food, that is, consumers have the option of going to a variety of outlets if choices at one do not appeal to them. This is the main characteristic of the commercial (free choice or non-captive sector) of the food service market. The commercial sector is profit motivated. There are commercial eating establishments that are operated marginally but these are run as a service or a labor perk and some person, company, or organization is subsidizing the operation. Examples of this type of operation that come to mind are: a government- or university-supported hotel/restaurant training school's demonstration restaurants; employee restaurants under-written by a company and part of a labor contract, and restaurants, often open to the general public, at private golf clubs and subsidized in part by members' fees.

- The noncommercial sector (the institutional or captive sector): here customers have no or only a very limited freedom of menu items and no choice of an alternative outlet. Diners in such situations are captive; they cannot go elsewhere to eat. Travellers switching travel plans to more accommodating carriers, riots, hostage-takings, and demonstrations due to poor quality meals in prisons, and angry parents complaining to elected school officials on the quality of food in school lunch programs are powerful incentives for change. While this sector is not profit driven, it is not meant to be managed at a loss. Purveyors of these establishments determine variety of menu items, quality standards and nutritional criteria of the food within constraints of budgets established by government, school, or hospital boards, etc. Diners are presented with limited or no-choice menus in the following situations:
  - Hospitals and convalescent homes, social services for homeless, institutionalized patients.
  - Military feeding for troops on maneuvers (e.g., arctic survival training), on air and naval patrols, or on active enemy engagement.
  - School meal programs in public schools as well as food service outlets on college and university campuses.
  - Company cafeterias in isolated suburban industrial parks or remote work camps.
  - Transportation meals served during air, bus, boat, or train travel.
  - Meals served in prisons and other rehabilitation centers.
  - Relief feeding in famines and other natural disasters, and in refugee camps.

Bolaffi and Lulay (1989) separate out the military food service as a distinct entity for food developers by virtue of the product standards, labeling and packaging requirements, and the bidding and tendering requirements demanded by governments. However, in this discussion, military feeding will be included with the noncommercial sector.

10.1.2 Customers and Consumers in the Food Service Industry

There are customers whose needs are to be met and consumers who dictate those needs to the customers and there are purveyors, a term I have introduced in place of retailer as a convenience in understanding this industry. The food service industry is similar to the consumer food products retail industry in this respect. The concept of the customer as the gatekeeper is also valid but requires a stricter examination. There are two distinct customers in the food service industry:

- The customer is the individual who orders from the menu, or who answers, "Chicken, please" to the steward on a plane, grabs a steamer from the street vendor, simply points to a steam table tray in a cafeteria line-up, or puts money into a coin slot at a vending machine. They eat the products either on the premises where they were purchased, munch on finger food on the street, dine in cafeterias at their work places, or pick up prepared food to eat at home. They are customer and consumer combined.

- The following are also all customers: executive chef of a restaurant or even a chain of restaurants, the owner of a restaurant or a diner or a *patates frites* stand along a highway, a dietician manager of a commissary of a hospital, the food-purchasing agent for government institutions, a food store owner with a deli bar, a franchisee owner of a franchise store of a fast-food chain, or the military quartermaster. They purchase food ingredients, prepared meats, *sous vide* products, and preprepared raw produce from suppliers. They are also the consumers (users) of the products purchased. They or their staff transform or assemble the basic food items into meals either through the artistry of chefs or by heating and plating *sous vide* products, by simply assembling prepared items, or merely displaying or dispensing items prepared by others. They have characteristics in some instances analogous to that of the gatekeeper.

- Purveyor is analogous to the seller or retailer in the consumer food product arena. They may have nothing, or very little, to do with the cooking, preparation, or assembling of the food. Possible examples are pub owners or franchisees of large fast-food chains. They have some characteristics of the assemblers above; they display or dispense items largely prepared by others.

This tiered customer, consumer, purveyor element introduces many novel problems. Developers of products for the food service industry are developing prepared or semi-prepared products for highly trained and skilled people (chefs, cooks, etc.), for people with rudimentary knowledge of food preparation, sanitation or hygienic handling, or for purveyors who may have assemblers trained by the fast-food chain. Thus, developers prepare products that must directly satisfy the needs and expectations of their customers and consumers (chefs, assemblers, etc.) and their customers' customers and of purveyors.

The complexity of the food service market is obvious. How does one screen a product such as a new product or ingredient when one does not know specifically how or in what finished product that ingredient will be used by a chef or cook or in what social setting those finished products will be served? How does one get market research, conduct a test market, or evaluate a test market?

10.2 Characteristics of the Food Service Market

Both commercial and noncommercial sectors can be discussed as one for the sake of simplicity, and any distinguishing characteristics of either highlighted. Some of these characteristics are presented in Table 10.2. Products designed for the food service sector must accommodate the characteristics of the sector they are to serve, that is, simple to prepare and present to customers, not conducive to waste production in a sensitive close-quarters working area, and not energy intensive in their preparation.

TABLE 10.2

Characteristics of Typical Commercial and Noncommercial Food Service Outlets

General Characteristics
There is small-scale preparation and assembling of a variety of menu items. Menus are cycled every 2–3 weeks; others have fixed menus with daily specials regulated by the day of the week or regulated by availability of fresh produce.
Food preparation area and social setting for eating may or may not be separated.
Uneven periods of food preparation and serving. Staff are often required to stand-by during lulls in activity. Military food preparation and feeding often conducted under stressful conditions.
Rarely are staff or outlet provided with a formal quality control and inspection procedure. Fast-food chains do have procedures.
Cook or chef has full responsibility for products and ingredients. Specification for purchase subject to cook's judgment or seasonal availability.
Preparation is labor intensive with variable skill levels required.
Storage of raw and prepared products, food preparation, waste disposal and utensil clean up often share cramped facilities. Often all activities proceed simultaneously and in hot, humid conditions.

10.2.1 Clientele

Consumers in high-class restaurants are looking for quality of taste, service, presentation, atmosphere, and relaxation. They are not concerned with price *but will demand quality and service for the price.* Consumers buying from street vendors are looking for cheap, good, quick, wholesome (safe) food while running errands on a lunch-break or between breaks in recreational play or at some outdoor event. Their only common characteristic is that neither the upscale diner nor the eater on the run are looking for a healthy nutritious meal; one is seeking gratification of the senses; the other eats to gratify a hunger and to get on with some activity.

Nevertheless, governments and groups concerned with health focus on nutrition of these meals with their concerns for obesity, fat calories, and high salt levels. An Associated Press report (May 28, 2009) reported that the New York City Health Department was meeting with food makers and restaurants to reduce salt content in foods; the Guardian News Service reported Russia published a "comprehensive meal planner advising Russians what to eat" to avoid fatty foods. Many jurisdictions have recommended, even compelled, restaurants to post calorie counts on their menus and to get rid of trans-fats. These demands and programs to promote better nutrition require developers to reformulate or offer alternative menu items. Popular chefs also influence restaurant menu items; Gordon Ramsay, a celebrity U.K. chef, advocates seasonal-only menus, maintaining that this would "cut carbon emissions as less food would be imported and also lead to improved standards of cooking," thus putting further pressure on suppliers and products (BBC News, 2009). Ramsay claims to have spoken to the U.K. Prime Minister to outlaw out-of-season produce!

Expectations of consumers in the noncommercial sector are quite different but just as diverse:

- Safety and nutrition are important in prison feeding, hospital, and long-term care feeding, military feeding, and school meal programs. A food intoxication running through these populations can have devastating effects. The presence of allergens in foods is of major concern in these categories of consumers.

- Quality, at the very least, must be acceptable to the majority and certainly not so bad as to be rejected. Poor quality or boring food variety and taste have caused dissension with diners. Transportation meals, for example, must not be so unacceptable that they are a factor in passenger's choice of carrier.

- Budgetary restrictions limit to a degree both quality and availability of choices. However, nutritional standards must be such as to prevent nutritional deficiency diseases.

Hospital care feeding presents a special case that will be discussed later.

10.2.2 Food Preparation and Storage Facilities

10.2.2.1 Equipment

Food service establishment run the gamut in equipment: Five-star restaurants, family-style restaurants, ethnic restaurants, hotel kitchens catering to two or three classes of in-house eating outlets, fast-food franchises, "mom and pop" diners, or bars serving hot hors d'oeuvres need a wide variety of food preparation or holding and serving equipment. These range from warming units, steamers, friers, wood stoves, and smokers to tandoors, woks, hot stones, 50-gallon steam-jacketed kettles with attached mixers, microwave ovens, potato peelers, grinders, slicers, walk-in freezers, refrigerators, air conditioners, and fume hoods (for odor control). Equipment in kitchens range from the well equipped to street vendors and hot hors d'oeuvres set-ups in bars at happy hour with less equipment than a well-run family kitchen might have.

When restaurant kitchens are run at one heating temperature for all menu items served, then preprepared products and ingredients must have attributes that can be prepared, assembled, and maintain their high-quality attributes at this operating temperature stress. Developers for the food service market design their products to withstand the challenges that these products might encounter in the stressful environments they will be assembled, finished prepared, and presented to consumers in whether this be a restaurant or a vending machine. Their design concerns must include safety features that meet these challenges.

Vending machines have become more versatile and from dispensing hot chocolate or coffees they now dispense flavored coffees, lattes, espressos, liqueur-flavored coffees, and some teas. Hot drinks can be brewed in machine or from sachets of instantized ingredients or ground materials or from liquid concentrates in capsules, this latter concept can provide either hot or cold beverages. Newer machines incorporate microwave ovens or deep-fat fryers making a wider selection of product offerings possible. Combination plates, sandwich, salad and fries or soup, sandwich, and dessert are now regular offerings (Saltmarsh and Hall, 2007; Williams, 1991). Such new menu items are much more sensitive to time and temperature abuse and require greater attention to stability, rotation of products, and quality.

In general, food preparation and storage facilities in noncommercial establishments are better than, or more adequate to their tasks, than those of commercial establishments; they were often built to specifications designed for the traffic. Hospitals and penal institutes planned facilities for their anticipated inmates. They are usually run by professionals trained in food service or are catered to by industrial caterers. While they are on limited budgets, their size permits them to purchase in bulk and obtain the reduced prices that quantity buying permits.

Food preparation systems in use in U.S. hospitals employ one of five common preparation systems (Matthews, 1982):

1. A cook–serve system, whereby food is prepared on-site, plated, and distributed on trays to patients.

2. A cook–chill system, in which the food can be prepared either on- or off-site ahead of use. After preparation, the food is rapidly cooled, held refrigerated for not more than a day (but see Mason et al., 1990), plated, and distributed to patients from specially designed heating carts.

3. A cook–freeze system that is similar to the preceding with the additional step requiring the thawing of menu items.

4. A thaw–heat–serve system, where the main menu item is precooked and frozen either on- or off-site. On-site, the product is thawed, heated, and plated.

5. A heat–serve system using thermally processed steam table tray containers. The main menu item can be heated and held hot before plating.

Not unknown are cook-to-order and room service systems in North American institutions. Assaf (2009) studied which systems were popular in Australian hospitals and found the cook–serve (cook–fresh) system still prevalent but cook–chill and the other hybrid systems becoming more popular. There was a definite trend away from in-house production and the use of outside commissaries was growing. These systems above are not unique to the health care food service industry. They are used in military feeding, especially the heat–serve system, and in many commissaries servicing a variety of school or industrial cafeterias. Developers must be aware of the rigors that such systems impose on food products and ingredients they are trying to develop.

Mason et al. (1990) and Livingston (1990) discuss commercial food preparation systems used in hospitals and other food service outlets and evaluate these with respect to their influence on the quality of the food. Quality of food, not its nutrition, has long been a complaint of patients and this is perhaps understandable as a review of the food systems above suggests the rigor that food undergoes.

An interesting combination of commercial and noncommercial food service in one facility is demonstrated by some U.S. hospitals that have rented out part of their food service facilities to quick-serve chains and developed a separate source of income while accommodating both their patients' needs and those of staff and visitors.

10.2.2.2 Storage Facilities

Limited storage in food service establishments requires frequent deliveries of produce and other nonfood items. Full-service restaurants will have adequate, but not excessive, storage facilities for supplies, food materials, and waste. Street vendors, on the other hand, have minimally adequate space for food preparation and storage for disposable flatware and must rely largely on municipal maintenance facilities for waste disposal. In happy hour setups in bars, there are no storage facilities so bar owners rely on same-day deliveries with the caterer bringing warmers and hot finger food and returning later to remove the dirty ware and warmers.

If storage for food supplies is limited, then storage for waste is even more at a premium. Products for food service must be designed to produce as little waste as possible. Prepreparation is a key added value feature. For take-away hot foods, fast-food outlets face another problem: packaging. It must be environmentally friendly and it must protect the purchaser from damage, for example, being burned or scalded. There is a growing uproar about the take-away–throw-away coffee cups of the major coffee chains despoiling the environments of cities. This has led to outlets offering discounts if customers bring their own containers, charging for the take-away containers, and making cups out of recyclable material.

10.2.2.3 Labor

Kitchen labor in either the commercial or noncommercial sectors is highly variable respecting skill and knowledge levels of food handling. To describe any one sector as characterized by unskilled labor simply invites a barrage of counter examples. Staff from chefs in upscale restaurants (where they are not owners) to temporary staff working their way through school are highly mobile or in transitional stages in their careers supporting themselves through school, or temporarily unemployed. (I had no idea until recently that chefs and sous-chefs had agents.) Their skills range from highly skilled and knowledgeable in food preparation and handling to rudimentary. Many of the latter are not careerists in food service; they want out at the earliest opportunity into other career paths. There was a recent promotional campaign by a quick-serve chain promoting their use of seniors in their outlets. Training such candidates can be discouraging and fruitless for food service employers.

To meet the challenges, therefore, design of food components used in preparation of finished products in many food service establishments:

- Must be kept practical, safe, and simple. Products for assembling or preparation must be "error-proof" in the hectic environment of a kitchen.
- Require the minimal amount of labor for their preparation. Ideally, they should require only assembly, reheating, or a finish cook.
- Produce the least amount of waste for disposal during preparation, and their preparation should not contribute unduly to the microbial load in the kitchen.
- Must be such that storage of unused portions will present no complications for their safety or quality.

Chefs and sous-chefs, today, are trained in a culinology, a combination of food science and chef's training; there is also another branch called molecular cooking that became all the rage, but mercifully, it seems to have died. They have the skill and knowledge to take unused portions and reuse these, safely, in other quality dishes. Yesterday's unused poached salmon can be today's salmon mousse or salmon cakes. Such prowess reduces waste and thereby minimizes cost overruns.

Nothing of chance respecting safety can be left to personnel in fast-food or more downscale fast-casual restaurants. Unused food or warmed and thawed portions must be disposed of. (These are often collected and used in hospices.)

The availability of skilled and trained labor is exacerbated by the rising costs of labor. A rather dated report by Pine and Ball (1987) found that wage bills (as percentage of sales minus pre-tax net profits) ranged between 8% and 38% in the U.K. industry depending on the class of the establishment. Those that are labor-cheap were largely in the catering-only end of food service where labor consisted of assembling preprepared items or they were businesses lacking a personal service aspect (transportation catering). These figures are still today generally applicable as well as in other countries.

The noncommercial sector faces the same labor problems as the commercial sector. Newly arrived immigrants with restricted communication skills are often forced to take work in kitchens; good communication is hampered by language barriers. In some penal and correctional institutions, unskilled inmates after minimal skills training may serve as help.

There are, however, wide variations in skills in the noncommercial sector. Dieticians required in hospitals and nursing-care facilities and military chefs are highly trained; school lunch programs are either catered or usually under the supervision of equally skilled personnel. Nevertheless, wages are at the minimum level for unskilled labor and at the bottom end of the wage scale, job security is absent. These are not elements conducive to attracting skilled people. On-the-job training is difficult in this environment. The exception is in the military where training programs have proven very effective in raising the skill levels of food handlers and cooks.

Engelund et al. (2009) describe the application of the Japanese *kaizen* and *kaizen blitz* to large-scale food service production. They call it "lean manufacturing principles" and describe the principles of lean manufacturing in detail. These are

- Determine value from customer's point of view
- Map the value stream
- Connect value-creating activities in a continuous flow
- Produce nothing upstream if it is not needed downstream (no waste production)
- Pursue perfection

10.2.3 Price, Quality, Consistency, Safety, and Sometimes Nutrition

Price, quality—more critically, consistency of quality—and safety from hazards of public health significance are necessary considerations in both the commercial and noncommercial food service sectors. Price, where it is an element in the noncommercial sector, is regulated by what the targeted consumer in these outlets (e.g., seniors' residences, long-care facilities) are willing to pay and governments are willing to budget.

Price is not only regulated by what the targeted consumer is willing or expects to pay but also regulated by other factors:

- Costs of raw materials and ingredients: Ethnic restaurants, upscale restaurants, product-oriented restaurants (vegetarian, seafood or local foods only [locavore movement]), or restaurants following religious observances will each have special product requirements that command premium prices both to purchase and to serve. Control of costs is partly obtained by adherence to a strict portion control program and skilled menu planning. Plate waste even if it has been paid for by the consumer is nevertheless an indication of prepared product losses that could have served more customers. Portion control, therefore, becomes a very important element of price and service provided by the supplier.
- Labor-related costs: The type of restaurant will determine the skill level, that is, a Chinese restaurant will require a cook/chef skilled in Chinese food preparation and Chinese servers familiar with the dishes and (often) the language.
- Site-related costs: These are costs such as real estate taxes, local economy, type of customer traffic around the area. The old real estate adage that "location, location, location" is a contributory factor to success and brings many cost consequences applies. Property that is in demand garners high taxes and demand by developers.

Quality must be consistent with the price the consumer is willing to pay and that quality and service must not vary from restaurant to restaurant in the same chain. This is a problem that has not been mastered in some doughnut chains. I have found wide variations in quality of doughnuts (from the same chain) within the same city, as well as variations in different states and provinces.

Food poisoning, intoxication, and allergic reactions of consumers (Gowland, 2002) are major hazards in the food service industry (Snyder, 1981, 1986). The culprits presenting the greatest safety concerns according to Solberg et al. (1990) in a study of meal items and food preparation facilities at Rutgers University's New Brunswick, New Jersey campus were protein salad foods. These are foods generally associated with "summer sickness" caused by improperly prepared and stored egg, chicken, or turkey salads.

Suppliers of prepared menu items (e.g., *sous vide* items) can only protect their new product introductions by designing a suitable protective constraint to food-poisoning microorganisms and by ensuring products leave their premises in excellent and safe condition with a reputable distributor. Suppliers need to provide clear instructions on the handling, storage, and any further preparation of sensitive prepared items respecting time–temperature tolerances, proper rotation of stock, zone isolation for the preparation of sensitive components to prevent cross-contamination in the kitchen. The attendant media coverage of food-poisoning outbreaks often destroys individual restaurants and seriously damages the reputation of fast-food chains. A close supplier/client relationship in which the supplier understands the needs and peculiarities of the client, and the client has clearly defined their needs, reduces chances of hazards occurring.

Allergenic reactions to ingredients in food products can cause suffering or death of consumers and certainly litigation. This is a major concern for the product developer, for the server who may not have warned consumers of the presence of possible allergens, and the food service outlets (Gowland, 2002). These are devastating incidents (Williams, 1992) since menus rarely have lists of ingredients to warn consumers (ingredient listing is now being required in many jurisdictions). Rarely do these incidents get the same coverage in the news since they only affect an individual whereas summer sickness at a company picnic affects many people and gets wide publicity. Allergic reactions are a major concern for caterers to school lunch programs. The offending foods in children appear to be age related: eggs and egg products, milk and milk-derived products, and nuts (especially peanuts) are often the source in the younger years. With age, seafood products frequently become problems. Children often exchange treats with classmates not realizing the possible dangers. Schools try to prevent this exchange by segregating children prone to allergic reactions in separate lunchrooms but little can be done when the exchange takes place outside lunchroom in the playground. Since only a few molecules of an allergen can trigger a reaction, developers must exert extreme caution to avoid allergenic ingredients in their menu components and provide truthful labeling statements.

Allergy specialists in Canada estimate there is at least one death a month due to allergic reactions. In the United States, with a ten-fold higher population, this suggests ten or more deaths a month due to eating prepared foods. Integrity of ingredients respecting the presence of allergens is a major problem of manufacturers; such integrity is very difficult to attain in multiproduct manufacturing plants.

To serve their customers (or shift responsibility from themselves) food service establishments, particularly the fast-food chains, have taken to posting allergy charts listing the ingredients they use in the menu items served (Anon., 1988b). The consequence of this precaution forces suppliers to food service establishments to ascertain that the composition of all ingredients they use in their products fit these charts. This requires vigilance on the part of manufacturers as they switch ingredient suppliers looking for better sources of supply and more effective ingredient cost control.

Bryan (1981, 1990) discusses the application of hazard analysis techniques to food service production. The earlier article treats temperature profiles of poultry and beef roasts while the later article presents a flow chart discussing hazards points and their controls for egg and potato salad.

Nutrition in the commercial sector was not a problem for product developers in the food service industry until just two or three years ago. The fast-food industry has been targeted with severe criticism of the high-fat and high-salt content of their meals and the lack of vegetables (beyond the ubiquitous french fries) and salads. The litigious nature of consumers and the growing problem of obesity and diseases associated with obesity have caused a welter of class action suits against fast-food chains; claimants accuse the chains of having caused their obesity-related ailments through the promotion of fatty foods. People eat out for a social event, for comradeship, work or business (networking) reasons, a celebratory event, conversation with good friends, and pleasure or as a convenience in traveling with children. Nutrition is not high on their priority list; good, tasty, safe, and filling food is their priority.

By contrast, food service facilities within the noncommercial sector have highly variable requirements for nutrition and quality of taste and flavor, which reflect the type of noncommercial food service outlet.

Meals served on an airliner are one-time occurrences; their nutritional content poses no risk to the well-being of the passengers whether the service be a snack, a cold or a hot meal. The long-term nutritional health and welfare of travelling passengers are not dependent on the food served; safety of the food, with taste and satiety close seconds, is always a major consideration. However, the pleasurability of the meal experiences may influence the choice of travel company.

Nutrition as a quality feature is two-faceted. There are both customer needs and consumer needs, particularly respecting health care feeding. Dieticians (customers) in the health care field require unique products to satisfy special dietary requirements of postoperative patients, patients undergoing cancer or other therapies, for children, convalescing patients or the elderly in their care. They need much more detailed, indeed, esoteric nutritional data than that required in most nutritional labeling regulations. Developers must be prepared to provide this information in order to assist dieticians in the preparation of the many different dietary regimens required for patients. Schmidl et al. (1988) provide an extensive review on parenteral and enteral feeding systems, a rather esoteric branch of hospital care feeding.

Is there a problem of nutritional quality with the food service most used by the consuming public, the fast-food sector? Yes and no. Health authorities are alarmed at the high concentrations of fat and salt these contribute or could contribute to the diet. A fast-food meal is usually fries and fried main course and rarely is there an accompanying vegetable or a salad. Certainly, the frequency with which this combination appears in one's diet could have a serious consequence over a long period of time. Fast-foods have not received good publicity concerning the nutrition they offer consumers. For example, Hibler (1988) reported the results of a sampling of burgers, fries, and chicken, all of which are highly ranked favorites among frequenters of these outlets. On a calories per gram basis, burgers were the least calorific choice (1.9–2.9 cal/g), chicken in a middle range (2.4–3.1 cal/g) and fries the worst (2.7–3.5 cal/g) but these figures are deceptive, these menu items are rarely the same weight in any serving. Hamburgers are rarely served plain and may have cheese, fried onions, ketchup (high salt) accompanying them or any combination of these. The calories from fat were highest in the chicken entrees (average 49.5%) and almost identical in the burgers (average 44%) and the fries (average 42%). However, if one analyzes the meal, a typical burger and fries or chicken pieces and fries has the percentage of calories from fat well over the recommended level of 30% (a figure now in some dispute). Ryley (1983) calculated that, even back in 1982, fast foods and snacks contributed over 16% of the daily fat intake per person in the United Kingdom.

The KFC Corporation of Louisville, Kentucky, to improve its healthy image, launched a new product Kentucky Grilled Chicken™, described as a slow-grilled chicken product. According to the company, the grilled chicken has 70–180 cal and only 4–9 g of fat: this compares to the original recipe with 110–370 cal and 7–21 g of fat (Anon., 2009a). This gesture must, however, be put in juxtaposition to KFC's Double Down chicken sandwich, which does away with the two bun halves and instead uses two pieces of original recipe fried chicken to enclose slices of bacon and two different cheeses and which was recently put in test market (Marco, 2009). Various newspaper reports confirm this test market (see, e.g., Parry, 2009). The response to this breadless sandwich is reported as strongly in favor.

There is a dichotomy that restaurants face described aptly in a news release issued by IFT at their annual conference (Klapthor, 2005). The release quotes Caty Kapica (director of global nutrition for McDonald's) as saying: "We don't even use the word 'healthy' anymore because our consumer research shows people don't understand it and it's actually a turn-off when it comes to food items." Yum! Brands, Inc. (owners of several fast-food chains, among them the KFC Corporation) have tried several healthy product introductions that have failed. On the other hand, Pizza Hut's double-stuffed crust pizza has been a success. The fast-food chains have to be practical and service their consumers with what they want.

To market nutrition, that is, healthy foods and hence new products, to people requires a dual-track marketing strategy. This was the finding of Lone et al. (2009). They grouped college student according to BMI as underweight, normal, overweight, or obese. In the campus fast-food outlet, they studied food choices. And determined interest in nutrition where nutritional behavior was available. Those with a high interest were largely female while those with a low interest were "overwhelmingly" male. There were twice as many males in the overweight or obese category. Lone et al. discuss marketing strategies to reach the low-interest group.

There has long been a concern with nutritional quality in the feeding of the elderly at home. Turner and Glew (1982) studied the nutrient content of meals delivered in Leeds (United Kingdom) and provided by six food service organizations. They found significant weight differences between the meals supplied by the different organizations as well as between the protein contents of the meals. Meals supplied between 20% and 48% of the recommended energy intake for elderly people, which represents under to grossly over the energy requirement for the main daily meal. Of particular importance to geriatric nutrition, the calcium content of the meals varied widely from inadequate to ample—dependent, primarily, on whether the dessert was milk based or not. Iron content was found to be just adequate for the elderly but ascorbic acid contents varied widely from providing more than 50% to less than 25% of the recommended daily intake. However, with ascorbic acid, significant losses were noted between the first and last meal deliveries as might have been guessed from the lability of this vitamin. Keeping meals hot during transportation and the duration of hot transportation itself were weak links in delivering vitamin C to the elderly. The rigors that the delivery (serving) system for meals will impose on meals must be considered by designers of products for such institutional feeding systems.

10.2.3.1 Standards

Some detail on the nutrition of food service meals has been recounted to point out a major quality problem for developers of products they supply. What are the nutritional standards or guidelines for these products? There are no nutrient standards or even nutrient specifications for products meant for institutional or military or school meal programs. And there certainly are no standards for the special diets required by health care establishments: What is a soft diet? How soft is "soft?" How does one measure soft? By fiber content? Developers have an opportunity to create added value menu items with established nutrient content or closely defined (by whom?) standards respecting such vague terms as low ash diets, low calcium diets, liquid or semi-solid diets, or semi-prepared foods with fixed soluble/insoluble fiber content ratios for health care establishments if they knew what the standards were. These are excellent opportunities where industry could assist these establishments and find good, and profitable, market niches for their products.

A much greater concern for health professionals is the natural desire that food manufacturers have to want to provide *nutrified* foods, that is, foods that have been fortified with nutraceuticals, to patients in hospitals, nursing homes, institutionalized care facilities and into retail outlets. To provide nutraceutical fortification of foods is a logical expectation if the benefits are true and able to be promoted. Such products would complement or substitute for the more expensive prescribed medicines and mood-altering drugs used with patients and perhaps reduce medical costs and reduce hospital stays. Nutrified foods with verified attributes show promise for those in the general population who self-medicate or follow preventive diets; they would not have to resort to pharmaceuticals.

There are hazards to nutrified foods. Would nutraceuticals be incompatible with, even antagonistic to, doctor prescribed medicines? There is concern—frankly there is as yet no firm knowledge—whether nutraceuticals, if used in food systems, would react synergistically or antagonistically with other phytochemicals or other components in the food system. Would the concentrated phytochemical in capsule form prove effective? And which phytochemical of the many hundreds in natural products might prove effective? Giovannucci (1999) in an extensive review of research work found that tomatoes, tomato-based products, and lycopene (a nutraceutical and an antioxidant in tomatoes) displayed an inhibitory effect on certain cancers. Boileau et al. (2003), on the other hand, found that tomato powder was effective against prostate cancer in rats but that lycopene in concentrated form had no better effect on cancer than did the placebo. Both Giovannucci and Boileau et al. stress that tomatoes are very complex mixtures and that attempting to single out one component as having the beneficial effect could be useless. The concern for all developers is that if varieties of tomatoes with greatly increased levels of lycopene, a major antioxidant, or products with added lycopene are developed it may be found later that perhaps it was something else in this complex of chemicals called a tomato that was effective. Food legislators want to curb false claims and antibiotechnology groups would be quick to damn the plant geneticists. Such adverse reactions would not benefit science, scientists, or developers.

Another hazard is that in the present state of knowledge, there is little or no policing of the claims made about herbs or herbal extracts, particularly in North America. There is no informed knowledge base of effective dosages of nutraceuticals. In practice, a slurry of the herbal preparation is sprayed onto snack items, often potato chips, but there is no information on what the dosage level is. Did the slurry contain those parts of the plant containing the active nutraceutical? Even a cursory glance through any modern herbal will reveal that it may be only the leaves or flowers or roots or stems or bark that may contain the active phytochemical to the exclusion of the other parts. Caution is needed by developers wanting to fortify foods with phytochemicals; they need to proceed with caution especially in the fortification of candies, drinks, and snack aimed at the school meal programs or leisure and discretionary food markets.

Kuhn (1998a) and Neff (1998) discuss at length some cautionary comments of health professionals and manufacturers respecting nutraceuticals and provide a good overview of products and activities in this market although not directed specifically at the health care sector. Schwarcz (2002), director of the McGill Office for Chemistry and Society, urges standardization of herbal preparations and an accurate statement respecting the dosage in nutraceutical fortified foods, along with information on which part of the plant has been used in preparing the consumer products. He describes several interactions of herbal preparations and prescribed medicines: ginkgo biloba, an anticoagulant, can exaggerate the blood-thinning properties of aspirin and coumadin; St. John's wort, a mild antidepressant, is antagonistic to cyclosporin, an immunosuppressant; and ginseng, ephedra, and valerian can interact with anesthetics (severe warnings against the use of ephedra have been issued). Kava kava has been banned as a liver toxin in many other countries including Canada yet its use is not actively discouraged.

The use of these supplements either as herbs or extracts should be approached cautiously for development of products intended for children (snack foods) or for the health care food service sector. A minor caveat for developers is that many herbal preparations and extracts are not pleasant tasting and flavor systems need to be developed to overcome their unpleasant taste.

10.2.3.2 Health Care Sector of the Institutional Market

This sector includes hospital feeding, convalescent care feeding, and feeding of the elderly in their own homes or in residences for the mobile elderly, or for those institutionalized in nursing and psychiatric homes and who are mainly bedridden. Most health care feeding presents two feeding opportunities. There is the public cafeteria for hospital staff, medical students, outpatients, and visitors, open often on a 24-hour-a-day, 7-days-a-week basis, and the institutional sector, where patients require special diets in accordance with their medical conditions as well as their personal, religious, or ethnic taboos.

The buying power of hospitals is governed by their regional health boards and can vary widely. Meal items are usually purchased from privately run food commissaries unless the institution prepares their own special dietary meals: many facilities prefer to prepare their own special diets because of the lack of accepted industrial standards (Burch and Sawyer, 1986; Matthews, 1982).

Considerations besides nutrition (digestibility, absorptability, and flavor to encourage eating) are important characteristics of foods needed in hospitals and long-term nursing care facilities, social agencies providing home delivered meals to the elderly or to the homeless or in other humanitarian feeding programs. Several qualitative characteristics of the meal have a direct bearing on the consumer's health. Such patients (consumers) are recovering from the trauma of an operation, undergoing irradiation treatment or have medical conditions causing malabsorption of nutrients or undergoing long-term convalescence require special diets. Food must be attractive and flavorful for those who have little appetite, and it must be nutrient dense. Many of these consumers have weakened immune systems and would be stricken eating foods with microbiological loads that would normally be tolerated easily by healthy individuals.

Confounding all the above is the lack of standards for special dietary foods. That is, there are no standards for nutritionally designed foods or for any of the special types of foods dictated by convalescent diets. Consequently, developers wishing to put products into the special market niches have no guidelines. They must therefore work closely with dieticians in development programs.

10.2.3.3 Military Sector of the Institutional Market

The dictum, "an army marches on its stomach," attributed to Napoleon Bonaparte, was never truer. Good quality, flavorful, capably prepared food keeps an army healthy and fit and its morale high. In military feeding, as in health care feeding, food safety is extremely important. An army cannot come down with some foodborne illness, which in emergency situations could incapacitate the troops.

Food products, besides meeting the above criteria, must be suitable for serving in highly variable conditions from action in hostile enemy territory in humid, tropical jungles, in dry desert conditions, in arctic terrain, or in quiet barrack life in peacetime. Conditions of serving or storing food make military feeding situations a difficult challenge for developers. Products must have a long stable shelf life, must be easily portable, hence, light in weight and small in volume for it may have to be carried by the individual combatant, flavorful, and contain all the nutrients for an active, stressful lifestyle. Variety of menus and ease of preparation with minimal equipment are also requirements. Mermelstein (2001) discusses some of the requirements that were required at that time for the basic combat ration, the Meal Ready-to-Eat (MRE):

- Shelf stable for a minimum of 3 years at 80°F and a minimum of 6 months at 100°F
- Able to be airdropped by parachute from 1280 feet
- Able to be dropped out of a helicopter from a height of 100 feet without the aid of a parachute
- Able to withstand environmental conditions of −60°F to +120°F during storage, distribution, and handling
- Packaged to be resistant to wildlife

In addition, it must be acceptable organoleptically to the soldier. Mermelstein presents further details regarding requirements for military rations as well as for rations for relief feeding.

The military market is also highly fragmented. Food service facilities and operations range from products and services for officers' messes to noncommissioned officer messes to combat field kitchens to vending machines and snack bars to hospital care facilities. Most of these facilities must be operated 24 hours a day to accommodate the shift work involved.

Ease of preparation and serving and consuming introduces problems not apparent in other food service venues. Food may be prepared or reconstituted in combat zones, in cramped conditions, in submarines, or on airplanes, and often by the unskilled personnel themselves. Packaging must be light in weight yet provide protection for the food from all environmental conditions (which are unspecified) and from any treatment (including air drops) it might endure. Yet the package must be easily opened and the packaging readily disposed of lest obvious waste disposal dumps be seen by enemy aerial observers as a sign of a field kitchen nearby.

As is the case with all government procurement purchases, products must adhere to rigid standards and specifications respecting ingredients, processing, and packaging. Suppliers are advised to obtain copies of them before attempting to manufacture products.

A possible cause for failure of new products, discussed earlier, could be the introduction of products into markets dominated by a single customer; if the major customer does not like the product and it is too customer specific, it cannot be retailed elsewhere. The military represents such a market. On the other hand, the military market and the food service market in general, do present unique opportunities to introduce new products within a select portion of the population who develop a liking for and a familiarization with—that is, an educational opportunity for—novel food products. Many returning veterans developed a taste for and liking of products encountered in messes (e.g., McIlhenny Tabasco Sauce™ packed in MREs) (Rothfeder, 2007).

10.3 Developing Products for the Food Service Sector

Diversity is the key descriptive characteristic of this sector (Table 10.3); there is diversity of facilities and equipment in these outlets; diversity of labor skills and wages, from chefs in top-of-the-line gourmet restaurants to street vendors in their peddle carts selling sandwiches, hot dogs and sausages, and chili sauces; and diversity in the consumers and their expectations.

10.3.1 Physical Facilities of the Customer

Products developed for the food service industry must be capable of being finish prepared or assembled with the equipment that food outlets have on their premises, which range from primitive (street cart) to sophisticated (centralized commissaries). Therefore, product developers design their products recognizing the limitations of the preparation equipment of the kitchen and skills of its staff. New products must suit chefs, cooks, kitchen labor, and fit in with the physical kitchen and ultimately satisfy its consumers as part of a menu item.

TABLE 10.3

General Problems in Food Service Establishments That Influence
Product Development

Problem Areas

Clientele

Runs the gamut of those seeking gratification of sensuous pleasure with cost no
concern to those eating of necessity in order to get on with something else to those who
have to be fed for humanitarian, military, or medical reasons.

Facilities

From well-equipped to barely adequate preparation equipment. Caterer often has to supply
equipment. Many are "heat and hold" facilities.

Skill levels

Skill levels are highly variable within the commercial and noncommercial sectors of the
food service industry; training levels are high in white table cloth restaurants down to
minimum in fast-food chains.[a]

Labor and labor costs

In general, labor in the food service arena is mobile or transient or both; chefs and
sous-chefs are highly mobile with their skills; summer help is transient. Labor costs are
high in such establishments. Labor in lower quality restaurants is usually lower paid and
is often transient. Food commissaries have labor problems similar to those of any food
manufacturing establishment; workers work in hot, dirty, noisy environments where
often little formal education is required and basic language skills are not necessary
(Fuller, 2001, p. 267).

Expensive real estate

Restaurants, company cafeterias, and fast-food outlets are sited in high traffic areas
(downtown areas, office buildings, busy streets or highways), and this results in high
taxation and high property values. For these reasons, work area to income earning area
is kept at a minimum.

Environmental areas of concern

Site locations make odor elimination and waste removal imperative; elimination and
removal is expensive. Removal and elimination can themselves be operations that irritate
neighboring residents or establishments.

Hours of operation (noise and traffic pollution) can result in local by-law regulations.

Energy costs

Restaurant operations are energy intensive.

Consistency, price, quality and safety

Quality and price must vary with budget restrictions established for raw materials and
ingredients. Consistency must be constant and safety is paramount.

---

[a] The skill levels in the central commissaries of fast-food chains are high but levels in the actual
serving establishment may be nonexistent and dependent on preparation and serving protocols laid down by the central commissary.

Settlemyer (1986) presents this situation perfectly with the development of a buttermilk biscuit for a chain of restaurants. During development, it was determined that ovens suitable to bake the biscuits were available in individual outlets but warming ovens to hold the biscuits were not. Thus, included in the development process was the need to research different brands of biscuit warmers. The needs of the kitchen have to be satisfied.

One fast-food multioutlet company solved the problem of compatability of products and equipment with a fully operational outlet at its development center, not merely a mock-up test kitchen. The sole purpose of this outlet was to provide a one-store test to determine how operationally compatible products were before any test market was attempted (Peters, 1980). In both Settlemyer's and Peters' examples, the needs of customers (chef/owner/managers) were catered to first.

But there is another element that must be satisfied: the franchisee. The Kentucky Fried Chicken Corporation developed a grilled chicken product with fewer calories than the traditionally fried recipe. This has angered franchisees who claim advertising is directed to the grilled product when the fried product, they claim, is 80% of their business. Thus, we see franchisees who regard a new product as cannibalizing their bell-ringer product and diluting their efforts. New products must fit and be profitable.

Availability of storage space whether for dry, refrigerated, or frozen products or ingredients varies according to the type of outlet but is generally limited. New products must not require special storage conditions and must be adaptable to available storage space. Unused product must be easily resealable and capable of being safely stored.

10.3.2 Energy Requirements

Energy usage and its associated costs in food service facilities varies widely with the type of food service outlet, within any country and from country to country. It is a major factor in the vending machine trade. Where prepared foods received from a central commissary are used, less energy is required in preparation and presentation. De novo preparation from raw ingredients requires more energy for food preparation and cooking. Energy is, along with labor, a major contributing factor to overhead expenses. Efforts to reduce the energy used in meal preparation would be greatly appreciated by food service operators. In military feeding, in particular, energy conservation would be very welcome in food preparation since energy sources must be moved with the marching, sailing, or flying consumers.

To design products that conserve energy in their preparation and presentation in kitchens requires a greater input of energy in the prepreparation of products; therefore, developers need to determine how and where energy is required in food preparation. If only the prepared food itself is considered, energy is absorbed by the following

- The average temperature increases in foods as the masses of foods are brought to the necessary preparation and serving temperatures, that is, in the heating of food to the presentation temperature.
- Phase changes in foods of mixed composition as solids are converted to liquids (dissolving and melting) or water to steam (concentrating). (Norwig and Thompson, 1984).

Norwig and Thompson demonstrate the necessary energy requirements using as examples the frying of french-fry-cut potatoes and the cooking of frozen hamburger patties.

Snyder (1984) suggests that if sauces and gravies had the correct amount of water incorporated initially to get the desired finished consistency, then the need to evaporate water by boiling reduces energy requirements. This eliminates a concentration step and saves both time and energy. An additional benefit would be the improvement in quality, since less heat damage to sensitive ingredients in the sauce would occur.

Many factors affect the energy absorption (and quality) of food products during heating: shape, thickness (McProud and Lund, 1983; Ohlsson, 1986), density (McProud and Lund, 1983), composition (Bengtsson, 1986; McProud and Lund, 1983), and method of heating; by microwaves (Decareau, 1986); by convection or radiant ovens (Skjöldbrand, 1986); by boiling (Ohlsson, 1986); by frying (Skjöldbrand, 1986); or by grilling (Bengtsson, 1986). Optimization of the entire production facility is discussed by Engelund et al. (2009).

Energy usage is not confined to food preparation in food service establishments. Energy is consumed in lighting, heating, air conditioning, and fans in both the dining and in the kitchen areas. In addition, refrigerators, freezers, serving cabinets, and warming ovens, all indirectly associated with food preparation, consume energy that must be dissipated. Vending machine operators use motion sensors to control their energy usage for display lighting.

Food service operators reduce their overheads by reducing energy consumption. The design of energy efficient products is an interesting and important added value for products for the food service industry.

10.3.3 Labor

Labor, its availability, its skill level and its cost, are problems for managers of any food service outlet. Upscale restaurants are the exception; their skill levels are high. Preparation and serving must be simple in, for example, fastfood outlets where skill levels are low.

Developers need to provide clear and explicit instructions for

- Storage of the product and unused portions of the product
- Preparation of the product itself and preparation of any of its recipe variations (multiple use products or ingredients)

- Safe display of the product
- Serving the product

Products must not be labor intensive for preparation in the rushed, hot, steamy, crowded, and temperamental atmosphere of the kitchen. Preparation may not always be by trained chefs or cooks but by young, minimally trained, part-time staff working at a job, not in a career as a cook.

In fast-food restaurants, preparation must be simple, uncomplicated, and require minimal preparation time—essential in periods of high-volume turnover—to produce menu items with uniform quality with tight portion control. Pehanich (2003) discusses working relationships using food chains in particular.

10.3.4 Waste Handling

The ideal new product for the food service industry needs no preparation and creates no waste to dispose of. With storage already limited for supplies and food, storage for waste must be held to a minimum. Unused portions must be easily and safely stored without special storage requirements.

There is another consideration: sanitation. Any product introduced into the food service establishment must not introduce any unusual hazards nor require unusual or extraordinary handling techniques with respect to hygiene, cleanup, and sanitation.

10.3.5 Customers and Consumers

So far, only the needs of the customer or purveyor (i.e., the food outlet's owner-manager) have been considered and these are very specific to the type of eatery. Development is directed to serving this market with products that answer to its specific needs and that provide added value that is valued by the purveyors. Conventional retail products served up in institutional-sized containers are not product development in this unique market. Product developers must think ahead to the varied needs of the multifaceted industry, see all its niches and design products that satisfy these needs. This is no different than the design and development of consumer products for the many niches in the food retail market.

Catering is unique in that in addition to providing an *eating occasion*, with the emphasis on occasion, there is a sociability aspect attached to that provision:

> Catering systems that are centred wholly on the technical aspects and ignore the social aspects will fail....production methodology or product formulation and presentation must recognise the social context within which the final product outcome is to be consumed (Glew, 1986).

Whether consumers "dine out" or "grab a bite," developers must recognize that these meals are being consumed in a social context. Food must please; food must entertain; food must satisfy; food must comfort. Product designers must fit their products into the social context of the food whether it is served in a fast-food restaurant or an upscale restaurant.

What does this conjure up for new product developers? Fun foods? Comfort foods? Entertainment foods? The context of food usage, the occasions on which it is consumed, must be understood as important factors in both development and marketing. Movie theaters have grasped this concept quickly; they have become a social and eating experience for the whole family. There is an entertainment-sociability-warmth factor that must be designed into these foods. Food retailers, for example, attempt to capitalize on this aspect of food with the in-store bakeries wafting the aroma of freshly baked bread or barbequeing ribs or chicken throughout their stores and providing an eating section and children's entertainment section. The mother with young family in tow has a place to sit and relax and have a bite after the rigors of shopping. The smell of freshly baked bread has a warmness to it, evokes memories. It is a smell that was most missed by soldiers in combat situations.

An interesting development in dining (perhaps not the appropriate word) has come about. Experiential restaurants have appeared and (perversely) become popular; these are experiential restaurants. I am aware of blind restaurants whereby the patrons eat in total darkness, served by blind servers. Another is prison-geared, with food served by prisoners and usually referred to as "slops." Another, obviously seasonal, serves food in an ice palace-cum-hotel with food served on tables of ice and chairs of ice. Drinks are served in ice glasses. How these experiential eateries might stimulate product development is left to the imagination.

Moskowitz et al. (2007) demonstrated a method, called consumer-driven development, to create new coffees for three different types of outlet.

10.3.5.1 Consumer and Nutrition: An Oxymoron

Nutritional concerns in the commercial and noncommercial sectors of the food service market have been discussed earlier. Today's consumer, as well as governments and health groups, is concerned with nutrition as it is reflected in the so-called obesity epidemic. Today's consumer is interested in foods that provide a health benefit (but how to explain the growing problem of obesity?). How deeply this concern extends when the context of eating is social as with the candlelit dinner or even noshing in the campus cafeteria after a heavy lab or strenuous badminton set, is very difficult to assess. Nevertheless, customers (food service outlets) recognizing the consumers' desire for cutting down excess fat, getting plenty of good fiber in foods and wanting minimally processed foods were driven to devise products that satisfy these needs. They came up with reduced-fat burgers, veggie burgers, and provided vegetarian menu items and described menu items as "heart friendly." This fizzled (see, e.g., Lone et al., 2009). Reduced fat-burgers have disappeared from the marketplace (veggie burgers have, surprisingly, quietly gained wide acceptance). Is the entertainment/sociability/warmth factor (comfort foods) associated with eating occasions in consumers' minds greater than their nutritional concerns? Most food service outlets have found nutrition to be not a major concern in a social context.

Many fast-food chains have extended their menus to present healthful variants of standard products or to present new eating opportunities, for example, breakfast menus have been developed; soup and a variety of salads are being offered as calorically lighter fare. Extended menus require new ingredients and new food products. Where new products can serve a multitude of uses in a food service outlet to satisfy the consumer with a variety of tasty products and the customer by reducing the variety of ingredients carried, everybody wins. But, it must be asked, is this satisfying the consumer's wants?

Nutrition, price, taste appeal, convenience, and gratification of the gustatory senses influence acceptance of new products in food service markets. Nutrition is a factor in acceptance but not in my opinion, the most important one in "eating out." The trend to healthy eating will remain fed by government and medical authorities but this market must be allowed to develop at its own pace and perhaps, as Lone et al. (2009) suggest with a more directed marketing approach to the nutrition-interested and the nutritionally uninterested groups.

10.4 Quality in the Food Service Market

There are two judges of quality: customers (chef/cook/owners/purveyors) and consumers (diners/eaters/users). Quality is first determined by the management of the food service outlet, be it a commercial or a noncommercial outlet. If quality is to be understood as the satisfaction of needs, then product developers must first meet the needs of the kitchen staff. Here, quality attributes must include characteristics discussed in previous sections. When these are satisfied, then perhaps, the product will be put to the consumer for final judging using the skills of the chef or cook.

The ultimate judge of quality is the consumer. Criteria for assessing this judgment are the trash bin (plate waste), non-ordering of menu items, loss of sales and complaints.

10.4.1 Safety

Concerns for safety with respect to both hazards of public health significance and to hazards of economic significance are paramount in developing products for food service. Any hazards with products introduced into this marketplace are multiplied many times with the many customer contacts involved. Programs (e.g., HACCP programs) used to resolve these concerns do not differ greatly from those discussed for product development earlier (see Bryan, 1981, 1990).

The stresses unique in the normal food service kitchen operations and the potential for widespread food intoxication require extra consideration be given for the safety and stability of menu items. The limited storage area, the frenetic activity in the preparation area, and the need to display or hold product hot in serving areas present extraordinary challenges to the quality characteristics of menu items. Health care feeding presents further problems of safety; there are patients with compromised immune systems, and special care must be paid to the microbiological safety of ingredients and products intended for their use.

10.5 Development of Products for the Food Service Market

Challenges found in the diverse subsections of the food service industry have been discussed in the foregoing. Product development for the food service market parallels a pathway similar to that for the food retail sector. Because of its highly segmented nature, the food service sector must be researched for what customers (both purveyors and diners) in and frequenting food service outlets want (Peters, 1980; Settlemyer, 1986). Developers wishing to bring new items onto its menu first develop objectives that they wish to attain and ascertain how these objectives will be reacted to by both company- and franchisee-owned and operated outlets. (This is something the KFC Corporation failed to do; they saw their future in what they perceived as the healthier grilled chicken product and not the fried product. This was not clearly communicated to their franchisees, who make their profits from the fried item.) From these, there should arise strategies and then tactics to reach these objectives within the time and financial constraints that the management desires. A take-out establishment may wish to expand to table seatings or a chain may wish to go upscale trying to attract a more sophisticated crowd with a more sophisticated menu. Clarity of purpose is necessary to target diners' needs.

To enhance chances of being successful, product developers (if not already in the business) need to speak to food service operators, discuss with them who their target market is and what their requirements for new products are, and be guided by these in new product development (Pehanich, 2003). Such communication with purveyors permits the developer to better focus research to design products more suited to the physical and operational needs of purveyors. Without this communication, and for external developers not in the corporate chain, the cooperation of corporate head office, developers will have difficulties designing suitable new products not knowing what the ultimate preparation in the kitchen and presentation to the diner will be.

To become successful suppliers to the food service industry developers must be very knowledgeable about this market, each segment of which represents a marketing niche in itself. Development teams require

- A clear, specific statement of what menu item their new product is going to be part of, that is, dessert, main course, side dish, etc.
- Identification of the targeted consumer.
- Information on which meal (breakfast, lunch, dinner, snack) the product is intended to be a part of.
- The budgeted allowance of the product—that is, what cost is that menu item going to contribute to the entire meal? What is the diner willing to pay? Is it inconsistent with the quality of food and class of diner?

Prices of menu items stick out like the proverbial sore thumb. Comparison pricing of menu items is much more obviously done on a menu. Price in a food service outlet influences a hedonistic choice. In a grocery store, the customer is seeking an item that is necessary, a staple. Prices of items on a menu are a more important factor than they were in the retail food market. Prices of new products for use with menu items must fit into the price structure of a whole meal.

Companies that are serious about developing products for the food service market no longer rely on food technologists but now rely on chef technologists. These are people who not only are skilled food scientists but are fully trained chefs. They combine those skills to be able to work knowledgeably with the food service outlet and the laboratory and the manufacturing plant through the routine of consumer research with focus groups, questionnaires, and interviews to get a clear and comprehensive reaction by the targeted consumer to the product concept. They understand the need to design meal occasions and menu items to meet the needs of the kitchen and the diner and to fit the many styles of menu types and varied customers.

Competitive products on the market are audited more competently by these chef-technologists than food technologists to provide some idea of the quality levels in the marketplace, the pricing structure, and expected consumer reaction to them. Criteria in screening are

- Price and profitability—Do the products provide labor-savings in preparation, rapidity of preparation, and serving time as well as portion control?
- Do the products answer to the constraints imposed by preparation and display equipment?
- Are the products compatible with the menu items and well received by the kitchen staff and diners?
- Do the products add any additional strain or hazard to the already stressed atmosphere of a kitchen?

Many products are prepared in a central commissary situated anywhere from several floors away in a hotel to a few miles away for health care feeding establishments to several hundreds of miles away in the situation with fast-food chains and the military. An obvious criterion of screening is whether the product provides better protection for both nutrition and safety against any abusive mishandling that distribution might impose. West (1994) reviews the similarity between good manufacturing practices for the food manufacturing and food service sectors. She emphasizes the need for designers of prepared foods and food components to know these.

Developers of food service items need to provide a very clear list of instructions for the storage of the product, its preparation, whether it is capable of multiple uses, for preparation of all its variants, its display, the method of serving, and storage or treatment of unused portions. This requirement is less essential in retail food product development but most food manufacturers do provide recipes, product information, and preparation advice on Web sites for good consumer relations.

At this stage of development, a consumer test can begin. This can be a small or large test of the product involving a few units in a local area to a dozen or more food service outlets spread widely. The KFC Corporation chose to test market their bunless sandwich in the city of Omaha, Nebraska. Marketing support can vary from simple table tents to TV, radio and newspaper advertisements supported by coupons, in-store displays and free sampling.

10.6 Criteria for Evaluating a Test Market

Reaction to new food service products can be found in the following ways:

- Consumer intercepts (questionnaires) permit the developer to determine both the food preparers' and the diners' reactions to the new product. Results permit, if evaluation warrants, a refining of the product to better adapt to both the kitchen's (purveyor's) and the diner's needs. How the intercept is carried out depends largely on the product under study—is it one that only the food preparer or the diner would appreciate? The social context of the meal often precludes the use of diner intercepts.

- Cash register data can provide information on the new product as a percent of sales. The impact of the new product on improving sales or the impact of its introduction on the cannibalization of some other menu item can be determined.

- More dramatically, analysis of the amount of new product found in the trash containers at test sites provides ample evidence of acceptance.

A careful interpretation of all data will be necessary to provide information that may herald a successful introduction or prevent a commercial disaster.

There are three sources of waste in any food service establishment. There is preparation or kitchen waste that can broadly be classified as food purchased for kitchen use but discarded during preparation or spoiled in storage. Kitchen waste will, of course, be higher in establishments doing their own preparation work and not relying on preprepared foods. The second source of waste, service waste, is prepared food left in warmers or steam tables and not chosen by consumers. The amount of kitchen waste and service waste is largely, although not entirely, a measure of the management skills of the establishment. Finally, there is consumer waste: food purchased by the consumer but discarded. This is a measure of product rejection.

How waste is to be measured and assessed presents some problems. Banks and Collison (1981) studied the problem of waste in 39 catering establishments in the United Kingdom and discussed the factors affecting waste, not the least of which is the size of the meal. Lack of attention by the establishment to portion control increased waste but the amount of convenience food used by the establishment decreased it.

Kirk and Osner (1981) agree that consumer waste can be a sign of poor portion control:

> It may be thought that plate waste does not represent a financial loss to the establishment since the food has been paid for by the consumer. However, poor portion control can lead to more food being produced than is required or to a loss of potential sales.

Another factor contributes to plate waste. People have varied attitudes toward the edibility of particular food items, for example, the skins on vegetables such as potatoes, cucumbers, or zucchini; and meat items such as giblets, fat and connective tissue, or chicken skin.

Examples of two different diner assessments done by questionnaires are those by Lülfs-Baden and Spiller (2009) and Porter and Cant (2009). Lülfs-Baden and Spiller were studying school meals (in Germany) and evaluating students' perception of these. Porter and Cant were evaluating hospital patients' satisfaction with a cook-chill food service system. Watson et al. (2008) went to social networking to evaluate the sharing of restaurant experiences using online communities. Their paper is an interesting analysis of the application of a modern tool to sharing food experiences of foodies: this may have deeper applications for evaluating restaurant experiences. As Watson et al. point out, consumer interfaces, diaries, and focus groups produce results that can be criticized for respondent inhibition—the researcher influences either in person or by the questionnaire and a personal bias. Blogs, they claim, are freer of these inhibitions. Foodies are defined at some length but simplistically they can be looked upon as not gourmets but collectors of food experiences.

Research on diners requires careful assessment. Does the data represent only the regional preferences of the test area selected for introduction or can the data be extrapolated to wider market areas? These are problems no different than those found in consumer food items. However, acceptance of regional dishes in one area of the country may not be equally well accepted in other areas. There is a growing body of evidence that food preferences may not be entirely cultural but may have a genetic element. Impartial answers to these and other questions are required.

Questionnaires (consumer intercepts), in combination with analysis of consumer waste, are useful tools in assessing consumer acceptance of newly introduced menu items. Interpretation, especially of waste, must be used cautiously. Cash register receipts provide an indication of purchase but the garbage bin audit can tell of the acceptance of the new product. The intercept gives further evidence of degree of acceptance.

The introduction of any product, even one so seemingly simple as a different style of hamburger, into a fast-food chain can involve several unexpected, interwoven variables which need to be assessed. One fast-food chain spent $1 million on taste tests to develop a better hamburger (Anon., 1986). This chain experimented with 9 different buns, over 3 dozen different sauces, 3 types of cuts of lettuce, 2 sizes of sliced tomato not to mention 10 colors of 4 different boxes and some several hundred different names. It was even determined that the order of the condiments was important to consumers. All in all, this new product introduction represented a formidable task in market analysis!

11

Product Development in the Food Additive and Food Ingredient Industries

They say everything in the world is good for something.

John Dryden

11.1 Additive and Ingredient Market Environment

Dryden's words hold true for the ingredient and additive industry. Ingredients and additives have been made out of plant, animal, and mineral materials through extraction and concentration, physical or chemical modification of materials, or through de novo synthesis of natural ingredients. Old herbals, old recipe secrets, and the recipes of other cooking traditions have been sought out for possible sources of new ingredients. Suppliers of food additives and ingredients supply materials that enhance or create desired qualities or give added value to consumer products that developers are attempting to create.

Definitions of additives as found in the food legislation of many countries have been given (Fuller, 2003). An overly simplified definition for an additive could be the following: a substance not normally consumed as a food used in a food that causes it to become part of that food or to alter the food's characteristics. Two exceptions to the U.S. definition (21 CFR 170.3) of additives exist. These are so-called prior sanctioned items and GRAS (generally recognized as safe) items [21 CFR 170.3 (k, l, n, and o)]. The prior sanctioned items (42 categories of products) can be described roughly as ingredients (as is commonly known) and the GRAS items as processing aids.

For textual simplicity, both additives and ingredients will be designated ingredients, but it must be understood that this use is merely a convenience. Suffice it to say, ingredients are not additives.

11.1.1 Characteristics of the Food Ingredient Industry

11.1.1.1 Chain of Customers and Consumers: A Welter of Identities and Needs

Developers of products for the food service or food ingredient industries have one thing in common: they may never see nor know how their product was used. They may also never get any credit for their creation except perhaps in a list of ingredients on the label, and even there, it may be hidden under some generic label or even a code number.

A distinction between *customer* as one who buys a product to be used and ultimately consumed by a *consumer* has relevance in the ingredient industry. Customer and consumer may be one and the same entity. The chef in a restaurant purchasing a soup base and then using it to add value to a menu item is, in effect, both a customer and a consumer. That chef has produced a new menu item that is then purchased by a customer (the diner, also a consumer). In the ingredient industry, the customer and consumer roles are often blurred, and the needs of each in the chain must be identified for a series of new products to be developed. The chain can be very long.

For example, in scotch whiskey manufacture, barley is sold to a malting company (the first customer and user) for soaking, malting, and drying over peat fires; the malting company then sells malt, smoked and dried to a specific brewer-distiller's specifications; and a brewer (the second customer and user) steeps and ferments the brew. The brewer sells the spent malt, called draff, to an ingredient manufacturing company or a feed manufacturer (either of which is the third customer and user). After suitable modifications, this spent malt may be sold to a bakery where it is used in a baked finished product which is sold to consumers or it may be sold to other food and feed manufacturers. The chain is clear: each customer and consumer in this chain has needs that must be identified in order to be satisfied.

As another example: flavor extractors purchase spent seeds and skins (called chaff) expelled from the finishers from hot pepper sauce manufacturers for extraction of the color and heat principle for sale to confectionery manufacturers and pharmaceutical companies to be used in their products; this chaff is purchased for enhancing spaghetti sauces; or the chaff is dried and sold as a hot pepper sprinkle-on product. Ingredient suppliers must understand who their potential customers are, what their needs are, and how they use the product. They also need to look ahead at trends in food products and determine how to provide products meeting the ultimate consumers' expectations.

The general public seldom notes ingredient developers' products except in a generic way as *flavors* or *thickening agent* or *chicken stock* in the list of ingredients. Some exceptions here are those diet products containing artificial sweeteners, fat substitutes, or specific source of fiber where the ingredient's brand may be named for the particular cachet it carries. McIlhenny's Tabasco

Sauce® is often named for the cachet it brings to products it is partnered with. Most ingredients, however, are not identified by the original manufacturer.

Some ingredient manufacturers have, in addition, a profitable retail market niche for food ingredients, for example baking powders, flavors, food colors, various types of flour (e.g., stone-ground flour), meat and vegetable hydrolysates, etc. These markets are subject to all the pressures of the retail marketplace, and their development has been treated in the previous chapters.

11.1.1.2 Similarities and Dissimilarities to the Food Service Industry

New product development for the ingredient industry presents interesting differences from and similarities to the development in the retail food market and in the food service market. Similarities of product development in the food ingredient industry to that in the food service industry are very close, indeed startling. The similarities are threefold:

1. Ingredient development tends to be reactive or crisis-oriented development. A company has a waste by-product. What can they do with it? Or a demand originates primarily with consumer product manufacturers (customers) who want some ingredient to create a particular property in a food; can they supply it? These manufacturers believe their customers have a need; therefore, they approach ingredient developers for problem solving.

2. A corollary of the preceding; that is, ingredient suppliers go to manufacturers with ingredients they have developed that possess unique properties. Ingredient concepts are obtained by a thorough understanding by ingredient suppliers of their customers' markets. Ideas originate with customers with whom the ingredient supplier works closely often in a partnership arrangement. Where are the customers going with new products and how best can the supplier aid their customer? Food service providers do much the same: they follow the tastes of their customers with various cuisines (nouvelle cuisine, ethnic cuisine, and fusion cuisine) that suit these changing styles, which are communicated to the ingredient supplier.

3. Ingredient suppliers serve at least two masters: the immediate customer who purchases ingredients for use in a product and the succeeding customers down the chain who use that enhanced product and rely upon its unique properties. Each customer in the chain puts in added value; each customer has different needs, which have to be catered to. This mimics the food service industry.

There is one difference, a difference of degree perhaps, rather than a difference of substance. Development of products for the food service market is focused towards the customer or consumer or both. That is, products must contribute to the over-all aesthetic characteristics of a product without being obtrusive but also must provide convenience in the work environment of a busy kitchen at a price the customer is willing to pay.

The situation in food ingredient development is slightly different. Here, the efforts of food ingredient manufacturers are still largely concentrated on developing ingredients that contribute added quality attributes to the finished product or that provide some processing advantage, thus satisfying the technical demands of the product rather than satisfying the aesthetic tastes of the end consumer. The ingredient should not contribute to any undesirable changes or cause an undue increase in product cost.

Ingredient developers are caught up in technology. They have shown great skill and versatility in making fat replacers from so many different substances at such a rapid rate that the casual observer gets the impression that there is a contest to see how many different substances can be developed into fat replacers. Here, developers have produced products that substitute for oil with most of the characteristics of oil except oil's essential fatty acids.

This is similarly the situation with fiber for which commodity processors have used their particular fruit, vegetable, nut, cereal, or by-product thereof and tried to develop it as a fiber or to emulate the properties of fats and oils.

11.1.1.3 The Ever-Present Government

The likelihood of government intervention is high in the developmental activities of ingredient suppliers. This interest takes the form of safety testing and approvals for the many novel ingredients developed from new nonconventional food sources (e.g., phytochemicals from nonconventional plant sources), from genetically altered organisms, or from chemically modified conventional food sources. Indeed, the whole question of calling such ingredients natural or even that such products will be termed environmentally friendly will be challenged not only by governments but also by consumers.

The introduction of novel and imitative products faces hurdles. Righelato (1987) put this succinctly:

> Regulations exist primarily to protect the consumer, but they are necessarily concerned with existing products and hence serve to maintain the status quo. In doing so they protect the existing producer, who, in fact, probably helped frame the regulations.

In short, the introduction of novel ingredients faces challenges imposed by legislation and from the suppliers who provided opinions and advice to the government on the formation of the legislation in the first place. The opportunity for bias and opposition to novel ingredients is obvious.

11.1.1.4 *Proliferation of New Ingredients*

New ingredients and ingredient technology, a technology whereby food scientists can develop ingredients for specific applications or functions, have grown at an amazing pace. A classic example of this growth is the many products that can be derived from milk, each with its own unique flavor or functional property that it contributes to foods in which it is used. A quarter of a century or so ago, milk was added as an ingredient, and a dairy product tree based on milk would have numbered only a handful of products:

Ice cream	Evaporated milks	Cultured milk
Market milk	Cream	Butter (salted and unsalted)
Cheeses	Dried milks	

Today, milk constituents have been prepared into a wide variety of ingredients (Kirkpatrick and Fenwick, 1987):

- Whole milk products: pasteurized milk, powdered milks, sterilized milk and flavored UHT milks, and milks from other species such as goat, sheep, camel, and buffalo (Nuttal-Smith, 2009)
- Products derived from compositionally altered milks such as fat-reduced milks, protein-enriched milks, lactose-reduced milk, sweetened milks, and reduced-mineral milk for infant foods
- Milk powder products from modified milk with heat stability properties or high dispersibility
- Products based on milk fat ranging from the common cream, butter, and anhydrous milk fat to compositionally modified milk fats with better spreadability or altered fatty acid composition to fractionated milk fat with controlled and defined melting ranges
- Products based on the proteins isolated from milk; whole protein coprecipitates, components of proteins such as casein rennet, whey protein derivatives (whey protein concentrates), lactalbumin, milk proteins combined with non-milk proteins, and modified protein fractions
- Whole cheeses, cheese powders, sprinkle-on cheeses, reduced fat cheeses, processed cheeses, flavored cheeses, and modified cheeses
- Products based on lactose where this sugar's low sweetness can be utilized and on the other end of the sweetness scale where its enzymatic conversion to glucose and galactose can produce sweeter products
- Biologically active materials that can be obtained from milk

These dairy ingredients find uses in dietetic foods (McDermott, 1987), in meat and poultry products as calcium-reduced binding and emulsifying agents (van den Hoven, 1987), in confectionery products (Campbell and Pavlasek, 1987), and in bakery products (Cocup and Sanderson, 1987).

Plant materials such as sea weeds, fruits, herbs, and spices are being similarly purified, fractionated, extracted, and blended to produce functionally important fiber ingredients, viscosity-adjusting agents, and antioxidants. Underexploited plants and underutilized fish caught as a by-product are finding uses as new ingredients (fish protein concentrates) or sources of flavonoids or other polyphenolic compounds for use as natural ingredients.

IFT's newsletter for the period February–July 2008 addressed beetroot juice reducing high blood sugar; grape skin resveratrol protecting against diabetes and obesity; tagatose, a reducing sugar with probiotic properties; properties of caffeine; use of extract of Chinese red yeast rice reducing risk of heart disease; orange terpenes as an antimicrobial against *Salmonella* spp.; vitamin D and heart attacks; vitamin K_2 and its relationship to bone strength; and Truvia™ introduced by the Cargill Company. When functional foods are considered (nutraceuticals, probiotics, and prebiotics), the list of possible ingredients with processing, health, or quality attributes is indeed huge.

11.1.2 Focusing on the Customer Who Is Also the Consumer

All product development gets its impetus from the customer or consumer or both; they have some perceived need that can be profitably satisfied. The ingredient developers challenge themselves by asking, "How can we convince potential customers that our products will improve their products?" or "How can my ingredients be modified to enhance the high quality characteristics of my customer's product?" Many ingredient developers are a step ahead of their customer developing prototype products to demonstrate to a customer what their ingredient can do. The possibilities for the development of close supplier/client relationships are apparent as the special needs of customers are catered to this might account for the diversity of ingredients.

The technological ability of an ingredient developer to transform some raw material to simulate the textural properties of fats, for example, is unimpressive unless this altered product contributes a desirable added value to the manufacturer's product that can be sold profitably to other customers. Ingredient suppliers must adapt the properties of their ingredients to the food systems of their customers' products and demonstrate to their customers the value of their ingredients. Ingredients are designed to complement the quality and stability the customer has designed into their finished product.

11.1.2.1 Customer Research

The ingredient company's strength is in the ability of its technical sales force to articulate the needs and desires of customers back to technologists in their research and development department. It is here that multidisciplinary teams prove their value: sales personnel sell with their applications technologist at their sides. They as a technical sales force are the primary route to customer and consumer market research information.

The usual market research techniques, focus groups composed in the usual manner with a company's customers, questionnaires for use in individual contacts, and mail or telephone surveys have no practical use. Of the surveying methods, only the Delphi technique (see Chapter 3) of querying company executives appears to be of any general help to ingredient developers; it is looking ahead too far to be useful for innovation. Demographic and psychographic data about industrial customers are simply nonexistent.

Trade, business, or commodity associations such as the American Association of Meat Processors, the American Spice Trade Association, the Chocolate Manufacturers Association of the United States, the Milk Industry Foundation, and the International Ice Cream Association can provide general information of ingredient trends and industry needs. These associations hold exhibitions and conferences where ingredient suppliers can develop excellent contacts networking with potential customers to discuss needs. *Prepared Foods* as well as several other national and foreign food trade magazines issue an annual index of trade associations and their exhibits (Anon., 1991).

After a contact is made, the ingredient suppliers' technical sales personnel will interview their potential customers' technical staff to communicate back the problems encountered by customers. Market research is largely customer intercept with a hands-on business with the ingredient supplier concentrating on the specific needs of one client. That is, the ingredient supplier's technical sales team works one-on-one with its client discussing that client's needs and developing, blending, or otherwise adapting ingredients to that client's needs. While an ingredient may serve several different clients, each customer's needs are different, and each customer's product is a different food matrix. Ingredients must provide their industrial customers with finished products that have distinct points of difference. Ingredient users cannot always rely on off-the-shelf ingredients to obtain this point of distinction.

11.1.2.1.1 Partnerships

Partnerships develop between ingredient suppliers and their customers and both will work together toward a common goal. These can be formal with contractual agreements or remain at handshake seller–buyer agreement. In partnerships, each customer of the ingredient supplier then is unique. The distinctiveness of an ingredient must belong to that customer and that customer alone, but it is not necessarily the property of that customer; much depends on the details of the partnership agreement; thus, we see the need of lawyers.

Flavor houses have developed this art of focusing on the needs of the customer to a high degree. They work closely with customers to produce any flavor sensation their customers want. They can blend natural flavors, create flavors imitative of natural ones, or create unique flavors not found in nature. Ingredients and the technical service supporting the use of the ingredients are designed to satisfy the customer's needs.

The ingredient developer's goal is to provide a quality service with a product distinctively designed to meet the needs of the added value manufacturer who, in turn, wishes to have their product satisfy the needs of their customers. Development is directed to these needs and their gratification. Ingredient suppliers certainly do sell ingredients, but they are just as likely to sell services.

An ingredient developer cannot, after creating a family of ingredients each with slightly varying properties, approach a potential client with these samples saying "Try these; one of them should work." The ingredient supplier must know that their ingredient works before it is passed on to the customer. Perhaps it will not work to the full satisfaction of the customer, but if it works well enough that a partnership relationship between supplier and customer can be developed, some success has been achieved. Hence, the need for a close working relationship between the ingredient supplier and customer through technical sales personnel.

There is an analogy with the food service industry; suppliers to the food service industry must adapt their products to conditions in the kitchen, to the skills of the kitchen's labor force, and to the style of the food outlet itself and its clientele. Similarly, developers of food ingredients must adapt their products to satisfying the equivalent requirements of their customers.

Customer intercepts with the intercept being made with the ingredient supplier's technical sales team is the preferred potential client researching techniques. This provides the necessary information that permits selected targeting of customers and consumers or the development of specific niche markets for ingredients. Heavy promotion through cooking schools and cooking demonstrations in schools, church basements, carnivals, and agricultural fairs, plus recipe booklets and free samplings usually accompanies retail sales. Feedback from these promotional tools provides its own consumer research information.

11.1.2.2 "Consumer" Research: "Yes" and "No" Possibilities

Is there an opportunity for consumer research in the food ingredient field? The answer is an ambiguous "yes" and "no." There are two consumers that the ingredient manufacturer treats with: first, there is the supplier's customer with whom the supplier works to develop an ingredient specific for this client's needs but this customer uses (consumes) this ingredient.

The supplier's client has customers and consumers that both the ingredient supplier and the client are trying to please. This latter consumer is indirectly the ingredient supplier's customer/consumer and can be found in the public marketplaces somewhere. The ingredient manufacturer cannot be unaware of trends in the public sector.

11.1.2.2.1 The "Yes" Side

For the yes side: ingredient developers can make themselves aware of all activities in the various food marketplaces. They should be aware of the trends in health foods and the nutritional concerns of the public, those touted by the medical profession, and those supported by government, for example, that low calorie, low fat, high fiber, low or no cholesterol, and no salt foods are in favor. Therefore, they develop suitable ingredients for different food systems that satisfy these niches. Nutritious (good-for-you) and diet foods, once relegated to the slow-moving section of the supermarket, are now mainstream and prominently displayed. Green and natural are in, and food manufacturers are attempting to draw attention to the natural ingredients in their products. Natural ingredients can be prominently displayed in the list of ingredients for their image of purity and wholesomeness. The consumers' desire to self-medicate and a need for foods fortified with ingredients to combat disease have caused suppliers' rush to develop nutraceuticals (functional ingredients) for manufacturers of foods. All the foregoing researchable trends provide consumer research for ingredient suppliers.

So, yes, ingredient manufacturers can indirectly research their client's consumers, determine consumer trends, and fabricate ingredients that incorporate these desirable characteristics into their products. In a sense, they leapfrog their clients, the consumer product manufacturers. This leapfrogging can be used for what Lee (1991) describes as "proactive product development." In proactive product development, ingredient manufacturers bring to fruition product concepts that if adopted as products in the marketplace by the consuming public would result in heavy usage of the ingredient supplier's newly developed ingredient. The ingredient makes possible products (for which, it is understood, there was an undiscovered marketplace need) that food manufacturers could neither produce previously without this ingredient, nor produce at a reasonable cost, nor produce at an acceptable quality.

Textured vegetable proteins, surimi-based products, and the mycoprotein product, Quorn, developed by Rank Hovis McDougall, are typical examples of such ingredients. All find wide use as analogues in various engineered consumer products: see, for example, Duxbury (1987) and Brooker and Nordstrom (1987) (surimi); and Godfrey (1988), Best (1989b), Bond (1992), and Wilson (2001) (Quorn). A consumer desire for new texture is identified and manufacturers can satiate with new products using the textures these new ingredients provide.

11.1.2.2.2 The "No" Side

For the no side: ingredient developers find it necessary when they have created a new ingredient or when they want to capitalize on the unrealized potential of a hitherto poorly used ingredient that they must develop a need for it by demonstrating new and exciting innovative products; this is expensive. They must have a very accurate and intimate knowledge of the general population of consumers and their needs: for example, if an ingredient is "in" for any reason, (cf., Oatrim™), then there is a scurry to make a variety of fat replacers from a multitude of sources of fiber and demonstrate the new fiber's uses in products. Currently, a wave of interest in natural preservatives and antioxidants has spawned a wave of research into açai (Tonon et al., 2009), Bahraini date palm (Allaith, 2008), fruits and leaves of the olive tree (Silva et al., 2006), marine alga (Abd El-Baky et al., 2009), and various Turkish plants (Kirca and Arslan, 2008) for natural-looking, minimally processed foods.

Such products require extensive development resources (surimi, an exception, had a long history of development and application and was readily adapted but required extensive education; Quorn is much slower to be adapted). Many of these natural preservatives bring with them strong flavors characteristic of the sources from which they were derived. For some ingredients, safety testing may be required and government approval needed before selling into the trade. Panickar et al. (2009) in a study of cinnamon polyphenols caution that compounds of cinnamon may accumulate in people and that cinnamon should not be eaten on a regular basis over long periods of time as more than an occasional spice. Beneficial phytochemicals used in flavors or preservatives as spices or herbs are far different in potency of side effects when extracted and concentrated. Spice and herbal extracts and concentrates require education of clients to their use or to learn how to adapt them to local food traditions or to their brands.

The major way for ingredient developers to convince their client food manufacturers of the opportunities of their new ingredients is by developing novel products and recipes themselves. If this is successful, then they must rely upon manufacturers to develop, market, and promote their own branded finished products. When the advantages of their ingredients are noted in finished products on shelves, ingredient suppliers hope new clients are attracted.

11.1.3 Development Process

The development of new ingredients differs from the development processes discussed previously; it is complicated (a) by the need to establish the efficacy of their new ingredients (does it do what the supplier hopes in food matrices it was designed for?), (b) by the need to determine effective usage levels for the many different types of products and processes of their clients, and (c) by having to determine the safety or possible side effects of any new ingredients developed as concentrated extracts of nonfood source phytochemicals.

11.1.3.1 Development Process and Food Legislation

Legislated standards and regulations for ingredients respecting safety and permitted usage levels in products are more significant factors in ingredient development than they are in the development of products for either retail food outlets or for the food service industry. There is currently no rationalization of food legislation among nations; consequently, manufacturers have to constantly reformulate products when ingredients are proscribed or permitted usage levels differ in an importing country. Attempts at harmonization of food legislation among countries are proceeding but are far from completed. One attempt is through *Codex Alimentarius*, but not all countries are signatories to this; unfortunately, many blocs develop harmonized regulations but as a protective trade barrier.

Obviously, food legislative activities occurring worldwide affect ingredient developers and their clients in many ways. A change in food legislation literally anywhere in the world can have a devastating effect on the acceptability, or not, of ingredients, on which foods and at what levels the ingredients are permitted; changes in trade barriers or legislation may affect not only the costs of ingredients but also the availability of the raw materials from which they are made. For example, many lesser developed nations have raised grave concerns over the rape of their traditional heritage flora by large industrial powers to produce and patent products to which these nations feel entitled. Time and money spent problem solving their customers' product development must not be wasted by experimenting with new ingredients that do not conform to local, national, or international laws or do not meet or respect religious customs.

11.1.3.2 What Are the Criteria for Screening?

Screening in the ingredient development process includes determining the safety and permissibility of the ingredient (in partnering with a client, this includes acceptability of the ingredient at the intended levels of usage in the client's products) and functionality of the ingredient in the client's products. That is, does the ingredient do what the client wants at a price the client is willing to pay? Herein lies a financial constraint.

Ingredient suppliers use basic research and applied research teams complementarily to do this. The basic team studies the physical and chemical properties of the ingredient and modifies the base ingredient to obtain desirable properties the client wants and the applied team studies specific food formulation reactions with the ingredient.

11.1.3.2.1 Financial Constraints as Criteria

Financial criteria for new ingredient development have different time horizons. Return on investment can be accepted over a longer period of time, measured in years rather than in months—that is, ingredient developers do not expect a payback in 3–6 months as might be expected in the retail food market. Standard ingredients are sold to many diverse clients such that profits are maintained as long as demand continues. Ingredients designed specifically for a single client bring in a higher return, but this may only last until the client's product loses popularity. Ingredients do not require the same level of the expenses of advertising and other promotional gimmickry that consumer food products do. Advertising is directed at specific targets, that is, it is narrowcasted rather than broadcasted.

> An ingredient which has been well researched and developed should be filling a market need and will sell itself to some extent, whereas it is often necessary to create a market for a new consumer product by intensive advertising (Lee, 1991).

There is no easy way to predict or evaluate potential financial returns for a new ingredient. Financial criteria as discussed under retail food product development cannot be applied to assess the financial success of ingredient development. Difficulties arise when trying to assess financial criteria in a partnership arrangement between the ingredient supplier and the consumer product manufacturer. Here, there is only one customer since such arrangements often involve exclusivity agreements; the two parties must enter into some legal agreement.

Our company had developed an Italian herb mix for breadcrumbs. We blended this in-house. Thinking it would be cheaper and easier to have this made by a spice house who had more business savvy than we did in the buying and blending of herbs we approached a spice house. Over a handshake the breadcrumb mix was handed over. After a few years of successful operation we decided to have the herb blending back in-house. We no longer owned the herb mix. The spice house had, over the years, improved upon it. We tried to make our original again but the breadcrumb mix was distinctly different. We received complaints.

A novel criterion is introduced similar to that encountered in the food service industry. The cost of an ingredient and its modification to meet the client's demands must be balanced against the customer's financial criteria. The required usage level of the ingredient must not force the client to price their products out of competitive ranges. When ingredient costs rise too high, those exclusive users must seek alternatives. For example, marketing boards may set artificially high commodity prices in order to establish good prices to farmers. These prices may compel consumer product manufacturers using these commodities to find cheaper alternatives. This sharing of costs is a hazard of partnership agreements that must be worked out.

11.1.3.2.2 Foreign Market Constraints

Ingredient suppliers are able to distribute their products more widely than do consumer food product manufacturers since ingredients, generally speaking, carry no ethnic, cultural, or nationalistic biases although religious concerns do arise regarding whether ingredients are of animal origin or are kosher, pareve, or halal. Consequently, ingredient suppliers usually have greater export market potential with only legislated barriers but without cultural or traditional taboos. Three major criteria for screening are unrelated to the specific ingredient itself:

- Can ingredient suppliers adequately service foreign markets and provide the technical sales support with its associated research and development, and marketing activity required to support these market niches?
- Do ingredient manufacturers understand the foreign market and its local food customs sufficiently well to identify potential consumer product manufacturers and to establish a rapport while recognizing the difficulties that language barriers and different business customs bring?
- Are these suppliers aware of national, regional, and local food regulations that must be observed?

Obvious solutions are the use of local agents familiar with local conditions in the foreign markets or the establishment of satellite operations in the foreign country. Both solutions have shortcomings:

- The use of agents interposes one more hierarchical level between client and supplier through which communication must be filtered. Ingredient suppliers must rely on their agents both for reliable market information and for providing competent technical support to clients.
- Satellite operations, unless they have all the research facilities of the parent company, must send samples from the foreign customers back and forth to head office for experimentation. Delays and inconveniences for customers result.

Both avenues represent added costs: agents want fees, and satellite facilities are costly to maintain with the double teaming of technical and marketing staff that they require.

11.1.3.2.3 Applying the Client's Criteria in the Development Process

Clients have criteria by which they evaluate ingredients, and these impinge on the ingredient developer's criteria:

- Does its use reduce ingredient costs?
- Does its use reduce production costs or increase production efficiencies?
- Does it provide the promised advantages the supplier indicated? Is there a satisfactory quality vs. cost ratio that justifies using the new ingredient?

- Are the promised attributes of the ingredient's use valued by or obvious to the consumer? Will the use permit the manufacture of more environmentally friendly products with a clean, green label?
- Does this ingredient permit the development of products previously impossible to manufacture?

The client requires affirmative answers to some of these questions if the supplier's ingredient is to be successful.

11.1.3.2.4 *The Ultimate Criterion: The Client's Test Market*

There is no formal test market for ingredients. Neither ingredient suppliers nor their clients select a geographic area representative of targeted consumer product manufacturers and proceed to launch new ingredients supported by advertisings and promotions. They are entirely at the mercy of their customers' success using their ingredient.

What is more likely to happen is that ingredient manufacturers, after extensive business and customer research, will target potential high-volume users of their newly developed product. They will conduct carefully rehearsed and well-researched individual presentations with demonstrations to show each candidate the values that will come from using their new ingredient.

New ingredient launches are usually heralded by announcements in trade magazines and technical journals or demonstrated in the carnival atmosphere of food ingredient trade fairs. A common routine for any ingredient suppliers at a trade show is to hand out free samples of the ingredient or demonstrate it in some prepared food handout in which the new ingredient has been used. Admittedly, this is primarily a gesture of goodwill, but it is frequently a hazardous thing to do: products are presented under less than ideal conditions and are not examples of "best foot forward" perceptions of what an ingredient can do.

11.1.4 Looking to the Future for Developments in Food Ingredients

Assessing the future is always fraught with difficulties (see Chapter 12). The future, as described by Best and O'Donnell (1992), Dahm et al. (1999), and McKenna (1999), has changed somewhat. All described and reviewed a number of future new ingredients and their applications. Many are products of biotechnology—new ingredients and biotechnology appeared to go together:

- Bacteriocins of microbial origin
- Transgenically altered milk production to produce natural preservatives, pharmaceuticals, altered milk-fat fatty acid profile or to remove a milk allergen producing off-flavors in UHT milk
- Genetic manipulation of poultry to improve texture and water-binding properties of the meat

As usual, when prophecies are made, the prophets hedge their predictions with several cautions that are just as apt today as they were nearly 20 or even 10 years ago. Their first caution: what will governments do? It is a certainty there will be intervention by governments concerned about the safety of and usage levels of ingredients that are GMO derived, and especially from plants with enhanced phytochemical concentrations. Certainly, there is disagreement among nations regarding the need to label transgenetically derived organisms or ingredients in foods.

The second caution concerns the economic impact such novel techniques may have upon some commodity industries. The dairy industry, for example, already faces two conflicting government interventionist policies. On the one hand, in many countries including the United States, the industry is encouraged to produce milk with a high milk fat content, while there is an obesity epidemic and government health policies exhort the public to consume less fat; children and teenagers are the prime consumers of milk and milk products.

Finally, the desire for new ingredients with functional properties, that is, with disease prevention properties, puts enormous pressures on ingredient manufacturers to develop products with these enhanced properties or separate, purify, and concentrate these beneficial products for the consumer product manufacturers who will want them and the multitude of customers who want to self-medicate. This will be an expensive gamble for ingredient suppliers to develop products with as yet, only anecdotal or less than authenticated medicinal value.

Consumers' desire for naturalness in products, that is, for products manufactured using natural ingredients or for minimally processed products, provides impetus to ingredient suppliers to explore the use of ingredients derived from a variety of natural sources and, concomitantly, the use of novel, minimally processed (i.e., modified) derivatives from these natural ingredients. Manufacturers of consumer products want to present to customers and consumers a label with an ingredient list that appears less chemically and synthetically derived and more natural, that is, a green label (Fuller, 1993).

Natural ingredients are currently "in," and a scramble is on to explore the world's biota for natural ingredients with preservative properties. The legal question remains whether those ingredients from plants biotechnologically enhanced to contain something they would not naturally contain or to contain that something in biotechnologically enhanced concentrations will be considered "natural"—hence, "in"—by customers and consumers or are even safe. With the consumers' fears of big science so prevalent, a massive educational program will be required to allay these anxieties. Ingredient manufacturers must use caution.

This intense interest in naturalness in colors, preservatives, and antioxidants, indeed anything natural that can replace "chemicals," is attested to by the abundance of research papers and reviews of literature in these areas: natural colors (Engel, 1979; Francis, 1981; Gabriel, 1989; Shi et al., 1992a,b);

antimicrobial agents, (Barnby-smith, 1992; Baxter et al., 1983; Beuchat and Daeschel, 1989; Golden, 1989; Shelef et al., 1980; Zaika and Kissinger, 1981; Zaika et al., 1983); and antioxidants (Kläui, 1973; Pokorný, 1991). The pharmaceutical and ingredient industries believe as Dryden stated at the start of this chapter, there is good in everything. The need for natural ingredients with functional properties such as preservation, thickening, emulsifying, coloring, flavoring, taste modifying, etc., that have nutritional and pharmaceutical properties has driven and will continue to drive future new ingredient development. Today, blueberries, grape extract, and açai berries have become popular because of their antioxidant properties, and vitamin D, to read various articles in the press, has become a panacea for many health issues.

A major caution to the use of all new ingredients in general, and ingredients of biotechnological origins in particular, will be their safety and acceptance by customers and consumers and their legalization by governments. A major requirement for acceptance of any ingredients for a food is that they must serve a useful purpose; they must create something desirable, whether that something is good color, flavor, texture, or health-promoting properties in the product for the buying public.

11.1.5 Meeting the Challenge: New Ingredients

As stated earlier, no clear distinction will be made between ingredients and additives (for a discussion of additives, see Fuller, 2003). Food legislation and regulations respecting the definition of and the use of additives and ingredients vary from country to country: acceptance in one country does not guarantee that the ingredient will be accepted in another. The mention of ingredients here does not imply they are permitted but merely illustrates their potential applications as ingredients. Even when the ingredient is permitted, there are restrictions regarding usage levels and types of foods in which the ingredient may be used. Food products that have standards of identity—in the United States, more than 300 foods have standards of identity—the addition or substitution of ingredients may be restricted.

11.1.5.1 *Marketing's Impact on the Direction of Research and Development*

Marketing personnel, especially those associated with technical sales, influence, based on their contact with clients and at trade fairs plus whatever market research has been done, the direction of research and the kinds of ingredients that are developed. If marketing personnel's research sees a need to pursue a program of development of ingredients with a health promotional benefit, for example, "lite" or low-calorie, or low- or no-fat, or high-fiber, or with high antioxidant content, then ingredient developers have direction; they research sources of ingredients that provide the promotional characteristic wanted. Where the relationship is a client/supplier one, obviously the direction is dictated by the client's needs and not necessarily the ingredient manufacturer. Nevertheless, marketing's research either directly or indirectly points the direction toward targeted clients and hence for development.

Export markets are always a profitable goal, and ingredient suppliers must explore the requirements of permissibility, of usage levels, and of labeling laws as well as advertising and promotional regulations in the importing countries. Even culturally close countries have very different food laws and regulations (e.g., between Canada and the United States). Consumers' interest in green labels (their desire for natural ingredients) and marketing personnel's obvious desire to please consumers will limit the ingredients that can be chosen in formulations. Development teams need to consider early and carefully all the implications that the use of any particular ingredients may have for labeling and for promotional claims for products for environmentally concerned consumers and for export markets that they may consider.

11.1.5.1.1 *Fat Ingredients*

Obesity has reached epidemic proportions in many countries, and governments of these countries have recommended nutritional guidelines or enacted legislation to advise (Anon., 2009b; Karlin, 2009) or penalize their citizens (Rawls, 2008) to reduce the amount of calories consumed. This usually targets those derived from fat. Thus, today's consumers are spurred on to the desirability for reduced calorie foods with their government's endorsation, with their own awareness that obesity is related to many disease conditions, and with the strong suggestion reported in the media that high fat diets and breast cancer as well as many other disease conditions may be directly associated with obesity (Falagas and Kompoti, 2006; Larsson and Wolk, 2008). Because weight for weight, fat contributes more than twice as many calories as protein or carbohydrate, it is the easiest food component to target. The only techniques to eliminate or lower the fat content of food are as follows:

- Reduction or removal of the fat by trimming, skimming, pressing, rendering, or solvent extraction.
- Reduction or replacement of the fat in the food by substituting, wholly or in part, with some less calorically dense material (fat extender). Fat extenders still provide calories, but weight for weight they provide fewer calories than the fat they replace.
- Reduction or replacement of the fat with some nonabsorbable substance (fat mimetic or replacement) with fat-like properties.

The first solution is not a path to tasty food as fat provides many desirable organoleptic properties to food nor is it one spurring ingredient technology. The remaining two solutions require a substance, a fat-extender or mimetic, that provides all the sensory properties of fat and is nontoxic. It is this area in which developers of new food ingredients have concentrated their efforts.

Fat replacements or mimetics have been derived from a number of products. They can be very roughly grouped into three classifications:

- Replacements based on physically or chemically modifying a polysaccharide fraction such as rice bran, pea fiber, oat fiber, corn starch, or tapioca dextrin.

- Replacements based on altering the physical character of a natural protein, e.g., Simplesse™, a trademark of The NutraSweet Company (Bertin, 1991). It is a suspension of protein microparticles, which has the mouthfeel characteristics of an oil.

- Replacements synthesized from natural foods such as sucrose or other sugars and esterified with fatty acids; these have fat-like properties. Silicones and long-chain members of alkane hydrocarbons with characteristics resembling fats have been used as fat substitutes (LaBarge, 1988) and can be included in this category.

Synthetic or engineered fats (Singhal et al., 1991) are somewhat different alternatives; they are derived from modifications to the basic skeleton of a fat molecule itself by

- Replacement of the glycerol backbone with other polyhydroxy compounds (sucrose, raffinose, and amylose) esterified with natural fatty acids. Propylene oxide can serve as the backbone. Sucrose polyester (Toma et al., 1988) is nonabsorbable and also able to reduce the cholesterol level in the body. Other replacements provide fewer calories per gram than do fats and qualify more as extenders.

- Substitution of acids with branched carboxylic acids, dicarboxylic acids, or shorter-chained fatty acids for some or all of the long-chain fatty acids of the triglyceride structure. Medium-chain fatty acid triglycerides are specialty fats prepared by the hydrolysis of vegetable oils, the subsequent fractionation of the fatty acids, and re-esterification of these fatty acids to a glycerol backbone (Megremis, 1991).

- Substitution of a polycarboxylic acid for the glycerol backbone and then esterifying suitable long-chain alcohols to the acid (referred to as reversing the ester linkage). Common acids used for the backbone are citric acid or tricarballylic acid, the tricarboxy acid of glycerol. LaBarge (1988) discusses these and feeding trials using them.

- Reducing the ester linkages of the triglyceride to ether linkages, and thus changing the properties of the fat. LaBarge (1988) describes these ingredients as much more slowly susceptible to hydrolysis than the esters but more prone to oxidation.

(Mention of these ingredients here should not be interpreted as meaning that all, or any, countries have sanctioned the use of these products. Developers must verify whether the ingredients they use are permitted in the class of foods they want to use them in and in the amounts they wish to use.)

Medium-chain triglycerides have limited value as fat replacements where caloric reduction is the major requirement. However, they do serve as processing aids as flavor carriers and in confections where their low viscosity is used to prevent sticking and provide gloss (Megremis, 1991). Medium-chain triglycerides are absorbed and metabolized in the liver as rapidly as is glucose (LaBarge, 1988), so their usefulness is in special diets providing a rapid and concentrated source of energy to people with intestinal malabsorption problems (Crohn's disease, colitis), a potential niche market (Babayan and Rosenau, 1991; Kennedy, 1991). Babayan and Rosenau (1991) also describe the use of a medium-chain triglyceride oil (chiefly caprylic and capric acids for the triglycerides) in Cheddar- and Fontinella-type cheeses.

A major problem exists with fat mimetic and replacement fats as some long-chain fatty acid triglycerides are essential fatty acid and needed requirements for the body. Therefore, lipids interesterified from medium-chain triglycerides and long-chain triglycerides (structured lipids) provide unique benefits summarized, in part by Kennedy (1991), as

- Improving immune function
- Lessening cancer risk
- Lowering cholesterol

Jojoba oil, a natural oil from the seeds of a hardy bush of the southwestern United States (Anon., 1975), has also been explored as a possible substitute for fat (Hamm, 1984). Its oil is a liquid wax comprising esters of fatty acids and alcohols. Its use in refrigerated foods is limited as it solidifies at refrigerator temperatures. It is used in such alternate medicines as aromatherapy and in cosmetics.

Gums aid in fat-reduced foods by stabilizing emulsions and suspensions and providing viscosity, thus simulating some properties of oils and permitting the reduction of oil in the formulation. Dziezak (1991) describes the properties of gums and reviews their applications in emulsification.

The fat replacers and extenders are not suitable in many food systems, for example, the alkoxy citrates are not thermally stable (Singhal et al., 1991); some interfere with the absorption of other macro- and micronutrients (LaBarge, 1988); and some may result in anal leakage when consumed (LaBarge, 1988). The entire food matrix in which the fat replacer or extender is to be used must be understood before the ingredient supplier can advise the client. All this on top of knowing whether the ingredient is permitted in the matrix.

Dziezak (1989) provides a very readable capsule review of some of the properties of natural fats and oils for technologists unfamiliar with fat chemistry.

Her accompanying treatment on fat substitutes is now dated. However, as she pointed out two decades ago, this area of ingredient development is still undergoing tremendous change due to the rampant obesity epidemic.

Product developers rapidly reformulated products containing saturated tropical oils to remove these ingredients from products. This occurred when the prevailing wisdom declared that these saturated vegetable fats behaved as saturated animal fats in the body. Berger (1989) describes concisely the situation concerning the use of tropical oils in the United States, presents a cogent argument for their continued use, and disputes attacks against their bad nutritional value. Vegetable fats and oils contain plant sterols that play a role in improving plasma lipid profiles in men (St.-Onge et al., 2003, but see Patel and Thompson, 2006 and Weingärtner et al., 2009, for side effects) and in benign prostatic hyplasia (Ishani et al., 1999). The last cited references point to the problem with ingredient developers regarding nutraceuticals, in this instance phytosterols, that is, they may not be all they seem, and products based on them may prove unreliable or give false hope to users.

11.1.5.1.2 *Sugars, Sweeteners, and Other Carbohydrate Ingredients*

Sherman (1916) in his classic *Food Products* (1916 printing) describes only cane sugar, beet sugar, molasses, refiner's sirup, maple sirup, open kettle cane sirup, and honey and their manufacture. Wiley (1917) included sorghum sirup. In the past, these were the sugars in common use; sucrose from sugar cane or sugar beets was, by far, the main sugar in food processing.

There are many reasons for the popularity of sucrose in world economics and manufacturing (see, e.g., Mintz, 1985). It is readily available and is, therefore, cheap, and in addition to sweetening foods, it has many functional properties in foods. Unfortunately, raw materials from which sugar is derived are used to produce ethanol used as an additive to gasoline; this has raised food prices. Sucrose is cariogenic, and with consumers' concerns about obesity and related health problems associated with high caloric intake from refined sugars typically found in North American diets, it, like fats, has become a target for substitutes. This awareness has led to the search for substitutes.

Alternative sweeteners to reduce or eliminate the need for sucrose in food products fall into two categories:

- Those that are intense sweeteners in their own right but contribute no or very few calories to the food. These replace a sugar *where sweetness is the only function demanded of the sugar.*

- Sweeteners that are caloric alternatives, that is, on a weight basis, they are more intense sweeteners than is sucrose. They also contribute other properties, including calories, associated with sugars. These are frequently referred to as the bulk sweeteners (Giese, 1993).

A partial (this group is growing as new sweeteners are being found) listing of the low (or no-) calorie group includes the following:

Cyclamates	Saccharin
Sucralose	Aspartame and other peptides (e.g., neotame)
Glycyrrhizin	Miraculin
Monellin	Thaumatins
Stevioside	Rebaudiosides
Neohesperidin	Dihydrochalcones
Acesulfame K	L-sugars

Caution must be taken in choosing intense sweeteners since some alternative sweeteners are unstable in certain food systems; some produce a slow onset of sweet sensation that peaks late in the mouth. Others have bitter or undesirable aftertastes. There is also the mistaken belief that since many of those listed are natural, that is, from plants found in nature, they are safe or without side effects. Any sweetener chosen must be safe, without unpleasant side effects or aftertastes, and complement the food system where it will be employed.

The alternative group, by far the more important for developers, comprises crystalline fructose, high fructose corn syrup, mannitol, sorbitol, xylitol, hydrogenated sugars (e.g., maltitol), hydrogenated starch hydrolysates, and isomalt. Members of this group possess, in addition to sweetening, many of the functional properties of sugar, including the following:

- They provide mouthfeel (viscosity) in some beverages and control texture in baked goods.
- They regulate a food's water activity and hence the food's stability (in jams and jellies).
- They serve as bulking agents in hard candies and other confections and control graining in soft-centered candies; they assist along with pectins in the control of gel texture for jams, jellies, and marmalades.
- They stabilize color in some food systems.

The uniqueness of sugars, in particular sucrose, is best exemplified by a simple demonstration described by Chinachoti (1993). In what she called her "magic trick," she mixed corn starch and water to produce a wet, unpourable powder. Next, she added another dry powder, ground crystalline sucrose. Mixing produced not an even drier mixture but eventually a flowable slurry.

The corn starch/water mixture despite its dry appearance has a high a_w (approximately 0.98). With the addition of the sucrose, the dry appearance disappears after mixing. The seeming solid becomes liquid with an a_w that has dropped to 0.93. As magical as this may appear, Chinachoti (1993) went a step further. One would expect the corn starch/water mixture to spoil faster microbiologically than the corn starch/water/sucrose mixture if one's prediction is based solely on the two water activities. Chinachoti demonstrated that this was not so. The mixture with the lower a_w spoiled more rapidly. The addition of sucrose had indeed lowered the a_w, but she demonstrated it had also increased the amount of mobile water. As Chinachoti explained, "Sucrose undergoes a phase change from crystalline to dissolved sucrose upon hydration....solvation of sucrose promotes mobility and helps facilitate the spoilage."

The lesson for product developers and spoilage modelers is this: sucrose and other crystalloids lower water activity, but stability and predictions of stability should not be based on water activity. Other factors such as water mobility and phase transitions play a more important role than does water activity alone in product stability.

The importance of phase transitions and the physical state of food components, with some references to carbohydrates, are treated in depth by Slade and Levine (1991), Roos and Karel (1991a,b), Noel et al. (1990) and Best (1992). Noel et al. (1990) and MacDonald and Lanier (1991) discuss the importance of phase transitions and the glassy state in the storage of frozen foods, freeze dehydration, and as a cryoprotectant for foods. If the aqueous system of a food can be formulated to maintain the matrix in the glassy state, then water crystallization and damage caused by crystal growth are minimized. The drying of pasta products is greatly improved by close control of humidity, temperature, and drying rate to control glass transition; this maintains pasta in the glassy state.

The quality of extrusion-puffed snack products owes much to control of phase transitions. Roos and Karel (1991a,b) suggest that glass transition temperatures may be a factor in browning. Grittiness of lactose in ice cream, caking in milk powders, and stickiness of hard candies are all related to the behavior of ingredients in the glassy state (Noel et al., 1990).

11.1.5.1.3 Fiber Ingredients

Dietary fiber is an all-encompassing term for a group of poorly defined, natural plant components including polymeric carbohydrates, cellulose, hemicelluloses, pectins, gums, and lignin, a noncarbohydrate component. None of these are themselves compounds of fixed composition; hence, fiber refers to an ill-defined mixture of plant materials. Historically, fiber has been referred to as roughage or bran regardless of source. This diversity of characteristics is partially overcome by purification through a variety of extraction procedures, enzymatically "chopping" the polymers to desired chain lengths and blending to produce fibers of specific, uniform properties.

Cellulose is a linear polymer of varying lengths of glucose units. Hemicelluloses are a complex mixture of linear and branched polysaccharides with side chains composed (most often) of other hexose and pentose sugars. Pectins are equally complex mixtures of polygalacturonic acid units as are many of the plant gums and mucilages. The seed gum of mesquite is a polymeric galactomannan (Figueiredo, 1990). Lignin is a polymer of units of phenyl-propane.

Dietary fibers have been derived from many plant sources, often as by-products of processing the raw plant: prunes, corn, oranges (Hannigan, 1982), soy beans, peas, oats, rice, spent barley, sugar beets, tofu processing, various vegetable gums such as gum arabic, guar gum, locust bean gum, and mesquite (Figueiredo, 1990), wood, potato peelings, carrageenan, and psyllium grain husks (Anon., 1990a), and the list grows longer each year. Fibers come from marine and microbial sources; chitin from the shells of crabs and chitosan prepared from chitin are biopolymers with promise as non-plant sources of fiber (Knorr, 1991).

Each source produces fibers with unique properties and when used in combinations, many interesting gelling, thickening, or texturizing applications in food systems are possible. A particular fiber can frequently be used synergistically with other fibers or with starches. Arum root from which konnyaku is made, a vegetable jelly for many centuries a mainstay of Japanese cuisine, is the source of konjac flour, a glucomannan (Downer, 1986). Tye (1991) reviews the synergism of konjac flour with kappa carrageenan and starches in maintaining the structural integrity of shaped foods during thermal processing and to simulate the textural properties of fat and connective tissue in sausage-like products.

The use of any particular source of fiber or combination of fibers presents technical problems whether the fiber is added for health benefits or to serve some functional purpose in food. Fiber will affect color, flavor, oil and water retention (hence the a_w of foods), rheological properties such as colloidal/emulsion stability, texture, gel forming properties, and other viscometric properties such as thickening and mouth feel of products (Penny, 1992). They are used to control crystal formation in sugars or to act as cryoprotectants in freeze/thaw food systems. The versatility of one fiber, cellulose powder, is demonstrated by the many products in which it is used (Ang and Miller, 1991):

Bread	Rusks	Cakes
Cookies	Pasta	Cheese
Soups and sauces	Yellow fat spreads	
Comminuted meats	Meat and fish pastes	
Slimming foods	Dietetic products.	

Cellulose powder has been used in stabilizing frozen surimi analogues. Ang and Miller (1991) and Penny (1992) describe the many applications of fiber in foods.

Heightened awareness of the importance of fiber in the diet dates back to the mid-1980s when the National Cancer Institute (United States) promoted the potential cancer prevention benefits of a high fiber diet (NCI, 1984). This campaign alone gave a boost of an estimated 30% in sales to one well-known ready-to-eat bran breakfast cereal (Anon., 1988c). However, the beneficial use of high fiber diets in aiding bowel regularity has been promoted for many years, and its laxative effects have been recognized for many more years. Background information into fiber as an ingredient in diet is found in the following articles: Rusoff (1984) and Schneeman (1987) on physiological responses of soluble versus insoluble fibers; Wood (1991) on the physicochemical properties of oat beta-glucan; Ripsin and Keenan (1992) on oat products and their value in reducing blood cholesterol levels.

Claims that certain sources of fiber such as oat, rice (Normand et al., 1987), pectin (Reiser, 1987), and guar can clear cholesterol from the circulatory system suggest a role for fiber in lowering the risk of cardiovascular disease. Furthermore, fiber may play an as-yet-undetermined role in the prevention of some cancers, particularly colorectal cancer. However, Fuchs et al. (1999) found no evidence of any preventive effect of dietary fiber against adenomas or colorectal cancer in several thousands of women over a 16 year period.

Goodlad (2001) raises the possibility that the value of fiber in the diet rests not with the fiber itself but in the consumption of an unrefined diet (one that is high in fiber and therefore high in other congeners such as antioxidants, vitamins, and other phytochemicals). Naturally, high-fiber diets have a low caloric density and are low in fat, therein may be the advantage. Regardless of the doubt respecting the efficacy of extracted fiber, demand for products with high fiber content has grown as food companies try to satisfy this increasing awareness by consumers that food, and in particular certain constituents in food, and health go together. Fiber is now advertised in a wide variety of foods from breads, muffins, and cookies to pastas, beverages, and processed meat products for its purported health benefits.

The function that fiber is to serve in a food system determines which fiber with the desirable characteristics to select. The two most important characteristics are particle size and shape for texture considerations and soluble-to-insoluble fiber content ratio. (See Olson et al. (1987) for a discussion on analytical procedures for soluble dietary fiber and information on total dietary fiber and insoluble dietary fiber.) Soluble fiber has excellent water-binding properties and as such will have a pronounced influence on the texture (moistness) of baked goods in particular, on the stability of products, and on viscosity in beverages. In general, soluble fiber is highest in vegetable gums and lowest in cellulosic fractions of plant materials. From the natural sources available, high to low ratios of soluble to insoluble fiber may be obtained to suit whatever purposes developers have in mind.

Crude fiber should not be equated with dietary fiber; crude fiber consists largely of cellulose and lignin remaining after the treatment of a food material with sulfuric acid, sodium hydroxide, water, alcohol, and ether.

11.1.5.1.4 *Proteins as Ingredients*

Proteins as ingredients have suffered from a lack of attention as most interest has centered on fats because of their high caloric density and carbohydrates because of their contribution to sweetness and other functional properties in foods. Substitutes for both fats and sugars lead to low calorie products.

Vegetable proteins, textured to simulate the fibrous structure of meat, have been available for several years. They were popular as meat extenders when meat prices soared during a scarcity of beef in the United States during the latter part of the 1970s. When beef became available and prices fell, many of these ersatz products failed, partly because these products did not satisfy consumers as did the real thing. Consumer demand was transitory. Quality of textured proteins has improved considerably, and these are now widely used in dried soup mixes and vegetarian dishes simulating meat dishes.

How strong is the growth in the vegetarian food section is a moot point; suffice it to say, it is there in the marketplace and is gaining prominence. Textured vegetable proteins have had a reasonable success with main dish items for vegetarians. A mycoprotein, developed by Rank Hovis McDougall, has been successfully texturized and used by the Sainsbury company (J. Sainsbury plc, United Kingdom) in a series of meat-flavored products. In test market locations, these have been well received (Godfrey, 1988). These products have a dual advantage over meat, which marketing personnel can promote: they contain no animal fat and they contain fiber. The mycoprotein has been tested by Sainsbury for vegetarian (i.e., non-look-alike meat) dishes.

Surimi, a textured fish protein, is fresh water-leached fish muscle. It is hardly a new development—indeed, it is a very old technology. Lee (1984) has described its processing in detail and its application in some products. Lanier (1986) discusses the functional properties of surimi, particularly its excellent gelling properties. Surimi has been used successfully with crab meat, shrimp meat, and smoked salmon in fabricating substitutes for, respectively, crab legs, butterfly shrimp, and lox. Sausages made with surimi, very popular in Japan, have not been successfully received in North America.

Soy beans have become a popular source of ingredients with a wide range of applications in ethnic cuisine. Their popularity is supported not only by the taste sensation of umami that they contribute in the various forms (Marcus, 2009) but also by their value for the cancer preventive properties it is claimed that they have. Soy milk has been used with various flavors as a non-carbonated drink. Tofu has been used as an ingredient in many vegetarian dishes. Japanese miso and Chinese chiang are both fermented soy bean curd products used as flavoring. Tempeh, mold-fermented soy beans of Indonesian origin, has been used as a meat substitute. With these ingredients, developers can create a wide range of added value products with unique textures and forms, especially ethnic dishes, and in the growing vegetarian niche.

11.2 Ingredients and the New Nutrition

Food was good for one's health; certainly, lack of food led to starvation. This had been known for millennia. Then, it became apparent that some foods were better for one than others. Then, it was discovered that food supplied things the body could not make, and the lack of these things in one's diet resulted in deficiency diseases. Then came the dawning that components in foods could ameliorate or prevent certain disease conditions in the body, but they themselves were not nutrients; fiber may have been the first of these discoveries. Some other examples follow:

- Kritchevsky (1991) reviews the effect of garlic on cardiovascular disease and discusses its hypolipidemic, hypotensive, anticoagulant, and fibrinolytic properties.

- Mills et al. (1992) discuss work on the biochemical interactions of peptides derived from the breakdown of foods and the possible influence these may have on food intolerances—work of immense value to consumers suffering food allergies. Two possible spin-offs from this work are the screening of foods for precursors causing intolerance and the possibility for developers to remove such precursors from new foods.

- Bioactive peptides derived from milk proteins, for example, opioid peptides, immunopeptides, and mineral-binding peptides, were discussed by Meisel and Schlimme (1990). These authors speculate on the role of these derived dietary products as food hormones and "natural" drugs.

- Bifidobacteria have received prominence, and bifidus milk is promoted as a treatment for various intestinal ailments. O'Sullivan et al. (1992) and Hughes and Hoover (1991) review the subject of probiotic bacteria and describe products already on the market shelves in various countries.

- *Aphrodisiacs* is the provocative title of an article by O'Donnell (1992) that reports on the ability of certain foods to affect human reactions; that is, they can act on sensory perception (cf., alcohol or hallucinogenic mushrooms); these "mood foods" can irritate or stimulate and soothe (Hozawa et al., 2009).

- The antioxidant hypothesis of cardiovascular disease (oxidized cholesterol is the risk factor and not cholesterol) was reviewed by Duthie (1991). In this, insufficient antioxidant intake to prevent the oxidation of cholesterol and associated free-radical activity are the problem. This clearly suggests a tie-in of vitamins E and C, beta-carotene, and selenium, all food nutrients. The antioxidants in green tea have been noted as reducing plaque in Alzheimer transgenic mice (Rezai-Zadeh, 2005).

- The February 8, 2008 IFT *Newsletter* reported beetroot juices reduce high blood sugar; the March 19, 2008 *Newsletter* had a note that grape skin resveratrol protects against diabetes and obesity and aids metabolism. Every two or three *Newsletter*s finds another beneficial component of some food.

As responsible data accumulate, the justifiable evidence for these health benefits will mount, and new food product development will be influenced as people will demand these products. Governments will be forced to provide the legislation and inspection services to protect consumers with respect to the safety of these products and the truth of the claims made for them. Certainly, there should be a large demand for the "deep-sea protein" food supplement made by a European company and claimed to banish wrinkles (Cremers, 1993).

11.2.1 Opportunities Provided by the New Nutrition

Nutrition as a science, as a public policy to fight obesity, and as a lifestyle to be undertaken by the public is in a state of flux with the public not understanding nutritionally labeled foods, not understanding government's food pyramids, and bemused by what appears as "good-for-you" foods only to be contraindicated shortly thereafter (cf., the disastrous history of contradictions that has plagued coffee and alcohol—both are presently "good for you"). Initially, health concern had been with diseases caused by the lack of certain nutrients, and great efforts had been made to optimize nutrient intake (fortifying foods or calculating required daily allowances) to prevent deficiency diseases. Then, there came a move to remove things from foods or advise against eating too much of certain foods, for example, reducing fat, salt, or refined sugar from foods to produce foods reduced in these unhealthy items. Today, the major thrust of ingredient development is to develop new ingredients from foods (phytochemicals, functional foods, nutraceuticals, etc.) to prevent specific health problems, to improve effects of stressful situations, to improve physical prowess, and even to improve or recover failing cognitive and memory abilities. The nutritional sciences have become interested with how diet and specific food components operate at the physiological and genetic levels (genomics) to influence health and disease (Young, 1996). The age of food as medicine is upon us, and consumers are interested in self-medication.

11.2.1.1 Biologically Active Nonnutrients

Health professionals are still learning (and unlearning) that some foods contain biologically active nonnutrients that have an as-yet poorly understood connection at a physiological and genetic level to mechanisms of disease prevention. This understanding started through research of early herbals and through an accumulation of anecdotal tales and scientifically grounded evidence from controlled feeding trials. The published data, whether from information in medicinal lore found in herbal books or research data in technical literature, have provoked food manufacturers to develop new foods exploiting, sometimes wisely and sometimes with too much hyperbole, this disease prevention connection.

11.2.1.1.1 Functional Foods, a.k.a. "Nutraceuticals"

Throughout this text, these nonnutrients will be referred to as nutraceuticals or where appropriate, phytochemicals (pharmafoods and medifoods are terms devised by the media that have also been used). It has been suggested that the term *functional food* be used only for foods containing prebiotics or probiotics that provide a health benefit; the International Food Information Council defines functional foods as "foods that provide health benefits beyond basic nutrition" (Deis, 2003). *Nutraceuticals* is the popular term found in newspapers, in food trade magazines, and is used extensively in *Food Technology*; the term "nutraceutical" has been used by many scientists themselves; it is descriptive of the value that a nutraceutical is purported to have in nutrition without the need of any quantifiers or qualifiers. More on the definition of functional foods and problems respecting procedural rules for food producers can be had in Culhane (1999), Eagle (1999), and Deis (2003). This text uses the most descriptive term, *nutraceuticals*; functional foods serve a function in foods as colorants, sweeteners, emulsifiers, enzymes, stabilizers, preservatives, and so on.

11.2.1.1.2 Prebiotics and Probiotics

Prebiotics and *probiotics* are terms used to characterize nutraceuticals. Both are nonnutritive elements in foods that provide health benefits to consumers of these products, but here the similarity ends. Prebiotics are an uncharacterizable grouping of chemical entities found in a wide variety of foods; those of plant origin are better known as phytochemicals. Readers should note that some prefer that this term be applied to those nonnutrients present in plants in very small amounts, thereby excluding soluble and insoluble fiber.

Probiotics (Table 11.1) are active microbial cultures, or foods containing these, for example, yogurts and krauts that act on the intestinal microflora in a beneficial manner (Salminen et al., 1999). Microorganisms suspected of involvement are described by Hoover (1993), Lee and Salminen (1995), and Knorr (1998).

Prebiotics run the gamut of chemical structures, which can be guessed at from the nomenclature in Table 11.1. The curative properties of plant parts

TABLE 11.1

Some Biologically Active Nonnutrient Factors Determined to Have or Believed to Have Beneficial Effect against Some Disease Condition When Consumed

Classification	Category and Food Sources
Probiotics	Bifidobacteria
	Fermented milks; yogurt
	Lactobacillus species
	Fermented milks (acidophilus milk); yogurts
	Streptococcus species
	Fermented milks; yogurts
Prebiotics (phytochemicals)	Fatty acids
	α-linoleic acid (canola and flaxseed oils)
	Conjugated linoleic acid (safflower, sunflower, and soybean oils)
	γ-linoleic acid (evening primrose oil)
	ω-3 fatty acids (various unsaturated oils, e.g., salmon and tuna oils)
	Lecithins
	Phospholipids (various oils especially soybean oil); phosphatidyl serine, phosphatidyl choline
	Unsaponifiables of oils
	Phytosterols (canola and soybean oils)
	γ-oryzanol and ferulic acid (rice bran oil)
	Organosulfur compounds, especially plants of the cruciferous family (broccoli, cabbage, and cauliflower) and Allium (garlic, onion, and leek) family
	Isothiocyanates (mustard oils of cruciferous vegetables)
	Sulfides (e.g., diallyl disulfide) and oxides (allicin) from garlic and onions
	Terpenes
	Monoterpenes: limonene, perillyl alcohol
	Tetraterpenes: lycopene, β-carotene
	Polyphenols (including flavonoids and catechins)
	Anthocyanins (blueberries, cranberries, tomatoes, red wine, tea, onions, kale)
	Various other phenolic compounds
	Nonsteroidal phytoestrogens (isoflavones)
	Genistein, daidzein, biochanin, and formononetin (soy bean, whole grains, berries, flaxseed, licorice, red clover)
	Fiber (and associated material)
	β-glucans (oats, barley, wheat, rice)
	Lignans (flax)
	Other soluble and insoluble fibers (fructooligosaccharides, e.g., inulin in Jerusalem artichoke)
	Saponins (derivatives of pentacyclic triterpenes)
	Ginseng, soybeans, grains
	Herbal components
	See, for example, Tyler (1993); Anon. (1998d)

Source: Adapted from Fuller, G.W., *Food, Consumers, and the Food Industry: Catastrophe or Opportunity?*, CRC Press, Boca Raton, FL, 2001.

have been known for centuries. There are books on biologically active plants ranging from ancient texts to present day self-medication books:

- John Gerard's *Historie of Plants*, 1597 (*Gerard's Herbal*, edited by Woodward, M., Senate, Twickenham, United Kingdom, 1998.)
- N. Culpeper, *Culpeper's Complete Herbal, and English Physician*, J. Gleave and Son, Deansgate, Manchester, 1826 (reproduced from an original edition 1981, Harvey Sales©)
- A. Boxer and P. Back, *The Herb Book*, Octopus Books, Ltd., London, 1983
- V. E. Tyler, *The Honest Herbal: A Sensible Guide to the Use of Herbs and Related Remedies*, Pharmaceutical Products Press, Haworth Press, Inc., New York, 3rd edn., 1993
- L. Bremness, *Herbs*, Stoddart Publishing Company, Toronto, 1996

and many articles on biologically active nonnutrients: Caragay (1992), Ramarathnam and Osawa (1996), Kardinaal et al. (1997), Garcia (1998), Katz (1998), Ohshima (1998), Zind (1998, 1999), Anon. (1998), and Sanders (1999); a series of columns in *Food Technology* by L. M. Ohr, particularly Ohr (2003) on fats, Clifford (2003) on dietary phenols, and Ray and Sivakumar (2009) on fermented tropical root and tuber crops. Even the much maligned caffeine has a protective effect against irradiation, at least for mice (George et al., 1999).

Archeological and biblical records have shown that plants and their various parts have long been used both to flavor foods and to cure various ailments. Each culture has its store of traditional medicines based on extracts or brews of different parts of plants. The herbal medicinal practices of native cultures, folk literature and oral traditions describing the use of plant materials, old herbals, and ancient medical records have been treasure troves for pharmaceutical, chemical, and ingredient companies for phytochemicals to use as the basis for new product ideas (Wrick, 2003).

The "self-care shopper," as Hollingsworth (2003) calls these customers, has become an obvious target for consumer products containing nutraceutical ingredients. This target is one that should be aimed at with caution not only for safety concerns but also for presenting them with a confusing brand image: food or medicine?

11.2.1.1.3 *Prebiotics and Phytochemicals*

Prebiotics and phytochemicals are terms that are often kept distinct by some scientists; the terms differ by the observation that phytochemicals are found in smaller amounts in plants than are prebiotics (insoluble and soluble fibers, gums and mucilages, and fatty acids). Phytochemicals act differently in the body than prebiotics. In this text, no distinction will be made.

Prebiotics (Table 11.1) include the plant-derived vitamins and a variety of phytochemicals from flavors, colorants, antioxidants, noncaloric sweeteners, narcotics, texturizing agents (pectins, starches, and other polysaccharides), and so on—many of which have beneficial physiological effects in the human body (Zind, 1998). Food manufacturers are interested in those with both desirable physiological effects and functional properties in the foods in which they are used, such as

- Antioxidants: vitamins C and E have been used as such in foods; β-carotene, long used as a food colorant, also has antioxidant properties (Ahmad, 1998; Astorg, 1997; Elliott, 1999).
- γ-Oryzanol and tocotrienols: plant sterols found in rice bran oil (McCaskill and Zhang, 1999); tocotrienols are also good antioxidants.
- Lycopene and carotene: the carotenoids are found in a variety of plant sources, in particular tomatoes (Astorg, 1997; Nguyen and Schwartz, 1999).
- Isoflavones: these are a class of nonsteroidal phytoestrogens; genistein, daidzein, and glycitein are found in soy-based foods (tofu, soy milk, and miso) (Zhou and Lee, 1998); genistein, daidzein, biochanin, and formononetin have been extracted from red clover. All have beneficial physiological properties in human beings. Protein extracts containing these phytoesyrogens have been prepared from these sources.
- ω-3 fatty acids and the related ω-6 type found in the fats and oils of both plant and animal sources (Ahmad, 1998). Examples are as follows: palmitoleic, oleic, linoleic, and arachidonic acids.
- Lactulose is an interesting prebiotic in that it is synthetic derived commercially from the isomerization of lactose. It is gaining interest as a food ingredient, besides its usefulness in alleviating constipation and as a stimulant for intestinal flora (Olano and Corzo, 2009).

Drug, food ingredient, and consumer product-manufacturing companies see these products as potential ingredients to entice their clients with prepared products with physiologically beneficial products: Block (1985) on garlic and onions; Kinsella et al. (1993) on antioxidants in wines and plant foods; Gehm et al. (1997) on resveratrol in grapes and wine; Katz (1999) on European trend in nutraceutical products; and Petesch and Sumiyoshi (1999) also on benefits of garlic. Potato and corn chips, corn puffs, candies, and other snack foods have been "laced" with ginkgo biloba, St. John's Wort, kava kava and many other herbs, spices, and essences of flowers in the hopes of relieving tiredness, calming mood swings, improving memory, and combating depression. Green tea has been found to relieve depression in Japanese community-dwelling elderly (Niu et al., 2009), and Hozawa et al. (2009), in the Ohsaki Cohort study, found an inverse relationship between frequent drinking of green tea and Japanese respondents experiencing psychological distress, which probably goes far to explain its resurgence by ingredient suppliers as a flavorant in ice cream and other confectionery products. Cocoa powder flavonoids reduce the cardiovascular inflammation caused by atherosclerosis (Monagas et al., 2009).

The list of phytochemicals and references to them that food scientists have found with beneficial, nonnutritive properties goes on, and Dryden's words prefacing this chapter makes one believe there is, indeed, some benefit in everything, and ingredient developers keep looking for these. However, as is ever the situation, contradictory evidence is usually found. Fuchs et al. (1999) could not find evidence of any protective effect of dietary fiber, a prebiotic, against colorectal cancers or adenomas in a study of 88,757 women between 34 and 59 years of age over a 16-year period (see also Goodlad, 2001). Kava kava (in Canada, the government has recalled all products with kava kava but appears to look the other way regarding products still on the market) and St. John's Wort have both had suspicion thrown onto their safety, and even the efficacy of tea and ginkgo biloba has been questioned. So even with the nutraceuticals, there is confusion about their benefits in the diet. Developers are warned to proceed with due safety and legislative caution.

11.2.1.1.4 Prebiotics: Red Flags

The warning cannot be overemphasized; product developers in both the consumer products and ingredient industries should use caution in their development programs with nutraceuticals:

- The safety of all prebiotics, especially as isolated pure compounds, has not been established, particularly at the doses shown to be efficacious in vitro.
- There is still no reliable data on effective dose levels for all consumers, especially the very young and the elderly. Clifford (2003) presents challenging dietary data for polyphenols.
- Reliable information is lacking on interaction of prebiotics with other prescribed medication that consumers may be taking.

Methodologies for detecting and quantifying the presence of prebiotics in foods with accuracy are lacking. Two questions need to be answered: Where and in what form do the prebiotics in the body produce the beneficial effects they have? At present all that is known is that, for example, green tea calms people and that cocoa powders mediate atherosclerotic inflammation. Proper dosage levels and mechanisms of action in the body are not known nor are they clearly understood. Once this is accomplished, then isolated fractions containing the active components can be used as medicines; their presence in foods can be enhanced through conventional or unconventional breeding techniques, for example, carrots with more carotene (Zind, 1998); and concentrates of isolated prebiotics can be used as ingredients to make consumer products with physiologically beneficial properties.

With the availability of nutraceutically enriched foods and concentrated extracts of phytochemicals, self-medicating consumers run the risk of overmedicating to harmful levels, an occurrence that may provoke the introduction of strict legislation.

11.2.1.1.5 Probiotics in Health

Many cultures have used fermentation either for preservative action the fermentation conferred on the food or the ability of fermentation to convert a food into something more desirable (Fuller, 1994; Knorr, 1998). Only recently has it been learned that the microorganisms responsible for the fermentation also provided health benefits for consumers. Probiotics are added adventitiously with the fermented foods as the vehicle, or they are added directly to foods as live or dead cultures.

Typical food sources of probiotics are yogurt and cultured milk products (Speck et al., 1993), some fermented meat and cheese products, sauerkraut and other fermented vegetables, and selected kimchi that is considered to have both probiotic and prebiotic elements (Ryu and West, 2000). A major problem for developers of consumer products is not only to create new products but also how to popularize the enormous number of traditional fermented vegetable, meat, and dairy products available in other cultures, that is, to adapt these to new products suitable for a wider variety of food cultural traditions, particularly North American traditions.

The extensive development program in Japan starting in the 1950s is discussed by Ishibashi and Shimamura (1993). Problems in the design of novel products for both healthy and sick consumers are discussed by Brassart and Schiffrin (1997).

A growing demand by consumers for alternative medicines has led to research into medicinal folkloric herbal preparations that in turn has led to the production of extracts, tinctures, and tisanes of many herbs and plant parts. Today, these products are prominently displayed on pharmacy shelves. The popularity of alternative medicines has spurred the development of nonprescription medicines, food supplements, and ingredients as well as new foods.

There are now opportunities to develop food products directed to specific health problems (anticarcinogenic foods; foods to improve cognitive ability) or for use in particular stressful situations (space flight, athletics). There are already, for example, sports drinks designed to rehydrate the body rapidly, to provide energy, or to enhance energy metabolism during strenuous activity (Brouns and Kovacs, 1997).

Nearly 20 years ago, *Prepared Foods*™ (Anon., 1990b) reported that a significant percentage of research and development executives "expected their companies to increase efforts to reformulate ingredients with 'additive' label connotations out of existing product lines." This has certainly been accomplished with these nonnutritive beneficial and functional ingredients, thus doing away with the "unnatural" ingredient list on many products and providing a healthy connotation as well. Executives wish to promote their products with as "green" an image as possible, that is, as natural or made from natural ingredients.

11.2.1.2 Other Ingredients: Some with and Some without Nutritive Properties

11.2.1.2.1 Antioxidants

Historically, food preparers skilled in the arts of cooking and food preparation discovered early that certain herbs and spices could extend the shelf life of foods and make them taste better. Later as the science of food preservation grew, scientists found that some food components that occur naturally have antimicrobial and antioxidant properties. Thus, these natural additives in foods gave rise to the hope that they could be used to give foods a naturally extended shelf life. They can, but as might be expected, the very strong flavoring that herbs and spices contribute to a food limits their use.

The following herbs, spices, and other food materials have antioxidant properties, and some have beneficial nutraceutical properties (Kläui, 1973):

Oregano	Rosemary	Coffee beans	Wine
Nutmeg	Mace	Tea leaves	Onions
Sage	Turmeric	Citrus fruits	Buckwheat
Allspice	Cloves	Tomatoes	Carrots
Marjoram	Thyme	Beer malt	Heated gelatin and casein
Alfalfa	Oats	Rice bran oil	Rice germ oil
Pea flour			

However, the efficacies of these or of extracts of them as antioxidants depend on other components in the food matrix (e.g., the fat system) in which a particular material was tested. This is similar to the findings of Davies et al. (1979). They use

Cocoa butter	Cocoa powder	Cocoa shell
Oatmeal	Rosemary	Sage
Defatted cocoa powder		

or extracts of them made with different solvents in several fat systems. All displayed antioxidant activity, but they found that "the antioxidant activity of a particular material and its extract appears to depend upon the nature of the oil or fat used for testing and the type of solvent used for extraction of the antioxidant components." The type of solvent used for the extraction was an important factor; this suggests the antioxidant components in these plant materials were a more complex mixture not fully eluted by one solvent.

Pokorný (1991) reviews several sources of natural antioxidants: tocopherols from sesame oil; phospholipids from olive oil; phenolic compounds from cereals; flavonoids from herbs, spices and algae; carotenoids; polysubstituted organic acids; proteins, peptides, and amino acids; and Maillard products. He concludes, however, that the best approach for protecting foods is to avoid antioxidants altogether by removal of oxygen from the food, avoiding or eliminating the oxygen-sensitive substrate, or decreasing the oxidation rate by low temperature storage and avoidance of light.

In an undated report, the Specialty Food Division of Ingredient Technology Corporation (Woodbridge, New Jersey) presented evidence of natural antioxidants in molasses and attempted to characterize them.

In many instances, the source of the antioxidant activity has been identified and has been incorporated into some commercial antioxidants. The active component in cloves is eugenol. Unfortunately, eugenol's strong clove odor limits its use to baking products. Its isomer, iso-eugenol, has a more acceptable woody and spicy odor (Heath, 1978) that is also found naturally in nutmeg, basil, and other oils (Kläui, 1973). Gordon (1989) reviewed natural antioxidants and attempted to outline some mechanisms involved, so that developers' understanding of their use could be more directed.

A wide variety of natural substances have antioxidant activity, but not all can be used because of other characteristics that they may bring to the food. Much trial and error research must still be done to determine those best suited for further study.

11.2.1.2.2 Antimicrobial Agents

Some spices and herbs, as well as other natural materials, have been known through traditional food uses to have antimicrobial activity against a wide spectrum of microorganisms. Some microorganisms are surprisingly sensitive to these natural components, as Beuchat (1976) found between *Vibrio parahaemolyticus* and oregano. Shelef et al. (1980) screened over 40 microorganisms associated with foods for sensitivity to sage, rosemary, and allspice; both rosemary and sage demonstrated antimicrobial activity, while allspice proved less effective as an antimicrobial agent. Generally, the spices were more effective against Gram-positive microorganisms. The antimicrobial action was ascribed to cyclic terpenes present in the spices.

Zaika and Kissinger (1981) demonstrated both inhibitory and stimulatory factors (which they were able to separate) in the behavior of oregano against *Lactobacillus plantarum* and *Pediococcus cerevisiae* alone and in mixed culture. Low concentrations of oregano stimulated both microorganisms, but as the concentration rose, an inhibition of both acid production and viability was observed, a factor for developers to consider.

Zaika et al. (1983) extended their study to include oregano, rosemary, sage, and thyme against lactic acid bacteria. Again, microbial growth and acid production were retarded as concentrations of these herbs were increased. They observed that microbial resistance to a particular spice could be induced, another factor of concern for developers seeking natural preservatives. When microbial resistance to a spice was induced in a microorganism, that microorganism was resistant to the other herbs. This suggests that the mechanism inducing resistance in microorganisms is a broad-ranging one, a cautionary flag to be observed by developers. Antimicrobials in foods have been reviewed by Beuchat and Golden (1989) and the phytoalexins by Mertens and Knorr (1992).

The antimicrobial activity in these spices and herbs are most likely substituted phenols: oregano and thyme contain oils high in thymol and carvacrol, sage and rosemary are herbs containing cineole in large concentrations, and allspice is high in eugenol (Heath, 1978). Sage, rosemary, allspice, oregano, thyme, and nutmeg are the herbs and spices of greatest applicability as antimicrobial agents for foods, but all have strong flavor notes.

The growing popularity of ethnic foods, which are often highly spiced, combined with the trend to low salt foods and to vegetarian foods, both of which benefit from adding flavorful ingredients, is making highly flavored foods more acceptable. The high flavor impact from spices and herbs along with other flavorants such as vinegar, lemon, onions, shallots, and garlic can mask the lack of salt and the blandness of many products. The preservative action of more flavorful herbs and spices allows ingredient developers to provide their clients with products to prolong the acceptable shelf life of their products.

Bacteriocins are nitrogen-containing substances with potent antimicrobial activity; as such they are emerging as a new class of natural food preservatives.

Lactic and acetic fermentations have been well established as stabilizing processes for foods, as has acidification of foods with the addition of lactic or acetic acids. The lactobacilli, in addition to producing lactic acid, produce several substances that have antimicrobial activity against various microorganisms, such as hydrogen peroxide and diacetyl both with antimicrobial activity. Daeschel (1989) reviewed the antimicrobial substances from lactobacilli (e.g., nisin) and propprionibacteria. Some show promise against many foodborne pathogens including *Listeria monocytogenes*.

Another review (Stiles and Hastings, 1991) published just two years later demonstrates the amount of research taking place to discover new natural preservatives. Stiles and Hastings classify the bacteriocins as follows:

- Those produced by *Lactococcus* spp.
- Those produced by *Lactobacillus* spp.
- Those produced by *Carnobacterium* spp.
- Those produced by *Leuconostoc* spp.
- Those produced by *Pediococcus* spp.

As Stiles and Hastings point out, bacteriocins have little in common except their ability to inhibit microorganisms and that they are all proteins. They vary widely in everything else: the spectrum of microorganisms they are effective against, their molecular weight, how they perform, as well as their biochemical properties (see, e.g., Barnby-Smith, 1992; Juven et al., 1992).

11.2.1.2.3 Colorants

Safety concerns caused the removal of many synthetic colorants from permitted lists, a fact well known to all product developers. Equally well known is the importance of color to the appeal of foods. Natural colorants are labile and especially prone to destruction by the rigorous conditions frequently found in food processing, for example, in the manufacture of hard-boiled sweets. This spurred ingredient suppliers to search for more stable natural colorant and to research less strenuous processing conditions.

Engel (1979) examined factors that influence the stability of three classes of natural colorants:

1. Anthocyanins and anthocyanidins: As every first year student learns, anthocyanins are a group of pigments that are ionic, and their color is greatly influenced by the pH of food. They are unstable to heat, to oxygen, and in the presence of sugar and sugar-browning products. Sulfur dioxide bleaches them.

2. Carotenoids: There are over 300 carotenoids, most are only fat soluble, but some, bixin (from annatto; used in coloring cheese and baked goods), astaxanthin, and crocin (from crocus blossoms), are water soluble. Their highly unsaturated, conjugated double bond structure provides a range of red to yellow hues; blending increases this range of hues, but their unsaturated structure makes them susceptible to oxidation and light sensitivity.

3. Betalains: These are ionic, water-soluble quaternary ammonium compounds and as such are subject to changes in color with changes in pH. They are temperature, oxidation, and light sensitive.

Francis (1981) also reviews these colorants and turmeric, which contain the colorant curcumin that is of interest in its purported ability to reduce plaques associated with Alzheimer's disease. In addition, Francis notes some miscellaneous pigments such as cochineal and laccainic acid, which are of insect origin, a red colorant that has a long history of use in the Orient from the mold *Monascus purpureus*, and chlorophyll used in some specialty pasta products.

Gabriel (1989) and Shi et al. (1992a,b) worked on the separation and identification of anthocyanins from sweet potatoes and studied the stability of these anthocyanins in model food systems.

A major problem with natural colorants, in addition to their instability in many food systems, is the question of their safety. This is far from established for all natural colorants. In their natural, in situ state, safety concerns are perhaps negligible. When these are extracted from their natural source and concentrated, an assumption of safety is not justified. Natural colorants are expensive; first, they are scarce, and second, extraction and concentration of them cost money. As Francis (1981) points out, 75,000 hand-picked crocus blossoms are required to obtain a pound of crocin.

A new group of anthocyanins substituted on the B-ring were reviewed by Francis (1992). Acylated B-ring-substituted anthocyanins show promise as a new class of stable colorants because they have greater pH stability and produce brighter colors. They are, however, all from nonfood sources and would require extensive safety testing.

11.2.2 Challenges for the New Nutrition

The new nutrition, directed to disease prevention, opens up a Pandora's box of problems some of which have been alluded to. Food products directed to specific health problems become possible, but questions regarding the safety of these in concentrated form for consumers, effective dosage levels, standards of identity, and advertising ethics also have arisen. These questions will be a concern for ingredient manufacturers, the food companies incorporating these into their products, and for legislators.

11.2.2.1 Problems Presented by Enriched Foods

The public's quest for a long and healthy life free from debilitating diseases gave impetus for the development of products enhanced by or fortified with nutraceuticals. The scientific community and both ingredient and food manufacturers are financially interested in identifying these nonnutritive, disease-combating food entities and in developing products that contain them for the money-making potential they represent.

Manufacturers will want to inform customers and consumers of the benefits of their fortified products. Yet labeling and advertising regulations must protect the consuming public from any hyperbole and misrepresentation of health claims that may accompany the label statements and advertising claims. The Institute of Food Technologists' Web site at one time in the first half of 2000 displayed these announcements: "add more than value to your food and beverages...add life, with Polyphenols" to describe products put out by Templar Food Products; and "herbally-active, protein enriched frozen juice bars" contained chromium, manganese, 100% of the daily requirements of vitamins A, C, and E, as well as protein. These products were from Cold Fusion Foods.

With the wide availability of nutraceutically, phytochemically, or herbally enhanced products, there also came a warning from the health professionals of the dangers some nutraceuticals might pose for people on prescribed medications and for young children eating candies and snacks enriched with these ingredients.

Manufacturers want to inform the public of the benefits of a new product in a market niche where, metaphorically, shifting sands prevent the establishment of a firm product identity. Are these enriched products medicines or snacks? Is the company a pharmaceutical company or a consumer food product manufacturer? Management must decide its direction.

11.2.2.1.1 Delivery Systems

Delivery systems of these non-nutritive substances have taken many forms, and the ingredient companies have had to be prepared with neutraceutical ingredients that suit these food systems. The most used vehicles for delivery appear to be candy, snack food, and beverage products. Zind (1998) reported one candy manufacturer was adding phytochemicals purported to prevent cancer, bolster the immune system, and reduce cholesterol to the candies it manufactured. Fiber and calcium compounds have been added to fruit juice beverages—indeed, now orange juice can be fortified with fiber, calcium, or just plain juice. Iced tea preparations and ice cream have been fortified with polyphenols.

Potato chips, corn chips, corn puffs, and other snack foods have had added to them herbals and plant extracts claimed to have beneficial properties (Abu-nasr, 1998): some examples are ginseng (a long life promoter), St. John's Wort (a depression preventative), ginkgo biloba (a memory improver), or kava kava (an aid to relaxation). Each of the claims above should be preceded by "said to be or have."

Baked goods have also served as vehicles with ingredient suppliers developing muffin mixes with flax seed and blueberries. There are breads fortified with ω-3 fatty acids such as docosahexenoic acid.

Ice cream has become a vehicle for supplementation. It has been flavored with phytochemicals from green tea (Hozawa et al., 2009; Niu et al., 2009), ginger, avocado, sesame, wasabi, and capsaicins (hot pepper principle) to become nutraceutically fortified as well as flavored.

Old products have been given a new role by making them nutraceuticals with a role to play that confers some benefit, with added calcium to enhance calcium intakes added caffeine to combat drowsiness, creatinine as an aid for body builders, added fiber to assist in lowering cholesterol, and so on. Soft drinks with calcium, or protein, or vitamins, or concentrated herbal extracts (or all of these) have been moved from a refreshment role to a functional one as a nutrified beverage.

On the other hand, what are ingredient and consumer food product manufacturers to do when the general public, through the general media sources, reads that a natural polyphenol of blackberries kills leukemic cells—in vitro, but the public doesn't read that—Feng et al., 2007. The public wants this, and the manufacturers want to satisfy their customers and consumers.

11.2.3 A Cautionary Summary

That there are biologically active, beneficial materials in foods is a fact that ingredient manufacturers and pharmaceutical companies want to pursue and use in the preparation of beneficial ingredients and medicines. When these active ingredients are extracted from their natural sources, concentrated or modified, to make more efficacious materials, the public and the companies benefit. Their safety when structurally modified or used in prepared foods at higher concentrations than naturally found (see Panickar et al., 2009) may not be established for all segments of the population; in this respect, they are similar to many medicines where it has been found that one dosage level is not suitable for all ages, all ethnicities, or for both sexes. Extraction procedures produce a complex mixture of components, and which is the correct, active one with which to fortify the food or improve its health benefits is not known. Sophistication of foods in which they are not normally found can have serious consequences to some populations, for example, snack foods targeted for children or teenagers who are voracious snack food consumers may not be appropriate vehicles for supplementation with some phytochemicals.

Kardinaal et al. (1997), Chung et al. (1998), and Hasler (1998) all have expressed their concerns for the safety of consumer foods enhanced with phytochemicals as have been made throughout this chapter and have proclaimed a need for further research. These concerns are worth repeating here for developers:

- How these phytochemicals work in the body to prevent disease is not clearly understood. Are there any adverse side effects from their use in foods with enhanced concentrations and what is the nature of these side effects?

- Does the nature of the diet or the presence of other nutrients in the diet influence the activity of added phytochemicals? Do phytochemicals in enhanced concentrations influence the absorption or metabolism of other nutrients in the diet?

- There is no toxicity data on many of these substances, very little verifiable medical evidence of their effectiveness and many confusing studies. A large body of anecdotal information derived from traditional medicine exists, and some epidemiological studies (e.g., groups eating large quantities of soy products have lesser incidences of certain cancers). Tannins, for example, on the basis of epidemiological studies have demonstrated high levels of mouth, throat, and esophageal cancers in peoples using betel nuts and drinking the herb tea, *mate*, or in peoples whose dietary staple is sorghum (Chung et al., 1998); they display hepatotoxic activity and antinutritional activity by forming complexes with digestive enzymes and proteins. Chung et al. also found antimutagenic, anticarcinogenic, antimicrobial, and antiviral beneficial effects (e.g., tea drinking). Confusing evidence indeed!

- Physiological interactions of phytochemicals with other medically prescribed medications that consumers might be taking at the same time are a concern. Do they enhance or depress the activity of these medications?
- Information on what acceptable dose levels might be for all segments of the population is lacking.

Hasler, in particular, sees the need to balance benefits and risks with the use of foods containing physiologically active substances.

For herbs and herbal preparations, a different cautionary note must be sounded. Their long history in medical folklore as cures for many maladies has gained for them a veneer of grandfathered respectability such that they are readily available in health food stores, mainstream supermarkets, and drugstores where they are sold as preparations for teas or tisanes.

Tyler (1993) and others have suggested several dangers respecting the sale of herbs and herbal preparations that ingredient suppliers (and customers) should be aware of

- Are the raw materials what they are claimed to be? Not all parts of herbs contain the active ingredient and so are not effective. Ingredient suppliers need to be certain of their sources of raw materials. This is a poorly regulated area.
- Herbal preparations are not standardized with respect to the active ingredient(s) that they contain. From personal experience with fresh oregano used as a flavorant, there is a vast difference between geographic sources. Do they contain the active phytochemical in the concentration claimed and are these safe concentrations? There are no standard and authorized analytical procedures for quantifying the active principles in preparations.

The onus is on ingredient suppliers to be certain that they are using safe, sound, and authentic materials from which they wish to prepare their ingredients for food products.

Tyler (1993) does not give all the herbal preparations a clean bill of health. Ginseng has a very low level of risk even with excessive use. St. John's Wort also is relatively safe but may cause a photosensitivity with some people at high dose levels. Ginkgo biloba extracts do promote vasodilation with improved blood flow, but very large doses can have unpleasant side effects and may exacerbate the effect of aspirin should this also be taken. Already warnings have appeared from the scientists and medical profession about the overuse of products containing nutraceuticals, especially when these are used in combination with or in place of conventional drug therapies.

Many have thundered onto the market, and many with long history of use are slowly having their efficacy or their safety questioned by competent medical professionals. Taylor et al. (2003) found that echinacea proved of no value to reducing the number of colds in children 2–11 years versus a control group given a placebo. Kava kava, gingko biloba, St. John's Wort, green tea, fiber, and even ginseng have all received some negative publicity within the past years suggesting they are not as efficacious as some of the hype had promised (but see Hozawa et al., 2009 and Niu et al., 2009, regarding green tea). Caution is the by-word for developers in this somewhat chaotic, new, and promising field.

Ingredients used in the formulation of food products are in a delicate and carefully researched balance with each contributing some characteristic to the product. Whether used singly or in combinations, they play a role synergistically with the product as a whole. Altering either by removing or substituting or enhancing any one component upsets the balance in the formulation. These can have disastrous results on the safety, stability, or quality of a product.

I well remember the terrible taste in fried peppers when our supplier substituted a new and improved antioxidant into our fractionated peanut frying oil without informing us. The change had been meant as an improvement.

This latter point should raise concerns among ingredient suppliers; simply because some ingredient component is natural, it does not mean it will benefit a product economically or be safe.

12

Dancing but Uncertain of the Music

...There are known knowns; there are things we know we know. We also know there are known unknowns; that is to say we know there are some things we do not know. But there are also unknown unknowns—the ones we don't know we don't know.

Donald Rumsfeld, U.S. Defense Secretary, 2002

12.1 Introduction

Rumsfeld's tortuous statement describes, quite accurately, the environment and problems to be found in new product development. We know a lot technologically, and there is a lot we do not know socially and psychologically about customers and their buying habits and about consumers and their usage patterns. There is much to learn about the marketplaces of the future. What will they be? How do they, their layouts, their services, even their décor influence purchasing? The impact of social networking on searching customer and consumer needs is too young even for digital natives to assess its full impact on knowing customers and consumers; it is as yet an "unknown unknown."

In my years in the food industry, the tools of new product development have improved but not the predictability of new product successes. Somewhere the developers made an error, misinterpreted their findings, or, in some instances, chose to ignore the findings and chose to proceed on gut feel. Sometimes, but rarely, gut feel led to success. Was this due to an unknown unknown or a known unknown?

Life is the art of drawing sufficient conclusions from insufficient premises.

Samuel Butler

12.2 Looking Forward and Backward

> The future...seems to me no unified dream but a mince pie, long in the
> baking, never quite done.
>
> E.B. White

12.2.1 The Changed and Changing Scene

Four major elements exist within the food microcosm: the primary producer
or gatherer, the manufacturer (or assembler of food items), the purveyor
(a broader term than retailer), and customers with the closely associated consumers (Fuller, 2001). All others in the food microcosm are derivative of these.
All have shifted their positions of dominance with time, and such shifts have
influenced the direction of research and product innovation. These changes
in the power structure are contributing factors bedeviling the interpretation
of much market research.

The upheavals of power are described in Table 12.1. This table is divided
arbitrarily into three time periods; past or historical times, the recent past
and present times, and future times. The changes have not occurred simul-
taneously or uniformly in all geographic areas.

12.2.1.1 The Past

Historically, the primary producer or gatherer dominated, and development
in the food microcosm was driven by them. Farmers slaughtered and butch-
ered; made sausages and various krauts; and milled, brewed, and baked, and
they sold what was excess to their needs. Fishermen dried and salted their
catches. Olive growers pressed their olives for oil; grapes and dates were dried

TABLE 12.1

Shifts in the Power and Influence That Drive the Food Microcosm

Past Times	Recent and Present	Future Times
No. 1	No. 1	No. 1
Primary producer	Customer and consumer	Purveyors
No.2	No. 2	No. 2 (tied)
Manufacturer	Purveyors	Manufacturer
No. 3	No. 3	No. 2 (tied)
Purveyors	Manufacturer	Customer and consumer
No. 4	No. 4	No. 3
Customer and consumer	Primary producer and gatherer	Biotechnology company (seed, feed, and animal genetics companies)
		No. 4
		Primary producer and gatherer

or fermented by the producer and the products sold. These primary producers often were the sellers of their produce and finished goods in their farm markets.

With time, specialized crafts developed. Brewing and wine-making matured with specialists appearing; millers and bakers along with confectioners emerged as crafts people. These crafts people were the sellers, selling from their own establishments but at the mercy of the producers for their supplies. The food service industry, such as it was, grew their own foods or bought locally; the difference between farm and inn was not wide.

Customer and consumer duplicated, in part, the roles of the food manufacturer and the crafts people, that is, many baked their own bread, brewed their own beers and ciders, and "did down" by preserving and storing their own fruits and vegetables in season for over the winter months. Choices of available fruits and vegetables were not wide, and most were locally available produce. There was no market research to determine marketplace needs. No efforts were made to make shopping a pleasure or a convenience. Excess products were what was available and little more was. Retailing had not developed into the science it was to become.

12.2.1.2 Recent Times and the Present

A major shift of power and influence in the food microcosm occurred during the last 150 years or so. Customers and consumers began to dominate (Table 12.1); a middle class had clearly emerged. During this period, the purveying (retailing and selling) and the manufacturing sectors were developing better understanding of the eating and buying habits of customers and consumers and tried to capitalize on this understanding. Consumerism developed with the advent of women's magazines and guilds and eventually grew into the strong force it is today.

Improvements in transportation brought the availability of a greater variety of foods, and farms no longer needed to be close to cities. Advances in processing and preservative technology eliminated the seasonality of foods and allowed exotic foods from foreign lands. Techniques in market research became very sophisticated with the result that marketing and marketplaces became very complex systems with very intricate food supply chains.

As purveyors and manufacturers gained influence, they began to dictate to primary producers what they wanted, when they wanted it, and at what price they wanted it. Food manufacturers demanded products that closely fitted their processing needs, and many vertically integrated themselves with primary producers, thus gaining exclusive control over their suppliers.

Today, many towns and cities have reestablished a market day tradition prompted by a growing "locavore" market, one selling only products from farms within a local 100–200 mile radius. The locavore market is claimed to have fresher, better tasting, and more nutritious food than food transported from afar and thus to have a smaller carbon footprint. The local producer has regained the top position with the realization that customers' demands for freshness, nutrition, and taste command a higher price. The locavore market is more expensive. It has spawned a market based on a demand for manufactured produce produced locally—a *terroire* market. Its growth has not prevented and will not, in my opinion, prevent the ultimate dislodging of the customer from the dominant position nor reinstates the producer to the primary position except in small pockets.

There have been other challenges to the primary producers' position of dominance. Food manufacturers—and retailers (a subsection of the purveyors)—responded to demands of customers by giving customers products at a price customers could afford while making a profit for themselves. This, of course, squeezes the primary producers who in addition to all the vicissitudes of agriculture and fisheries also faced the anger of those opposed to the inroads of high-tech modern food production (factory farming, chemical fertilizers, hormones, etc.). Food manufacturers pressured primary producers, and large retailers pressured both the manufacturers and the primary producers to supply them with resources at prices retailers wanted for their store-wide promotions. Primary producers faced losing the business of both if they did not comply or could not supply on demand. For example, a major retailer might tell a chicken producer that they need chickens within a particular weight range at a given price per pound for a promotion: supply them or you lose the contract. Often on such sales, the supplier is expected to provide some advertising monies. Manufacturers, too, contract with producers for raw material of the size, shape, maturity, and condition that producers have to meet, and the manufacturers' field personnel inspect, grade, and set prices on the grade. Large manufacturers demand and get produce on their terms. Fast-food chains often have leverage to demand both quality and price from their suppliers, or these latter risk losing contracts. Power squeeze went down the chain because, at the top, purveyors had to be competitive with price and quality.

Primary producers, it must be remembered, are also at the mercy of seed, feed, and chemical manufacturers to whom these producers often owe large sums of money, which cannot be paid until harvest. The primary producers are truly between a rock and a hard place.

12.2.1.3 The Future

The purveyor (especially the large chain retailers) is now in the catbird seat (Table 12.1) with the most influential and dominating position within the food microcosm. Retailers need to make each foot of their shelves provide maximum profitability, and those products providing the best returns per measure of shelf space are displayed prominently. Unproven products (i.e., new products) or low profitability products are moved to low volume areas of the shelves or not permitted in store at all. New products per se are not in the retailer's picture unless their introduction means more *guaranteed* profits.

Purveyors mine their sales and marketing data obtaining information on their customers' purchases and shopping habits. They control this sales and market information that their suppliers value and often make suppliers pay for it. Such control of marketing data provides purveyors with dominance and influence over a major portion of the food microcosm. Their information is real-time information not based on mall surveys or on focus groups. It is based on real customers and consumers in real time at the cash register. Purveyors control

- Which products are displayed in the stores or prepared and assembled in restaurants or in fast-food outlets, all to maximize sales and revenues
- How prominently products are displayed and the number of facings to provide best exposure and the prominence and placement of in-store promotional materials
- The pricing of products and, hence, their attractiveness for purchase
- Food exhibitions, demonstrations, tasting events such as Taste of New Orleans™ or Taste of Toronto™ promoting restaurants, ethnic cuisines, etc.

They are gatekeepers for the introduction of new products (indeed gatekeepers for all products) onto store shelves or into kitchens. The retailing sector dictates to manufacturers through JIT (just in time) delivery systems and ECR (efficient consumer response) when and what they want delivered and through feedback to manufacturers what products they want developed.

How "tainted" is information gathered at the point of purchase? It reflects what items sold most, but it also reflects the disappearance of those items that were presented as sales features or were promoted by couponing. It is biased by how the purveyor has positioned or promoted products. Marketing personnel must study such marketplace data carefully and interpret it alongside their own market data. Purveyors knowingly or unknowingly influence product data in favor of what is profitable for their marketing plans. Are such data biased? Have purveyors manipulated the customer to maximize profitability?

Purveyors choose what products, menu items, or specials are displayed for sale based on what their sales data tell them and heaven help the customer (or consumer) who does not fit what their data tell them. This does have its amusing aspect.

In Quebec, liquor is sold only in government-run outlets. As a single malt scotch drinker, I repeatedly spoke to my local outlet about their poor selection of both single malt and blended scotches. The local store manager told me that I lived in—and his store was situated in—a wine, gin and vodka district and that to get a better selection of malt scotches I had to drive to a more distant store. This is hardly a convenience. My problem was solved by shopping on the Internet directly with the government warehouse; delivery is free, and I can specify when I want delivery.

Purveyors do preferentially stock their shelves with products they know suit the demands of the majority of their customers and hence provide products that are most profitable to their establishment. They ignore or pay only lip service to other market niches. My actions confounded local preference statistics and introduce another metric into the scene, Internet shopping.

12.2.1.3.1 The Tied No. 2 Position

Food manufacturers and customers with their consumers will in the future be tied for the second position of influence and dominance. The manufacturing sector is beset by pressures from the purveyors, and both give pressure to and get pressure from the suppliers below in the chain of influence. JIT and ECR both represent pressure put on the manufacturing sector. I cannot judge dispassionately either program; I have received mixed reviews and analyses regarding the value of these programs. They work well for some and have even been described as an aid in new product development as the retailer can feed back to the manufacturer information about what their customers want. I find this difficult to believe since retailers in particular do not like the uncertainty of new products. Distributors and members of JIT and ECR trade associations laud them; manufacturers and their traffic managers I have spoken with hate them or have great reservations about them. I am biased; it is naïve to think ECR or JIT work in anyone's advantage except the retailers. *If* it assists the customer and consumer, it is only peripheral to the interests of the retailer being satisfied. Retailers exert pressure on manufacturers.

12.2.1.3.2 Efficient Consumer Response

ECR is a more advanced version of the QR (Quick Response) tools describing close manufacturer and retailer partnerships. "Efficient Consumer Response incorporates the concept of customer-specific marketing into Quick Response" (Morris, 1993). Morris cites the Food Marketing Institute as defining ECR as "a responsive, consumer-driven system in which distributors and suppliers work together as business allies to maximize consumer satisfaction and minimize cost." My interpretation of this is that suppliers and distributors will see to it that there are no empty spots of the shelves; in some ways, this is a value to customers.

It has been suggested that ECR has improved the success rate for new product introductions by providing an early weeding out of "losers." Perhaps this is so, but I suggest that the weeding process strongly favors keeping those products that directly meet the needs of the purveyor and secondly the needs of the customer.

12.2.1.3.3 The Bottom of the Pile: No. 4 Position

Enter the biotechnology company. The biotechnology companies and many very large primary producers together occupy, for the time being, this bottom position. They represent a new entry to the tower of power. Chemical and pharmaceutical companies have moved into biotechnology and genomics either by forming new companies, buying companies actively in these new fields, or merging through partnerships with other like companies. These entities in turn have purchased many seed companies and are developing, producing, and selling new varieties of plant species developed through genetic research. In effect, these companies have strong influence, if not control, over the primary producer; they supply seed, feed, and chemicals (e.g., herbicides, fertilizers, and medicated feed concentrates) to primary producers who are often indebted to these companies until harvest time when they can pay back their loans. In time, this supplying of primary producers will likely include genetically modified animals, poultry, and marine species. They license the use of their seed, require that farmers do not carry over seed from past years for seeding (previously a common and accepted practice), and through the use of their seed limit pesticide use to their unique pesticide.

In time, this entity, the complex composing the chemical industry and primary producers, will exert greater influence—and power—in the food microcosm but at what price to the food microcosm? There has been an outcry from environmentalists, consumer groups, and lesser-developed countries that these large organizations are pilfering the crops of lesser developed nations, patenting them, and demanding royalty fees for their use. Others argue about the dangers of monocultural agriculture and the loss of varieties that the actions of the large biochemical companies might cause.

12.2.1.3.4 *Observations on the Future: Subject to Change*

This juxtaposition of influence in the food microcosm continues and promises, in the future, to influence the introduction of new products. Whether this is a negative or positive influence depends on which segment of the microcosm is in power for it is that segment that will direct development. The position of power may not be the same everywhere in a global market and the global marketplaces that will be developed; indeed, very likely it will not.

In any geographic area, the food microcosm is a complex mixture. The activities of purveyors have an impact on their primary producers, supplier–manufacturers, and on their customers and consumers; manufacturers, in their turn, command their primary producers, their ingredient suppliers, and their customers; and the assemblers (food service industry) impact on their supplier–manufacturers, primary producers, and their customers and consumers. The primary producers try to please everyone under pressure from environmentalists, animal rights groups, and Luddites denying advances in agriculture; primary producers, too, have suppliers, the seed, fertilizer, and chemical companies. Each of these suppliers, in their turn, influences their suppliers; it is like a row of dominoes in which tipping alters the whole mini-microcosm.

Market research data may mislead and the market researchers may be mis-led if the impact of these power shifts is ignored. The tools used to gather market and marketplace data are flawed, and the tools used to understand customer and consumer behavior are poorly understood, so that interpreta-tion of data obtained has often failed to fully integrate and understand the many roles of the different protagonists. Data gathered by the marketing people can be influenced by those in power or influential in the microcosm at any particular time and by any major development occurring in any seg-ment of the food microcosm.

Purveyors, especially major chain retailers, have no empathy for either encouraging a new product introduction as introductions are an intrusion into their routines, and unless they are assured of profits, introductions remain intrusions.

On the other hand, the developer has been intimately involved with the product. It is the developers' "baby." To avoid this lack of enthusiasm or even cooperation by chain store retailers, marketing personnel must work closely with retailers to get their assistance or at least to avoid their antipa-thy by demonstrating the value that would accrue with the introduction. A successful launch requires the active participation of the purveyor.

"The buzz" is an element for market research personnel to be aware of during an introduction, indeed to be aware of during any stage of a product's life cycle. It is unrelated to any impact that the retailing sector may have in the food microcosm. It may represent competitive activity. The buzz refers to the rapidity of communication, indeed basically to instant messaging. It is a tool of activist consumers and some consumer groups. Marketing and advertising people are frightened by the rapidity by which negative reactions to new product introductions or to newly opened food service establish-ments can spread. Bad buzz, instant feedback, can hurt new food products, especially those aimed at the teens and older groups of consumers who are active social networkers. The buzz is one powerful weapon that strengthens the customer–consumer segment of the microcosm.

Thus, in the future, there will be customers and consumers armed with the power of social networking and buzz. A biotechnology manufacturing-cum-producer segment joined uneasily with a food manufacturing tied into chain store retailers may try to generate buzz artificially, but the social networkers can fight back with very damaging buzz. All of this is closely watched by governments who look askance at artificially generated buzz as a tool, as well as environmentalists, and cultural and ethnic minorities.

12.2.2 Being Sure of the Concept

This book has been used extensively in universities and colleges. Its users, I assume, are digital natives, whereas I am a digital immigrant who frequently must get his son or grandchildren to help when in difficulties with computers and electronic equipment. This has given an interesting perspective that allows observations of the habits of digital natives. They can be both amusing and concerning.

Most software applications such as statistics programs are familiar to many only by plugging data into the program that then spews out a result. If the program is a statistical program and it says that the result is significant, then most accept this as gospel. Industrial food technologists, marketing individuals, or financial persons are often the pluggers-in of data. They do not understand what assumptions the software designers used in establishing their programs or what limitations there are to the programs: they only know data in, results out. Ergo, conclusion!

Product development is being flooded with software programs that model the development process, even model the test market and thus making actually undertaking one unnecessary. Data are entered and decisions come out, but who among the pluggers-in of data understands the machinations of the programs they use. They have a black box into one end of which they plug data, and from the other end, a result comes out. They know only that decisions regarding significance or irrelevance were made for them. The results are not challenged.

I do not decry the use of software programs for either statistics or modeling and certainly not as search engines. Much can be learned from what is obtained if wisely digested. I am appalled by the blind use of software without some understanding of pitfalls and limitations in its use. I decry the unthinking acceptance of the conclusions reached without questioning and examining these conclusions.

I attended a lecture given by an engineering professor who demonstrated the growth kinetics of a microorganism calculated using a software programs. His results were challenged by a microbiologist (familiar with the microorganism in question) who pointed out that the microorganism only grew in a very narrow temperature range and could not exist at either extrapolated ends of the growth plot and therefore all subsequent data were irrelevant. To which the lecturer answered in words to the effect, "I'm no microbiologist, but if the microorganism could grow at these temperatures, it would."

One must assure oneself the results "make sense": the results have to be logical and conform to the real-life situation they are meant to portray. The algorithms upon which statistical, decision-making, and modeling software are based must be understood for what many of them are: models and pretend playthings to be used to assist careful thought not to replace it.

The application of statistically based experimental designs in testing, such as rotatable designs (Mullen and Ennis, 1979a,b), reduces work, reduces the number (and costs) of experimental trials, and assists the interpretation of data. There are innumerable statistical software packages available to perform this design task. But the question must always be asked, "Are software programs applicable and appropriate and the results meaningful?"

These programs were not designed with a specific user in mind; indeed, they may be entirely inappropriate to a user's circumstance. Therefore, results and conclusions reached using these programs might not be pertinent to the user's real-world situation. "(T)he informed judgement of the investigator is the crucial element in the interpretation of data" (Cohen, 1990b).

> When the blind lead the blind, they will both fall into the water.
>
> **Chinese proverb**

12.2.2.1 Value of the Earlier Literature

The value of the older scientific literature should not be overlooked. I know of one professor who bragged to me he wouldn't let his students cite any paper older than five years. Despite the age of Cohen's article—it is nearing 20 years old—it still presents very sound thinking that extends beyond its statistical nature. Too often, the modern product developer rejects the older literature simply because it is old and may be in error (and maybe it isn't). Alongside Cohen's paper (1990b), I must also put Levitt's two classic papers (1960, 1975) on marketing that are older still in the must-read list for those contemplating product development. And then, there is Katz's paper on management (1955) that is older still than those by Cohen or Levitt. The above food science academic was so terribly wrong. His students would have missed Katz's, Cohen's, and Levitt's papers.

Technology, epitomized by the bells and whistles of the modern laboratory, may change, but sound thinking stands on its own merits—*plus ça change, plus c'est la même chose*. The earlier studies used equipment that is now dated, but the reasoning and thought behind the technology are still fresh and concise. There has been much earlier work lost or discredited because of the shortsightedness, indeed, arrogance, of scientists regarding its value, and many research workers are doomed to "reinvent the wheel" by not reviewing the earlier literature.

> What experience and history teach is this—that people and governments never have learned anything from history, or acted on principles deduced from it.
>
> **G.W.F. Hegel**

What values are to be found in the early food literature? There are many:

- The early scientific journals contain articles, reviews, and letters describing work from which students learn how ideas developed and progressed historically, not always moving in lockstep but often with abrupt changes from accepted dogmas to run off in new directions. They see how the modern equipment they work with grew from the jerry-rigged apparatuses that earlier workers had to build. Students can explore ideas that hadn't worked back then but might do so with more refined equipment and better understanding of processes.

- The early cook books are capsule social histories recording the changes in the popularity of particular foods, of food customs, of the availability and kinds of food ingredients, the appliances for the cook, and how preparation and cooking were carried out. One understands how people used food and were influenced in their usage of food by scientific and nutritional advances and how controversies as well as political and social upheavals influenced their daily lives. Parallels to today's situations can spark ideas for products.

- The pamphlets and brochures supplied by ingredient, food, and appliance manufacturers of long ago provide vivid illustrations of the food scene then, and one sees comparisons that astute developers can reinterpret to produce exciting new ideas.

- Textbooks on food processing, nutritional sciences, dietary guides, popular commentaries—often written by those now considered naïve or quacks but who at the time were revered authorities—are frozen capsules of the state of collective food knowledge. Parallels to the views of some of today's food writers are sobering.

This early food literature in scientific journals or in long-forgotten women's magazines is an excellent teacher of the progression of the laboratory bench into the home kitchen, from the beginning science of food and nutrition into the diets and well-being of people. There were no food scientists; the early workers in the food sciences were chemists, biologists, and physicists. The modern equipment in today's laboratory will be deemed historic pieces of junk 50 years from now if indeed they last that long. Students can learn how ideas and technologies develop.

We are bemused by the simpler beliefs, customs, and understanding of earlier food scientists and nutritionists; but we should realize that similar fallacies are rampant today. We quickly forget how our knowledge of diets and nutrition is changing, for example, our theories concerning fats, fatty acids, and cholesterol; how our knowledge of stomach ulcers has been reshaped; and how learned scientists tried to discredit the existence of prions put forward by Prusiner. As a young, fresh food scientist, I was asked by an older chemist what field of chemistry I was in. When I told him, he exclaimed rather derisively, "Oh, you're in Schmierchemie." Future generations will smile at us for our naiveté as we have smiled at those before us. It teaches one humility to read the early literature and see how wrong scientific experts can be and, often, how stubborn they were about changing their cherished beliefs.

12.2.2.2 What Customers and Consumers Want or What Purveyors Want?

Do all players in the food microcosm want the same thing? Do customers want what consumers want and are these wants different from what purveyors and manufacturers want? Do all want the same thing? And very simply, what is that "thing"?

Customers (the gatekeepers) want to be able to obtain those food items they have been sent to purchase for others or require for themselves. They want these with the quality suitable for their purpose, available within their budget, delivered in timely fashion, and with the assurance that these items are safe.

Consumers (the users) want the products they had asked, directly or through their gatekeepers, to get for them. They have the most complex needs that vary from utility and convenience, good taste, and a pleasant ambience (diners) to functionality to uniformity of quality at a price that is affordable.

Purveyors want to attract customers and their captive consumers, to encourage repeat purchases or visits, and to be profitable.

Manufacturers and the assemblers of foods want to satisfy all, customers, consumers, purveyors, and their suppliers.

There is a common theme: customers, consumers, purveyors, and manufacturers all want satisfaction, but what constitutes satisfaction is quite different for each. Gatekeepers are satisfied when they and their consumers' needs are met. The purchasing agent is satisfied when the required amount of produce with the required quality was delivered on time and within budget. Consumers have complex sets of needs; satisfaction is pleasure or gratification of the senses. Satisfying hunger contents the hungry teenager, and the successful provision of an added value characteristic to a product that encourages repeat purchases gratifies the developers and the retailer.

Purveyors know, from their studies, which products and product mixes attract local customers; they know the best sites to locate outlets to satisfy local population groups. Purveyors are satisfied if profits and a loyal repeat customer base results and they relocate outlets or reprofile their product mixes to advantage if profits lag.

In answer to the question put earlier, all want satisfaction of their needs, yet what is meant by satisfaction of those needs is different and can lead to a conflict of interests. Have all customers' needs been met? Yes, *if* needs are satisfied with products from which most profit is derived and that are most in demand.

A Montreal manufacturer of bariatric products that were sold only to doctors practicing bariatric medicine, required a special fraction of a dried milk powder. This was only available from Europe. No company in Canada or the United States produced it. Special permission from milk marketing boards was required to import it at great expense and with the requisite permissions and licenses required with each order.

Customers with special needs, with their consumers in tow, must go farther afield. I do not suggest that purveyors stock or assemble products or that manufacturers make products that do not sell or sell only infrequently or into small markets. I do suggest that market data of customer buying habits dominate the business philosophies and operations of purveyors and manufacturers, so that they often overlook potentially profitable niche markets in favor of big market share items. Profitability can be had with small volume items with small profit margins in many niche markets. Goodwill satisfying that small market segment may have inestimable value in the long run.

12.3 What Food Science and Technology Have Wrought

12.3.1 Impact of Food Science and Technology

> The old-fashioned housewife's menus were carefully thought out; the modern housewife's menus are carefully thawed out.
>
> **Anonymous**

12.3.1.1 How Food Savvy Are People?

Kesterton (2003) credits the above cynical aphorism to Harry Balzer, vice-president of the market research firm NPD Group of Park Ridge, Illinois, who commented that the majority of North Americans eat dinners that come prepared and need only to be reheated. Indeed, less than 50% of diners sit down to dinners in which at least one of the dishes was assembled. The description *assembled* applied to any food preparation activity, from putting together cheese and crackers or a jam sandwich to home-baked pies. Pyke (1971), writing some 40 years ago, startlingly suggests that the supremacy of the supermarket over the chain store over the corner store is the result of the impact of food science and technology and that this has had a profound effect on society.

Why is there this reliance on prepared foods? There are many causes:

- Foremost among these reasons is the simple fact that good tasting, convenient, prepared dishes that are quickly served are available at prices most customers can afford.
- The busy lifestyles of many people demand foods requiring minimal preparation in the home. The nuclear family sitting down together for meals has become a fiction. The family meal where social intercourse took place has gone except for festive occasions. Preparation time for meals has become scarce where both parents work. They need prepared foods for speed and convenience. Progress in food technology has been a socially divisive influence on home life (Pyke, 1971).

- People, women and mothers in particular, no longer feel guilty about not slaving all day in the kitchen or not laying before guests a home-prepared meal. They now take pride rather in greater participation in family activities, community affairs, or self-improvement courses.

- Many people do not know how to plan or prepare a nutritious snack or meal because they have never learned cooking skills or the basics of nutrition. They have no need to learn because they have a multitude of chilled, frozen, canned, or dehydrated prepared foods available for them.

A personal anecdote illustrates, somewhat amusingly, either of the last two above causes:

My wife regularly picked up our children plus a couple of others after school. For snacks, she invariably prepared carrot sticks by the simple method of cleaning, peeling and slicing carrots (this was before the days of the ubiquitous prepackaged baby-cut carrots). One young boy always devoured these avidly and one day commented to his mother, when she arrived to fetch him, "Why can't we have these?" To which she, a university lecturer, replied, "I've never seen them in the stores" and asked my wife where she bought them.

G. Woolf, editor of *The European Food & Drink Review* (issue 4, 2002), wrote in an editorial: "Many children have simply no idea that cereals have to be harvested to produce everything from bread to pasta, so what chance now has the communication of sophisticated concepts…?" Woolf's comments were not entirely out of context. He was commenting on how food technologists expected people to accept genetically modified foods based on how little people presently knew about the food they ate. Overcoming the ignorance of the general public about their food and about the food microcosm in general is a hurdle for food innovators.

To provide the convenience their customers want, retailers stock a wide selection of ready-to-eat chilled delicatessen items or they operate a hot deli-bar with fully cooked items for take-out or for eating in-store. They are thus competing with the fast-food chains. In the frozen food sections, easily prepared entrées vie for consumers' attention. Products are displayed to convey ideas for convenient—and only incidentally nutritious—meal combinations for hurried customers.

Unfortunately, often many gatekeepers do not understand any principles of meal preparation or basic balanced nutrition; hence, nutritious meal combinations have no meaning for them or their consumers. The August 20, 2003 IFT *Weekly E-Newsletter* provides a headline: "Consumers want healthy but buy convenience." This article describes consumers as confused about health claims and that they do not recognize which foods are healthy. Gitelman (1986) expressed similar views almost 25 years ago. Picard wrote about "Fat nation: heavier, wider weaker" (2010) describing a study of various obesity indicators showing Canadians to be getting fatter and weaker. The Associated Press, in a release, reported on the U.S.'s first lady, Michelle Obama, who is launching a crusade to wipe out childhood obesity. She faces much overt and covert opposition from vested interest groups. Governments are stricken apoplectic at the national obesity problem and the health issues caused by it, yet they have restricted funding for nutrition and household science education in public schools over the years. A "Healthy People" program launched some 10 years ago had a 2010 goal of just 5% of children being overweight. Latest figures indicate that 32% are deemed overweight. Governments now agitate for all kinds of legislation for funds either to educate the public, to prevent obesity in some unspecified way, or to apply a tax on fatty foods. The state of Alabama has given its employees a year to get into shape or face having to pay $25 per month for insurance that normally would be free (this in addition to having to pay a monthly penalty for being a smoker!).

At the same time, at the Florida State Fair in Tampa (February 4, 2010), items such as butter balls wrapped in dough and fat-fried, bacon strips dipped in chocolate and Krispy Kreme© doughnut burgers were on sale. These are hardly lean menu items, and one wonders at the motives of food service outlets in promoting such items to consumers. On the other hand, one wonders at the savvyness of consumers.

Wise development team members spend time watching and talking to people as they shop in the many marketplaces; here, they gather ideas to serve their customers better and see for themselves how knowledgeable shoppers are. Often, neither senior management who have the final say in the development process nor those responsible for the management of the new product development process ever venture into their marketplaces or question their sales forces who do talk with customers, consumers, and retailers and see the shoppers' knowledge displayed in their shopping carts. When those responsible for development see and talk with customers, consumers, and retailers and challenge for themselves their sophisticated market research data, then they can better evaluate their customers' needs and extent of savvyness. Researchers and developers must see the many influences and stimuli that affect customers and consumers at the various points of purchase whether these be the buy button on a screen or a cash register or an exhibitor's booth at a trade fair.

How food savvy are customers and consumers? If we consider food savvy as including a working knowledge of menu planning, food preparation skills, diet and nutrition, and health benefits, the answer is that the majority are not at all possessed of a working knowledge. Or else, why the increase in obesity?

12.3.1.2 *Impact of Technology*

Newspapers, popular science magazines, radio, and television hosts quote or interview experts from medical doctors to food journalists to dieticians to cook book writers on issues of food and food safety, health and disease, cooking, or nutrition. The nonscientific community has become confused by, frightened of, or suspicious of technology—and with good cause when such professed experts argue for their respective personal opinions, professional expertise, or downright quackery concerning food issues. It is good for readership or viewers if the journalists or the hosts of these shows garner interest and maximum publicity by stirring up sensationalism or controversy between opposing views and raise doubts in the minds of their readers and viewers. The more sensationalism the better the ratings are. The result for customers and consumers alike is not enlightenment but confusion and with confusion comes fear. Should food technologists and nutritionists wonder the public harbors a suspicion of science and technology? When the experts cannot agree on what appears to the public to be simple and clear-cut questions, such as "Is it good or is it not good for me?" they can only be expected to be confused. This is a question scientists either cannot answer or they waffle on their answers with abstruse, technical jargon.

12.3.1.2.1 *Big Science: Biotechnology*

When biotechnology shows promise to relieve pain and suffering, the public welcomes this advance. The public is not so welcoming when these same biotechnological techniques are applied to their food supply although this branch of science promises to have value for farmers, food processors, and the malnourished in lesser developed countries. Here, suspicion from many sources confuses the public. Luddites in our society, environmentalists and ecologists preaching caution, vested interest and consumer advocacy groups pulling political strings, science influenced by the support it receives from big pharmaceutical companies, and governments confused on the correct course of action to take all seem to cloud biotechnology's impact on agriculture and world food production. Many of the advantages in agriculture promise no benefits for the consumer but only for industry. Many in the public, therefore, want no messing around with their food.

Here are some issues confusing the public:

- Bovine somatotropin (BST)-derived biotechnology is used to increase milk production, yet the general public reads of gluts of milk on markets. The promised decrease in the price of milk and dairy products has not happened (indeed, prices have risen). If milk marketing boards control both the supply of milk and the availability of industrial milk, where is the need for more milk production from fewer cows? Consumer advocacy groups are suspicious of BST's long-term effect on people consuming products derived from milk of treated animals. Small dairy farmers are concerned that large dairy factories who can afford to use BST to increase milk production will force them out of business. Their dairy associations lobby government for protection. Who wanted BST? And why? are questions that confuse the public.

- Biotechnologists with leaf culture techniques can grow many plant products, for example, "natural" vanilla, in vats in factories situated far from Madagascar, the Comoro Islands, the Island of Réunion, Tahiti, and Mexico, the natural geographic sources of vanilla bean production (see, e.g., Knorr et al., 1990). The economic impact on these source nations for whom the vanilla bean is an important export commodity is a concern both for these countries and those activists against such technologies. Who does this technology benefit?

- Control of the senescence gene in tomatoes is of indifferent interest to the general public. Such tomatoes are for the salvation of shippers and distributors; they are not meant for consumers. Will there be a perceived advantage and will the customer, that gatekeeper, pay for it?

- Breeding and gene selection programs for sturdier disease resistant banana trees with straighter bananas are not advantages easily perceived by the consumer no matter how much they may be appreciated by the grower and shipper. "Will my bananas be different or cost more?" laments the consumer.

- The introduction of manufactured microorganisms into the environment for nitrogen-fixation or as a cryoprotectant for plants is an issue understood by scientists but not by the public. Any advantage to the public is seen dimly if at all.

These, all technological feats in themselves, are grist for the mills of the environmentalists who can readily shake the confidence of the public respecting the hazards, real or imagined, that such introductions may bring. The public has some good memories of bad decisions and advice:

- The introduction of DDT into the environment; they remember the assurances of scientists of the chemical industry who proclaimed its safety and who highhandedly attacked the naysayers.

- Chlorofluorocarbons (CFCs) that were touted by scientific experts as harmless, biologically inert, noncombustible substances. Then, the effect CFCs had on the ozone layer was discovered. Suddenly, CFCs, considered harmless, became dangerous when used in large quantities.

- The claim by leading experts on government-appointed committees that claimed lead in gasoline was safe until accumulations in the body fluids of children proved otherwise.

The key issue is convincing the public with tangible evidence that the technology will benefit them. So far the biotechnologists have failed to put this message across successfully. The public, through sometime sensational journalism, see only that the scientific community abetted by the large pharmachem companies simply proclaim that the genetically modified products are safe while they patent GMOs as quickly as they can. The general public too often sees these advantages as benefits through patents, royalties, and recognition to the scientists and the companies that supported their research. The potential value of biotechnology is lost through this inability to communicate value to the public.

Product developers must realize that raw materials derived from unproven technologies require careful investigation before they are incorporated into new products, for if they fail to be approved or accepted by the public, the work and effort, as well as the good name of their company, may be forfeit.

Harlander (1992) stated the problem regarding the acceptance of biotechnology succinctly:

> If biotechnology is to be used to ensure a safe, abundant and affordable food supply, it must be accepted by the public; therefore it is critical for us to come to grips with the scientific, as well as the social, moral and ethical issues that influence our thinking about the food supply.

For biotechnology to be accepted in new consumer food products, the products must be perceived as having benefits for the customer and consumer and as being environmentally safe or neutral *over the long term* and not merely as an advantage to the grower, shipper, distributor, or manufacturer or discoverer. (For a fuller discussion, see Harlander, 1989.)

Only when the public has been educated to know more about where their food comes from and how and why it is processed and they become better acquainted with its properties can their fear, suspicion, and confusion regarding science, politics, and big business be overcome.

12.3.1.2.2 *Consumer's Dilemma: What, Who, and When to Believe*

While with the H. J. Heinz Fellowship at the Mellon Institute for Industrial Research, I found an article entitled "Sufficient Challenge" (Smyth, Jr., 1967); its author, Smyth, was also at the Mellon, actually just down and around the corner from our laboratory. In the first line of the first sentence of this article was "You've gotta eat a peck of dirt before you die." This parroted my parents and grandparents, who told me similar things—"You'll eat a peck of dirt before you die," usually as we pulled a carrot out of the garden, brushed the dirt off, and ate it, or "A little dirt (or mould or bruising) never hurts anyone." It fascinated me because it seemed to go against everything I had been taught about hygiene, cleanliness, and sanitation since elementary school days: "Cleanliness was next to Godliness."

Smyth, an industrial toxicologist, named his hypothesis "Sufficient (but not overwhelming) Challenge," which earned him the sobriquet of "Dr. Smyth and his fellow poisoners" for suggesting that a little bit of poison was good for you—that is, some toxic material, but not an overwhelming doses fed to people (inadvertently) may have a beneficial effect: it stimulated the immune system and made people healthier.

The gist of "sufficient challenge" or the "hygiene hypothesis" that sufficient challenge seems to have morphed into is this: "In the past few decades, since World War II, we have antibiotics, vaccines, we live in insulated homes, and the cells of our immune systems have become uneducated and reprogrammed to react to allergens," according to Dr. Susan Waserman, president of the Canadian Society of Allergy and Clinical Immunology (2006). In short, are we figuratively killing ourselves with cleanliness and super-hygiene?

This hygiene hypothesis alone is enough to shake the public's faith in what the scientific community has been preaching for years, that is, cleanliness may not be next to godliness. They now question whether our fetish for hygiene has not upped the prevalence of autoimmune diseases such as diabetes and certain types of arthritis.

Tarnishing of the experts' image continued with the publication of numerous reports of scientists being supported by big pharmaceutical and chemical companies. Did such support influence judgment, bias results? Did freebies given by these companies starting with clipboards, pens, notepads and all with company logos, and then escalating to expenses-paid trips to conferences influence decisions regarding efficacies of drugs? Put another way, is research into the safety or nutritional value or the effect on the amelioration of some disease condition (high blood pressure, cholesterol blood levels, etc.) to be questioned if the work was supported by a vested interest group or company such as a milk marketing board, a confectioners' association, and so on? These are questions that new product developers must face. Have they communicated well and wisely to a skeptical public?

Some parallels in the early stages of the introduction of irradiation techniques to those of the introduction of genetically modified foods are apparent. It will be remembered that irradiation as a food preservation technique was originally touted as prolonging the shelf life of foods. This benefit was not perceived by customers and consumers as an advantage for them but rather it was perceived as an advantage for the processor (Best, 1989a) and the distributor. The public reasoned that the longer shelf life meant they would get old, stale product since it could stay longer on the shelves. Furthermore, they were convinced that processors would be able to skimp on sanitary procedures because they could zap the microbial contamination. Is this not logical reasoning on the part of the public? Irradiation for food products and the acceptance of genetically modified foods and food ingredients are both clouded with highly polarized views for or against. Some of the rhetoric and the extent to which apologists for irradiation will have to go in educating the consumer are reviewed by Pszczola (1990). It is good reading for proponents of the use of genetically modified foods.

"Geneticists' latest discovery: public fear of 'Frankenfood'" was the headline of an article by Molly O'Neill in *The New York Times* for Sunday, June 23, 1992. ("Frankenfood" is genetically altered food, a name coined by Paul Lewis in a letter to the editor of *The New York Times*.) This article highlighted developments of, in particular, a frost-resistant tomato incorporating a fish gene featured at the exhibition of the Institute of Food Technologists (1992). This development and the many others of biotechnology have advantages for a grower but unclear advantages for the public. It requires a bit of a stretch of the imagination for consumers to see the benefits. Such a headline and accompanying article do little to build public confidence in science.

This inability to appreciate the efforts of scientists is in part due to the public's ignorance of the activities in the food chain from grower to retailer. If the public cannot see how the modification benefits them directly or indirectly, it is, in their thinking, an unnecessary alteration of an already accepted and successful product. Many companies have backed away from the application of some of their genetic research; they fear the reprisals from educating the public to accept something should that something prove detrimental with later and newer knowledge.

Science has had a bad record with its pronouncements of what is good for the consuming public and of what is safe. Coffee and chocolate, butter and both highly unsaturated and saturated vegetable fats, cholesterol, fiber in its various forms (soluble and insoluble), red wine and its resveratrol, alcoholic beverages in general (beer is apparently good for women with osteoporosis) and foods high in β-carotene have had their ups and downs: all are apparently in the ascendancy now—at least until the next press release by a group of scientists saying they have no or bad effects. Kava and extracts from it were honored as a gentle stress and anxiety reliever. It was used on snack foods frequently eaten by teens and children; then, it was banned by many countries as a liver toxicant, but there is now a quiet movement to get it approved again, and some of those countries banning it simply look the other way at its sale. Of course, science always hides behind that excuse "best available knowledge at the time"—if they had only researched a little more deeply... The shape of hot dogs has been condemned as a choking hazard and some entrepreneur has already come out with a round (that is, circular) hot dog pattie. And recently, mother's milk was condemned as bad for infants!! What is the public to believe?

Expert (that is, scientific) opinion has been shaken with the cranberry and aminotriazole scare just before American Thanksgiving in 1959; the apple and plant growth regulator (alar) scare of 1986, the aforementioned lead in gas as not contributing to lead accumulation. Thalidomide was safe until it caused deformities in babies, and there are others. These create a degree of skepticism in the public's mind regarding both the scientific community and the government.

> If you consult enough experts, you can confirm any opinion.
>
> **Hiram's Law**

There was a delightful cartoon in a recent issue of *American Scientist* by S. Harris: A professor is sketched against blackboards filled with arcane mathematical formulae. She addresses her seminar groups seated around a table: "Today, class, we learn how to communicate with the outside world."

12.3.1.2.3 Agricultural Advances: More Dilemmas, More Confusion

Factory-farming techniques are perceived as cruelty to animals by many in the general public, by animal ethicists, by vegetarians, and by empathetic meat eaters. Odors from farms annoy suburbanites whose houses encroach on farmland. Run-off from pig and poultry factory farms polluting streams, rivers, and lakes angers environmentalists. Even a traditional practice such as force feeding of geese to produce foie gras has been declaimed. Governments have banned the manufacture of specialty cheeses made from raw milk and make it difficult to import such gourmet cheeses despite demand by consumers and the prevalence of these cheeses in other countries with seemingly no harm.

There have been violent demonstrations against both food retailers displaying factory-farmed poultry and veal and against farmers employing factory-farming techniques. These incidents plus car bumper stickers, T-shirts, and placards deploring certain agricultural practices (often displayed by those lacking any knowledge of farm animals and their habits) clearly indicate the existence of social constraints to the technologies and crafts applied to food production and processing. Spokespersons for the food microcosm are needed to teach the public about food, its production, manufacture, nutrition, and proper kitchen preparation lest many new product ideas using the promised benefits of biotechnology and new processes will die aborning. Developers, alert to the cachet that such novel product introductions carry, should have ready an effective educational campaign to accompany them.

Old food crops are being rediscovered, and developers have created new products from them to satisfy the consumer, for example, quinoa-based cereals, flaxseed bread, and amaranth flours. Such crops as millet, quinoa, and amaranth have long been favored in ethnic cuisine, and their rediscovery provides manufacturers with new product opportunities to satisfy the needs of peoples as they move around the globe. Articles on the history, properties, composition, and applications of both the grain and the vegetable amaranths plus information on its cultivation have been described by Brücher (1983) and Teutonico and Knorr (1985).

Organic farming and organic products have surpassed the multimillion dollar level but remain around 1%–2% of the market. Retailers are expanding sections of the produce and prepared food departments in supermarkets for the organically grown products. Restaurants make special mention of organically raised produce and meat from organically raised, free range animals. Many in the public now contract with market gardeners and farmers for weekly deliveries of organic produce to their doors. Processed consumer products using organic produce are already in production with some major food manufacturers going mainstream with such products.

There are questions for customers of organically grown produce that present them with a dilemma: Is it organically grown and by what definition of "organic?" It is unfortunate that confusion exists over this term—the public does equate this with "locally grown," and the two are not the same. There is also the question that many areas have banned the use of pesticides and the public has been astonished to see that many naturally occurring pesticides have also been banned. On investigation, the public finds that many natural pesticides (nicotine, vinegar, and pyrethrins) are as toxic if not more toxic than chemical pesticides. What is safe? The final question for the public is, "Do I want to pay the extra cost for organic food or local food?" A recent comparison of organic food prices (Canadian prices, March 2009): eggs at $2.99 per dozen while certified organic eggs were $5.19 a dozen; pork at $1.99 per pound for ordinary pork versus $9.99 per pound for organically raised ("…free range, of course, and happy pigs" [sic]) pork.

Customers and manufacturers who want to use organic produce are faced with extra costs for raw material and a controversy over what constitutes truly organic and whether organic produce and products are improved enough to justify the price. "Better" in this context has several meanings: environmentally (less of a carbon footprint), safer (free from chemicals), nutritionally healthier, and more humane. Farmers (primary producers) are equally confused; does the public want us to return to the farming practices of our ancestors? If so, yields will fall and prices will rise with the possibility of scarcities of some products.

Confounding the issue are those experts who are opposed to organic farming as wasteful of resources, unable to supply the world demands for food, and ultimately unsafe. These views are espoused by the very conservative think tank, the Hudson Institute. Their expertise has been seriously questioned by others. These matters present customers and consumers with more confusion about foods.

12.3.1.3 Trends as Social History

Many have tried to extrapolate trends as a means to predict the future. Predictions made in the past about what we would be eating today have been horrendously inaccurate. All I have learned is not to make predictions from trends as they are subject to change from many events that were themselves unpredictable. One can only interpret events in the environment of the times as they come using or relying on past events only as guidelines. One should learn from hindsight, but one should not apply it blindly to the prediction of future events too rigidly.

12.3.1.3.1 The Past and What Can or Cannot Be Learned

Market researchers are guided by the predictions of economic demographers (Foot, 2009; Foot and Stoffman, 2001), the agricultural economists, and professional market analysts (Sloan, 1999, 2001, 2003b) in their development programs. Reviewing earlier predictions of what was to have happened in food and agricultural areas is a sobering, educational exercise. Why have some forecasts come to fruition rapidly? Why are some still pie-in-the-sky dreaming? Why have some been slow to come? This analysis is the learning process.

Lawrie and Symons (2001) reviewed and commented on predictions made by Dr. J. G. Davis, then president of the IFST (United Kingdom), in his lecture *The Food Industry in AD 2000* presented in 1965 to the Royal Society of Arts. Davis' predictions are presented in Table 12.2. The right-hand column, headed AD 2000, has both Lawrie's and Symons' comments interspersed with mine.

Whitehead (1976), the American Society of Agricultural Engineers (Anon., 1977), Frost & Sullivan (Anon., 1980), and Ryval (1981) are amongst many who have predicted what and how we will be eating or producing food in the future. Time lines for the fulfillment of some of these predictions have passed, and for others, the technology is yet to come:

- Food through aquaculture: Seafood consumption has risen, as predicted. Wild fish stocks have been depleted through overfishing, and many governments have put a moritorium on the fishing of some species. Aquaculture for finfish, for example, salmon, halibut, tilapia, and catfish, and for shellfish, for example, oysters, clams, mussels, scallops, abalone, and geoducks, has grown in importance to meet the demand. Seafood farming has led to water pollution, introduction of diseases to the wild fish stocks, and concern for the genetic stability of the wild stock: all are ecological and environmental conflicts yet to be resolved. Underutilized species, previously used as bait, fertilizer, or dumped at sea, have received greater recognition as ingredients in fish sticks, fish sausages, and surimi-based products.

- Food from farm, field, and greenhouse: Future farms were redesigned and laid out in long narrow strips spanned by moving bridges in which sits the farmer, who, from his position on this bridge, programs all activities from cultivation, seeding, irrigation, weeding, harvesting, and primary processing of most crops. This never got anywhere and never will. Predictions had "farmers wearing cybernetic equipment may be able to control their machinery by thinking about what they want it to do." Practical thought control of machinery is years away and largely slanted to medical assistance to handicapped individuals. Hydroponic and greenhouse cultivation of crops are practicable possibilities.

It is interesting to note that neither of these predictions involved what we now call the biosciences—this had not been developed. So there are no

TABLE 12.2

Then and 35 Years Later according to U.K. Data Based on Excerpts from a Speech by Dr. J. G. Davis

1965	2000
Over 25% of personal income is spent on food, and the food microcosm employs more people than any industry.	19% of personal income is spent on food, and the number of people employed in the agricultural sector of the food microcosm has declined sharply.
The most important nutritional problem in 1965 and into the future is a sufficient supply of good quality protein.	FAO/WHO had recommended the amount of good quality protein to be 70 g per day. This has since been reduced to 53–55 g per adult male and 45–47 g per female.
The form of malnutrition in developed countries is overnutrition.	Obesity is a major health issue (American Obesity Association, [A.O.A.], 2002; Birmingham et al., 1999; Lachance, 1994; MacAulay, 2003; NCHS, 1999).
It will be necessary for a fourfold increase in the world's food supply by AD 2000.	There has been more than a twofold increase in world population. Population, though still growing, shows signs of slowing its rate. Food supplies have increased, but availability is uneven.
Qualified food technologists will control every aspect of food from production to selling.	There has been a great increase in the numbers of trained personnel within the food microcosm. Education of the public in nutrition and food hygiene has lagged (Fuller, 2001).
Protein extracted from green leaves could be used for human and animal nutrition (Pirie, 1987).	This has never been developed to the potential that Pirie and Davis anticipated partly because of the Green Revolution, the lack of a convenient, economical, and acceptable vehicle to use the protein in, and costs of extraction.
By AD 2000, waste and food poisoning will be a thing of the past.	1965: the innocence of the age of antibiotics! [author's comment]. Not only has the number of cases of food poisoning increased but the pattern of pathogens has also changed. Waste and environmental problems still confound the food industry.
Convenience foods, apartment living, and increasing numbers of the elderly and working women will lead to communal feeding.	Instead of communal eating, there has been a burgeoning growth of convenience foods, a wide gamut of eating establishments, and take-out food. Home cooking from scratch has declined to be replaced by so-called speed-scratch cookery.
Freezing as a means of preservation is the "least objectionable method" of preservation, and most people would own a refrigerator and freezer by AD 2000.	The availability of frozen foods has grown enormously, but so have many other means of food preservation.
Desalination would be economically feasible by AD 2000.	The availability of fresh water is a major world concern, and cheap power is as elusive as ever for massive desalination projects.
Africa will become a food-producing continent.	This has not occurred because of political instability, disease, and drought.

Source: Lawrie, R. A. and Symons, H., *Food Sci. Technol.*, 15(3), 50, 2001.

biogenetic developments predicted. Herein lies the shortcoming of predicting; what will be the next technological, sociological, or historical development that disrupts the predictability of anything?

Harvesting and infield processing of field crops are progressing slowly; field trials of hot-break processing of tomatoes that I have been involved with revealed too many insoluble problems—at that time (40 or more years ago)—to be resolved. Successful infield processing would permit the separation of the plant material into useful food and leave the waste material as either green compost or reserved for conversion into biodiesel.

Atomic energy plants as a cheap and safe energy source are many years into the future. Atomic power plants are being shut down or mothballed, and the public has been resentful of where any new plants might be sited, and new technologies have brought solar panels and wind power. Who could have anticipated these developments when these predictions were made?

Hydroponically grown fruits and vegetables raised in greenhouses near urban centers are possible, but they do require energy and produce liquid waste. Energy is costly and liquid waste is costly to dispose of or clean up. The locavore movement and environmental concerns (it is mistakenly believed that hydroponically or greenhouse-raised products have a smaller carbon footprint) made this prediction successful. Poultry and other protein sources will be raised in environmentally controlled high-rise or retrofitted buildings as part of urban areas with feeding and waste management automatically controlled and waste to be extracted for its energy. Animal rearing or hydroponic culture of vegetables in urban settings in retrofitted-abandoned office or factory buildings is simply too expensive to contemplate. Land is too valuable, and social activists and ethicists would surely object if potential housing was retrofitted for food production when housing for the homeless is lacking.

Other novel considerations that have been proposed are as follows:

- Unusual food sources: spun protein meat and cheese analogues as innovative as these products are have struggled for wide acceptance. Spun protein meat analogues have sophisticated canned stews, dried soup preparations, and pet foods. Surimi has become popular in its own right notably in simulated crab legs or other sushi dishes. Single-cell protein, particularly from yeast, has long been used as a base for food ingredients and in the fortification of foods. Insects (larvae, ants, grasshoppers, grubs, etc.) have been eaten raw, fried, boiled, and roasted in many cultures but are poorly accepted as food or feed by North Americans and Europeans. It will be the rare company that will risk introducing products based on insects. In Caucasian culture, insects are novelty foods—for example, chocolate-covered ants or grubs in alcoholic beverages. Much education is required before consumer activists and animal ethicists allow insect-derived protein or manure-derived feed to be fed to animals.

Taboos exist against many novel protein sources—for example, dogs and cats or horses and only with reluctance rabbits, reptiles, as well as many offal meats.

- New, or uncommon, cereal and legume sources are being rediscovered, for example, triticale, quinoa, millet and varieties of millet (e.g., fonio), amaranth, barley, oats (not rolled or instantized), rice of all types among them wehani, a patented variety, and the variety of legumes and products made from them is voluminous: their popularity is traceable to the cultural diversity of immigrant populations bringing their traditional foods and is partly driven by new foods with health benefits. This greater variety of plant protein sources complements traditional sources and invites a multitude of possible new products to satisfy the local population needs.

- Communal eating and eating out: this topic has cropped up often as a development to be watched. Architects have predicted that kitchens need not be designed into homes as separate single purpose rooms. By 2076, 75% of meals are predicted to be prepared at large food service commissaries and not in the home (Anon., 1977). This is an interesting prediction. Many hospitals, industrial and business canteens, and restaurant chains (including the quick-serve chains) are fed now from central commissaries delivering prepared and semi-prepared foods to them. Are these the communal eateries predicted many years ago but shaped by today's sociology? A majority of meal occasions are removed from the home. Are kitchens necessary as separate dedicated spaces in homes? The home kitchen is wasted space when people eat out, buy take-out food, or buy frozen prepared foods. Kitchen preparation is largely the putting together of meal components. This does not require a separate room dedicated to food preparation with built-in appliances and storage cupboards. Some city apartment dwellers do not have kitchens and roomers are, in effect, communal eaters.

However, kitchens are still built into houses, and today, there is a growing popularity of cooking classes as a desire for homestyle prepared foods grows; a decline in eating out and the closure of many French café-bistros in Paris were headlined in newspapers. Recessionary periods drive an increase in home-prepared meals and avoidance of semi-prepared or prepared components. Hard times bring back old favorites (cf., growing popularity of Spam™), comfort foods, or foods evoking memories.

On the other hand, the home-office and the office-in-a-car have become realities, thanks to improvements in telecommunications, the personal computer, and networking facilities; the distinction between home life and work life has blurred. Today's generation—that is, the New Busy—considers 9:00 a.m. to 5:00 p.m. a cute idea. How does this affect eating-out or eating-in?

With less time for food preparation, a greater impetus for home delivery of prepared foods may result; many restaurants now take part in meal programs whereby menu items can be ordered by telephone and delivered home with each member eating from a different restaurant's menu. What things have become known knowns and what are unknown unknowns?

- Meat consumption and vegetarianism present a dilemma. Meat consumption hasn't dropped as predicted but has increased as migrating peoples used to a more vegetarian cuisine have adopted a taste for meat. Vegetarianism, however, is slowly increasing because of personal health and humanitarian and environmental concerns.

- Fresh fruit and vegetable consumption has increased (as predicted) with a greater variety of fresh produce available year around, but the locavores protest at this variety of nonseasonal and exotic produce.

- Market segmentation in retailing and food service has made available a greater variety of ethnic foods to satisfy the tastes of migrations of peoples. This has fragmented the market with a broadened selection, as predicted. Variety rules despite the locavores' desires for only local and seasonal produce. Fresh herbs, herb blends, and salad greens all largely hydroponically grown have provided greater variety to produce counters. (But see previous bulleted point.) Where was hydroponic culture 100 years ago?

- The 65 and over segment of the population is rapidly growing, and this group is highly segmented. They present many challenges for marketers and food product developers. Couple this growing customer group with the growing nutritionally positioned products— nutrition being the "turn-on" for all future consumers but especially the aging but active seniors—that have partially displaced naturally positioned products (foods without additives, organically grown, and minimally processed). There are nutritionally positioned products to be found alongside food products enriched with, or genetically enhanced to contain, nutraceuticals and directed to prevention or amelioration of disease conditions. Many nutraceuticals, however, await confirmation and approval of their safety.

Food service menus have increased their menu selection under the influence of vegetarian and ethnic cuisines. Fast-food restaurants are certainly going upscale with white tablecloth seating, broader menu selection (Kentucky Fried Chicken now has grilled chicken), and expanded deli salad bars. These latter are found even in supermarkets. McDonald's and Burger King have both experimented with special dinner menus (*The Gazette, Montreal,* August 16, 1992), and some have described themselves as "adult-friendly" as opposed to "children-friendly" or "family-friendly." As said previously, I seriously wonder if this is not today's interpretation of communal eating:

it must be remembered that communal eating is not to be confused with a regimented form of eating where all file in to a communal kitchen and are served what is prepared.

- Waste and water recovery techniques: These, utilizing fermentation technology to extract valuable food and feed ingredients from the waste, are still highly experimental and facing the stigma of "food (or feed) from garbage."
- Technical developments: The retort pouch and sterilized milk were singled out by the seers of yesteryear. The retort pouch "spells the end of the tin can" because of the energy savings the pouch offers plus the improved quality: it hasn't been successful in North America and has had only modest success in Europe. Sterilized milk was predicted to move milk out of the refrigerator cabinet and onto the store shelves: it hasn't. The social and anthropological bases for these differences are not clearly understood, and it is rather facile to state milk has always been in the refrigerated dairy counter and not on a shelf, and the North American consumer has not found the retort pouch a convenient item unlike users in Japan and Europe. Developers are reluctant to go against the established preferences of their customers.

New ingredients such as whey-based ingredients, high-fructose corn syrups, encapsulated flavors, and savory flavors do not require education of customers; indeed, they add benefits, and their use will expand despite Pollan (2008).

Social, practical, ethical, environmental, ecological, economic, or legislative reasons have hindered the acceptance of many of the above predictions, the same reasons that squelch the development and acceptance of many new products and innovations. The unpredictability of the public's acceptance of technological breakthroughs and the unpredictability of scientific discoveries themselves, the vagaries of government activity within the food microcosm, and the volatility of customers and consumers and marketplaces confound the making of predictions or the discernment of trends that might lead to successful new products. Some predictions have come to pass, while others have sputtered out. Why? What don't we know?

12.3.1.3.2 Curbs to Predicting the Progress of Food Product Development

Are there natural curbs to an unlimited growth in new food products and, therefore, social, and anthropological barriers to predicting new product development events? The following observations give some credence to such a view:

- To support unlimited growth of new food products, old products must die. That is to say old traditions (cultural traditions?) must die. Food is such an ingrained cultural tradition I can't fathom its end. We are taught how we choose by parents and society, and, certainly, different cultures choose differently. Food may change to adapt to local resources, but many religious taboos regarding food and food's religious significance won't.

- Confounding this will be the reluctance of purveyors to see traditional, profitable, bell-ringer products die, and customers and consumers will dislike the loss of familiar products around which they have built their recipes, menu planning, food habits, yes, and cultures.

- The sciences related to customer and consumer research are not fully understood. Some understanding of individual responses to products is being made, but how these individual responses relate to the reaction of populations of individuals has not been successfully interpreted. Some claim these new techniques border on mind control, and the ethics of studies using these new techniques disturbs many.

- Customers, the gatekeepers, and consumers, the users, are difficult to educate to accept newer technologies associated with foods and its production or to accept changes to food customs and cultures. Altering these barriers (food cultures may have a genetic component) is a task neither scientists nor food companies want to undertake. Scientists do not consider it their concern, and food manufacturers fear reprisals if acting as advocates of such change.

- Events (legislation, governmental agricultural policies, international trade agreements and trading affiliations, stability of world politics, etc.) influence food companies' objectives and, hence, their new product development programs. Things happen. Predictability is lost.

- Projecting technological and scientific discoveries forward in a logical stepwise predictable fashion is a mug's game. Innovative discoveries do not proceed in that fashion but proceed most often with abrupt breaks from traditional dogma. Predictability is impossible.

- Diet and nutrition and their relation to one's general health are still not clearly understood by consumers or scientists. New health claims and relationships between nutrients and nonnutrients are discovered almost daily, and disclaimers from a rival group of scientists appear shortly after. Phytochemicals and their role in disease prevention is a field of study undergoing rapid rethinking and reinterpretation as new ones are discovered and controversies of effectiveness are argued. Because of the uncertainty in this area of study, developers are loathe to venture too deeply into new products to avoid costly misadventures and inciting government and the public's displeasure.

All the above serve to make the senior management and financial managers of food businesses nervous about success in uncertain—which new product development is—new product ventures.

The application of technology to food has outstripped the dietary and nutritional sciences upon which it should be associated. The technologies of stabilizing with high pressure, irradiation, water control, or pulsed electric fields still lack a theoretical and calculable basis for their effectiveness. This effectiveness is based on empirical evidence. There is no mathematical basis for irradiation or high pressure processing as there is for thermal processing. Development practices rely on trial-and-error and craftsmanship in equipment design, product formulation, preservative technology, and ingredient technology.

> "[M]an had always been concerned with the technology of food, because he needed food to survive and that his understanding of the basic principles underlying the art, or craft, had always been slower than its development" (Coppock, 1978; see also Taylor, 1969).

Without the science, applying technology for new product development will always be costly in time and money, and it will be hazardous.

12.3.1.3.3 *Consumer Responses*

Table 12.3, selected from Sloan (1999, 2001, 2003b) and now dated (although I see little has changed), illustrates the volatility and unpredictability of customer-purchasing trends in marketplaces over two-year intervals. The volatility of the purchasing habits demonstrates the danger associated with following trends: they are short range, often short-lived, and, by the time one is aware of them, they describe history. They are based on statistics from the past month's, past six months', or even past year's purchases. Introducing a new product into a trend may be jumping in at the wrong time in the life cycle of the trend. At best, a study of trends emphasizes purchasing volatility. If, indeed, as Sloan claims, the statistics show that children dominate food choice decision-making, who could be more faddish and volatile in food choices than the child, the tweenies, or the teens armed with their access to social networks? Trends do represent population dynamics in food purchases while neuromarketing techniques represent individual purchasing habits.

A more promising opportunity to understand the public is to study population dynamics. The problem is to segment the populations into recognizable subgroups with specific identifiable characteristics. For example, the traditional family unit has evolved: single parent families and same-sex parent families are common; two parent families now see both parents pursuing their individual careers or vocations; the proportion of active and healthy seniors in the population has grown and becomes a recognizable pressure group within society with money and influence. The yuppies and baby boomers have grown up to create new population dynamics causing their own problems. Teens, tweenies, and other groupings within which there are further subdivisions all represent separate market niches. With these changes, many traditional food habits have changed not the least of which is that the daily

TABLE 12.3

Top 10 Trends "to Watch for in the Future"

Food Technology, August 1999	Food Technology, April 2001	Food Technology, April 2003
"Americanization of flavor": plain American with just a twist of foreign	"Do-it-for-me foods": the use of pre-made or take-out meals	"Heat and eat": tasty, quick-to-prepare, no mess, no clean-up foods
"Super simple": simple-to-prepare foods with taste appeal for busy living	"Super savory and sophisticated": demand for more upscale and cultivated taste in foods	"Retro nutrition": renewed interest in sugar-, fat-containing foods but with reduced amounts of calories
"Street foods": fast, portable, tasty, and accessible	"Balance": extremism in food selection is vanishing	"Casual indulgence": upscale food for casual comfort
"Living foods": fresh, natural, organic foods	"Form follows function": growth of the "minis," e.g., appetizers	"Country charisma": ethnic and regional foods with an American twist
"Deliver me": carryout purchases and Internet food shopping	"A new kind of 'homespun'": developing foods for regional tastes; communal (one-dish) meals; homey, comfy, living room style restaurants	"Table talk": upscale, fashionability (more flavor) in favorite foods
"Eatertainment": family-style, communal "one-plate" eating; restaurant samplers	"Kid-influenced": catering to the influence children have on food purchases	"Simple solutions": single serve, combined (multi-ingredient, multi-nutrient foods). Mini-packs
"Freestyle": grazing, foods for all day, mealtime anytime eating	"Light and lively": fresher, healthier, and more attractive looking items	"Custom catering": catering to children as decision-makers in food purchases
"In-dull-gence": eating with moderation in all varieties of food	"Crossover meal patterns": mealtime is grazing time; out with traditional meal patterns	"Correcting conditions": foods to prevent, ward-off, or relieve unhealthy conditions
"Self-treatment and trial": do-it yourself health; eating foods to ward off something	"'Do-it-yourself' health": selecting food to improve or prevent something healthwise occurring	"Exceptionally pure": organic, natural, pure, no artificial additives foods
"Trusting technology": consumer willingness to try "genetically modified that tastes better" (cf., European attitudes)	"Clean, pure, natural, and safe": all natural, "green," non-factory-raised, organic foods	"Snacks and mini meals": snacking, grazing, and eat-all-day foods

Sources: Sloan, E. A., *Food Technol.*, 53, 40, 1999; Sloan, E., *Food Technol.*, 55, 38, 2001; Sloan, E., *Food Technol.*, 57, 18, 2003b.

family sit-down dinner is a rarity. Eating and, with it, meals have developed a complexity harder to understand and satisfy with suitable products.

The complexity of customers and consumers is being uncovered with new anthropological, sociological, and neurological technologies. There is a mixing of religious and cultural traditions with as yet no homogenization. There is a social networking communication that is not clearly understood as a power and influence in marketing or in commerce in general. The concept of markets and the many marketplaces in which markets find their placement is rapidly changing.

The global marketplace is here via the Internet, but there are not global tastes or foods. These are still, crudely put, tribal. Regional food tastes differ even within the same country. Even global brands change formulations to suit regional tastes. The response of consumers can be summed up as resulting in an almost chaotic segmentation of tastes to be accommodated. Attempting to get a majority market share with fewer but more profitable bell-ringer products seems hopeless, but getting good profits with smaller market share by satisfying some segments may be the best strategy. The president of Loblaw Companies Limited, a major food retailer in Canada, apologized to customers at its annual general meeting; they will return to the shelves the many products the company had considered less profitable and dumped in favor of those more profitable. The president said he heard the voices of the customers.

12.3.1.3.4 Politics and Government

Several factors have accelerated governmental intervention in the food microcosm. Two major concerns must be singled out:

1. Concern for food bioterrorism directed against crops and animals: Disruption of the food supply would create economic chaos and political instability. Stricter agricultural policies to prevent the deliberate or accidental introduction of animal and plant diseases have emerged.

2. Political instability or laxity of food controls in many food exporting countries: This has caused the illness of people, commercial animals, and household pets. The consequence has been delays at borders, increased costs of inspections, and increased regulations regarding animal inspections from exporting nations and by importing nations.

Governments' response to these concerns (not the only concerns governments have about the food supply and the health of its citizens) has been as follows:

- Directed an increasing body of legislation at national self-sufficiency by regulating the growing, harvesting, production, and retailing of food and increasing support for national agricultural and seafood policies. These directly influence regional agricultural and environmental policies.

- Established dietary food guidelines. There is major concern about obesity in children and its consequences in later life as well as obesity in adults. Other targets are high salt content in foods, high sugar foods, high fat content of fast-food restaurant items, and the presence of soft drink and candy dispensers in schools. There should be higher nutritional standards in school cafeteria foods. Special taxes on "bad" foods have been proposed.
- Have acknowledged that phytochemicals found in certain foods have important roles in the prevention of certain diseases. Governments are attempting to verify the claims and regulate to avoid quackery by manufacturers who fortify inappropriate foods (e.g., candies) and make exaggerated claims of yet-to-be-proven health benefits. This has alerted governments to the need to curb overhyped advertising with policies regulating safety, advertising, health, and nutrition.
- New agricultural and processing technologies have aroused passions in some citizens and have put agriculture and the pharmaceutical-chemistry industries on the defensive. Many in the general public fear the heavy involvement of the chemical, biochemical, and vested interest companies and trade associations in agribusinesses. Governments try to restrain these influences with legislation.

All of these arouse a profound interest by government in the food microcosm and confound prognostication in the food business.

12.3.2 Factors Shaping Future Product and Process Development

12.3.2.1 Influences: Known and Unknown

Which future developmental endeavors will be successful? There are too many unpredictable factors (unknown unknowns), and there are no algorithms yet to assimilate them all for accurate predictions. New food product development is an art with valuable inputs from both the soft and hard sciences; it is a very risky but necessary part of business—new products drive profitability. Developers can certainly study and apply those societal elements and current technological developments that are expected to influence the acceptance of new products or processes or conversely possibly avoid the rejection of them. These elements sort themselves into several broad areas, each with their own subsections:

- Social, cultural, religious, and ethical concerns of people as these relate to food that are hard to overcome through education. Food is culture.
- Public skepticism and fear, sometimes even apathy, about scientific and technological advances especially as these affect cherished food traditions. The public is confused by claims and counterclaims regarding the virtues of some foods.

- There is poor communication between agricultural, food, and nutritional scientists and the general public. Many in the public are ignorant of where their food comes from or how it is processed.

- The ability to reduce customers and consumers into more and more subcategories opens up more market niches for products satisfying these customers. The changing marketplaces in which these new market niches and their customers and consumers will shop in the future will change shopping habits of all customers and consumers.

Two disasters in 2010, one manmade and the other natural, have had an impact economically that could not have been foreseen. The Icelandic volcanic eruption was an inconvenience to travelers but an economic disaster to the shipment of fresh fruit and vegetables to northern European countries and for processors of these crops. The horrendous oil spill in the Gulf of Mexico caused by the explosion and sinking of an oil rig is an environmental disaster but will also have a terrifying impact on marine food products, tourism, and natural resources. These will have an as yet unkown impact on food production and marine product development.

One factor that does not seem to influence the future of new product development is the economy. Reference to Chapter 1 seems to indicate that the economy has no lasting effect on new product development over the short time studied.

12.3.2.1.1 *Social, Cultural, Religious, and Ethical Knowns and Unknowns*

Religion has a strong grip on the food we eat, on how it is processed, even on how it is served. Religion-based food laws fall roughly into two categories. The first category includes those laws that are primarily philosophical in nature; they prescribe a way of living. Eating, obviously an important part of living, is included in the laws. Where respect for life is part of the religious philosophy, this respect is practiced often by some form of vegetarianism. However, vegetarianism is not a key note aspect of this category. These religions, philosophically based, often lack both an orthodoxy and a ritual and are more flexible regarding food choices that are eaten.

The second category of religious laws has an essentially rigid orthodoxy from which come rules describing what is acceptable as food and acceptable for the preparation and serving of food as interpreted by shamans, religious or cult leaders, priests, and sages. It is imperative for the adherents of their religion to follow these rules. Such orthodoxy provides limited opportunity for product development except to alter established (bell-ringer) products such that they adhere to the preparatory laws, the processing, and serving rituals and to develop new products according to these laws. The developer aims at a narrower market with products prescribed by the religious laws.

The education of customers and consumers in North America was initiated by the writings of Catherine Parr Traill (1855) describing good household practices; by consumer-oriented magazines such as *Good Housekeeping Magazine,* started in 1885; and by the activities of women's associations such as, for example, writings by the Women's Alliance of the First Church of Deerfield (Anon., 1897). These were directed to women as the gatekeeper in a nuclear family and provided her with recipes and housekeeping advice.

Today, the Internet plays an increasingly strong role in informing and educating, for better or worse, customers and consumers. This tool lets customers and consumers comparison shop for the products they want and shop in a worldwide variety of electronic food marketplaces without leaving their homes. Information about food, nutrition, health, and cookery can be found. However, in searching for information, Web surfers often get well-meaning but unsubstantiated and perhaps dangerous quackery. For a layperson, distinguishing truths from half-truths from arrant mistruths is difficult.

The current "green" movement (i.e., one meant as having a smaller carbon footprint and not the green revolution of the 1950s) is a challenging influence. The movement is a mushy confusion of conceptual issues to which, nevertheless, companies had better pay attention in product and process development and in manufacturing in general (Mattson and Soneson, 2003). It involves opposition to free trade agreements that are unfair to third world nations, the exploitation of cheap labor in these countries (e.g., fair trade coffee), endorsement of environmentally and ecologically sound agricultural and fisheries practices—which means being actively pro organic or ecological farming, anti factory farming, and pro animal rights—and promotion of health issues, which mean additive-, pesticide- and herbicide-free food (i.e., pure food), availability of clean water and pure air (devoid of scents, perfumes, etc.).

12.3.2.1.2 *Healthy Eating with Pharmafoods: What Nobody Knows for Sure*

Knowledge of the interrelationships between nutrition and disease has progressed well beyond knowing the need for vitamins in the diet. Today, there is a growing awareness of the role that the foods people eat may play in preventing the onset of certain diseases (Jenkins, 1980; Taylor, 1980; Maugh, 1982; Ames, 1983; NCI, 1984; USDHHS, 1984, 1985; Cohen, 1987; Berner et al., 1990; Ishani et al., 2000; St.-Onge et al., 2003; Rajaram et al., 2009) and also in calming behavior (Barinaga, 1990; Erickson, 1991; Kolata, 1982; Wurtman, 1989).

Nutritional information for most people comes from newspapers, magazines, TV and its cooking networks, or via the Internet. Such information often comes through press releases issued by the university's or research institute's information technology officers; it is presented for publicity, seed money, and primacy regarding the discovery. As such, it can be overly exuberant in its as-yet-to-be peer-reviewed claims. Very little information comes through recognized educational sources or the original journals. Announcements of such advances or of newly discovered properties of some food are first digested by science journalists before being presented to excite the public imagination with their promises of benefits, usually some time in the future and on the day of the announcement the benefits only hold for mice. Then, later contrary opinions and refutations of the benefits are expressed by equally prominent scientists, well-meaning environmentalists, or other groups including the government. For example, Broihier (1999) highlighted beverage products containing functional foods among which were products containing kava. Later kava was pronounced a possible liver toxin (Branswell, 2002; Health Canada, 2002). Several countries ban kava outright even though it has been used for centuries in cultures in the south Pacific. Currently, its use is being rescued. Caffeine in coffee, cholesterol in eggs, and many other foods have been so treated.

A growing consumer mistrust of science within the food microcosm has developed aided by a consumer movement known as Naderism. Books such as Rachel Carson's *Silent Spring* (1962) and more recently Pollan's *In Defense of Food* (2008) have fed this skepticism. This antipathy towards science and its claims of authority have unsettling effects within the food microcosm. Sensational journalism sells, and some science journalists are popular and skilled communicators of their personal views and prejudices.

Confounding all this is the findings of genomic research. There are different reactions and benefits between males and females, different ethnic groups, and different age groups to certain medications, and nutraceuticals are a form of medication. Are developers now to develop self-medicating foods designed for middle-aged white men but not recommended for black men or women? Genomic research promises to put a new twist on product development.

12.3.2.1.3 Communicating Science: Educating the Public with What Nobody Knows for Sure

Science is not communicated well to the layperson, and scientists often are not the best communicators. Their public announcements are filled with wordy, technical jargon as if written for other scientists. Perhaps scientists should learn how to Twitter using only 143 characters! Where professional science journalists are intermediaries between scientist and layperson, communication can be quite good. When the intermediary is a public relations arm of the company or research institute, communication to the public can have ulterior purposes filled with hyperbole and promises of greater discoveries to come.

Good communication requires that a message be received by the target audience; in press releases, this is the general public. The message must be understood by the target: for this to occur, scientists must be careful to be clear. Finally, the message must be responded to in a way the sender, the scientist, desires. What then is the purpose of a scientific communication and what is its target? If it is honest education, it must be written in language the layperson understands and communicates in. If the communication is other than education, it may be written with some exaggeration to encourage support or investors. These two purposes often conflict and may confuse the layperson.

A healthy lifestyle plus good nutrition with disease-avoidance through proper diet with foods containing, or enriched with, prebiotics and probiotics able to prevent disease could benefit society and government in many ways. For this to happen, the public must be encouraged to eat or use supplements of extracts from prebiotic- or probiotic-containing foods or both. It is here manufacturers see great possibilities for new products. The communication problem is as follows:

- Education of the public to avoid, or avoid excesses of, those foods considered to be bad for health and thereby reduce one's risk of developing certain diseases
- Reeducation to adopt a healthier lifestyle and change one's diet to a more healthy one

Herein is the problem: for manufacturers, education promises no new products but only expenses with the possible admission that their products were formerly "unhealthy."

Enhancement of products with pre- and probiotics promises new products, hence, more profit. The public has rushed to self-medicate with foods with purported disease-preventing properties. What else were food product manufacturers to do but develop products such as power bars, sports drinks, teas, beverages, candies, and snacks fortified with prebiotic and probiotic substances of questionable efficacy? The onus is, then, on governments to educate the public to the advantages and at the same time to verify the efficacy, safety, and purity of these ingredients in order to protect their constituents public.

The result of better health through diet and lifestyle changes abetted by the pharmafoods is incalculable. The key, of course, is whether all the promises for the pharmafoods are scientifically true as opposed to anecdotally true: scientists are finding much contrary evidence.

Promotional activities to support these foods in diets blur the fine line between responsible claims ("good for you") and irresponsible marketing claims ("cures such and such..."). Most certainly, the food industry can expect, and is currently getting, government intervention in several countries in the form of advertising guidelines. Those developing such products for export need to be cautious as there is not unanimity on safety issues, claims, or permitted dosage levels.

My favorite black humor illustrative of this confusing and confused topic is the following:

1. The Japanese eat very little fat and suffer fewer heart attacks than the British or Americans.
2. The Mexicans do eat fat and suffer fewer heart attacks than the British or Americans.
3. The Japanese drink very little red wine and suffer fewer heart attacks than the British or Americans.

4. The Italians and French do drink red wine and suffer fewer heart attacks than the British or Americans.

5. The Germans drink beer and eat lots of sausages and fats and suffer fewer heart attacks than the British or Americans.

Conclusion: Eat and drink what you like. Speaking English is apparently what kills you.

Such humor illustrates a skepticism and cynicism toward scientific claims.

12.3.2.1.4 As Yet Unknown Markets and Marketplaces: Their Influences

Influences affecting customers and consumers within marketplaces have been described (see Figure 8.1). To introduce new products, some attrition of older established products is to be expected *if these latter do not satisfy the needs of highly volatile customers and consumers who, through new market research technology, have been subdivided by market researchers into new population groupings with unique needs.* Old, established, but less profitable products need to be culled unless maintained by developers with new upgraded characteristics and targeted to new users in new niches. An example of maintenance for a product is instant coffee: this first emerged as a liquid concentrate, then a powder, then a freeze dried powder, then liquid carbon dioxide extracted powders or crystals, and then versions with and without caffeine: it is a flavoring ingredient, and now there are liqueur-flavored versions. As an old product, it has, through maintenance, managed to stay on the shelves but not always by the original manufacturer.

Niche marketing can be seen to advantage in the brewing industry. For years, bigness was in, and big breweries amalgamated nationally and internationally to get bigger with promotion of a limited number of popular beers—an attempt at one beer fits all tastes. Then small, local breweries—microbreweries in Canada, craft breweries in the United States—started up. These kept away from the "popular" beer tastes of the large breweries and concentrated more on traditional beers filling niche markets by targeting beer lovers rather than beer drinkers. So successful has this been that big breweries have attempted a takeover of this market niche by buying the microbreweries.

Continued fragmentation of the alcoholic and nonalcoholic beverages into new niches has occurred: flavored beers, low- or no-alcohol beers, alcohol-free wines, wine coolers, carbonated and still fruit drinks, and flavored natural waters have all snatched parts of the beverage market away to create niches that have matured or are maturing into very profitable markets. Snow (1992) sees this move to niche markets as a result of the vicissitude marketing people feel to fragmentation of the retail marketplace by customers and consumers. In short, the buying public's volatility has fostered new and uncertain market opportunities that marketers have not quite come to terms with, an excellent example of the turmoil between developmental technologies in market research and the volatile consumer.

How does one target an elusive population in these new niches? Snow (1992) suggests that "instead of broadcasting adverts to the old 'admass,' the new buzz word is 'narrowcasting.'" Marketers must target niches of customers and blitz these niches with their highly targeted promotions and advertising. This can be done more easily via the Internet with social network sites and blogging.

Changes in the marketplace are legion. The warehouse outlet is climbing in popularity as customers buy in bulk to control food costs. Whether such purchasing really helps customers control costs or not is immaterial if customers believe they do. Which products could be adapted (and how?) to a bulk market customer is something developers must determine. But there are complaints that some outlets are too big with customers having to hike between departments to satisfy shopping requirements.

Communication is expanding at an enormous rate, in particular, interactive television. The impact of tele-boutique shopping on the food retail marketplace and ultimately on the introduction of new food products is as yet unknown. Developers faced with a heavy expenditure of television advertising monies can initiate a demand among customers and consumers through blogs for new products that retailers may have to stock (governments frown on this). Television and videoconferencing are costly for developers.

Conceptually the supermarket is changing. It is still a marketplace all under one roof, but now various departments may be privately owned or leased to specialist trades people, that is, the meat department may be owned and operated by a professional butcher; the in-store bakery run by bakers; and the fruits and vegetables run by a knowledgeable greengrocer. There is professionalism: these people know and care about their products. The new supermarkets have staff knowledgeable about the breads they sell, the sausages they make, the coffees they are roasting, and the tea blends available. In-store demonstrations provide customers and consumers with an opportunity to taste products. The whole effect is to produce a carnival-like atmosphere typical of a marketplace where people can eat, meet, socialize, and shop. This is the environment within which developers must fit their product introductions.

The marketplaces of the future will continue to evolve necessitating a change in food products. Supermarkets are challenging the fast-food chains with deli counters where customers can purchase prepared lunches, dinners, and even breakfasts. Street vendors—a growing food service segment in many cities—with their push-cart finger foods provide a direct challenge to the fast-food chains and stand-alone restaurants, each of which is saddled with high real estate costs. A colleague told me about a station restaurant in Westport, Connecticut, for commuters who place their dinner orders with the restaurant in the morning on their way to work and pick these up on their way home at night.

As traditional marketplaces have undergone and are still undergoing change, so have traditional markets been fragmented with greater opportunity for placing new products in markets that are either too small for large companies to fill or that provide margins that are unprofitable for them to exploit. Within these niches, there are very profitable markets for the right products.

12.3.2.1.5 *Economic Unknowns as Facts of Life*

It's the economy, stupid

Ascribed to James Carville, 1992

The theory, technology, and sciences upon which the development process is based need improvement. The odds of success are poor, and to improve the odds for successful development, developers need a better understanding of the interplay between the sciences, society, and societal changes that play an integral role in development process. Companies cannot continue to lose money on developing products that prove unsuccessful. Companies have to get smarter about carrying out product development. Management's strategy for its growth and criteria for evaluating new products in progress may have to be revised:

- How valid is measuring success by how much market share was obtained or taken from their competitor? Should partnerships or amalgamations be considered?
- Would establishment of a growable market niche of a group of faithful customers and consumers be an acceptable long-term alternative?
- Would the company accept a longer time frame within which to recover their costs of development and promotion?

These are management decisions, and if management sets its sights only on short-term gains, then new product development may have to remain a crapshoot for this ilk of management.

12.3.2.1.6 *The Unknown Impact of Government*

Government intervention into the food and agricultural industries will not stop. New reasons based on concerns for safety or politics will always appear. Intervention will be felt in areas as diverse as

- Patent protection for genetically altered plants, animals, and foods derived from these
- Toxicological testing of new ingredients and foods derived through biotechnology and submission of safety data to regulatory authorities for these and other new products such as degradable and edible films
- Development of a nomenclature for biotechnologically derived products
- Labeling regulations and guidelines to define or clarify such concepts as *natural, naturally raised, nature-identical, organic,* or *minimally processed*
- Advertising claims for the new ingredients and products with health benefits

Governments have already stepped in; the French government has attacked claims that special diet foods for athletes containing large amounts carnitine enhance athletic performance and reduce the amount of fat in cells (Patel, 1993); American and Canadian governments have warned against the use of certain herbal preparations (Clough et al., 2003). A caution must be given to those developers quick to leap in with pre- and probiotic enhancement of foods: because something is natural does not mean all plant parts, an extract of the plant, or concentrates of extracts are safe.

Political and economic factors influence the growth of new food product development. The connectedness of diet and nutrition on human behavior and diseases forces governments to adopt agricultural and food policies that promote healthy food production, reduce the risk of nutritionally related diseases, and provide for a secure food system. This has its problems. An attempt by the National Cancer Institute of the U.S. Department of Health and Human Services (1984) to publish a guide to healthy food choices to reduce the risk of cancer caused a furor from vested interest groups, mainly meat producers and sugar producers but also some vegetable growers. Those whose ox was gored, the meat industry, have found reputable science to make their point (Kabat et al., 2009). An example of the truth of Hiram's law: "If you consult enough experts, you can confirm any opinion."

New trade patterns emerge as political and economic alliances change. If nothing else, these agreements will usher in new competitors with new products to satisfy niches in ethnic markets. There will be new customers and consumers but also new legal requirements to adhere to. The impact on the movement and regulation of raw materials and added value consumer goods will be immense both within the established zones and also between the new trading blocs.

These alliances and free trade zones bring their own more subtle forms of protectionism, and they will most certainly introduce more food legislation. Agricultural powers within these alliances often find ways to protect their farmers. Vulnerable elements within the agricultural and food manufacturing systems will always clamor for protection, and governments pay them some lip service with the imposition of nontariff trade barriers as impediments to the free flow of products.

There will eventually be a global marketplace, but a global marketplace does not mean global food customs and traditions—these are too ingrained with religious and cultural traditions. Food products will have to be designed to satisfy each of the traditions, habits, tastes, and customs within this world marketplace. The successful food developer will adapt products, their shapes, textures, flavors, and colors to the needs and expectations of consumers in the particular targeted geographical marketplace that the developer wants to penetrate. The truly innovative developer will know that in established markets, the taste and flavor preferences of consumers are being challenged by the greater varieties of ethnic foods available.

No single product will be universally successful in all the marketplaces and market niches of the global community: there is no best product. The big breweries tried that and lost to small breweries. Developers must target specific niches with products designed for specific customers and consumers. Even the ubiquitous fast-food chains alter menus to meet local tastes in countries where they have penetrated. The success of any product in any given marketplace will always be limited by the social and political upheavals that will plague the world, either in cataclysmic fashion or in more subtle fashion as styles and eating habits change as populations shift.

12.4 What I Have Learned So Far about Product Development

12.4.1 My Mentors

Those familiar with Jacob Cohen's article "Things I Have Learned (So Far)" (Cohen, 1990b) will recognize this section's title. Cohen summarized his "things learned" as "less is more"; "simple is better"; and "some things you learn aren't so." These three pillars of wisdom serve me very well in new product development.

Butler's aphorism after the chapter's introduction provides another pillar of wisdom I have learned: in new product development, one must draw sufficient conclusions from insufficient premises. I might add to that: "from insufficient premises based on inconclusive data produced by overly self-assured people." An aphorism used very frequently in today's computer work, "garbage in is garbage out," can be traced back with many twists and turns to Butler: therein are many errors in development. And it is with these things experientially learned that much of successful product development must be viewed and judged.

Alongside Cohen's paper, I must also put Levitt's two classic papers (1960, 1975) on marketing that are older still than Cohen's and Katz's paper (1955) on management in the must-read list for those contemplating product development.

12.4.1.1 New Food Products of the Future

New food products of the future are impossible to guess except in very broad terms. In short, future new products will meet the diverse requirements of a highly mobile society of diverse food cultures and ethnicities, provide customers and consumers with confidence in their safety and nutritional value, be available in the many marketplaces that customers and consumers will be found, respond to the economic reality of society, and meet the increasing demands of government. All that at a price customers will pay.

12.4.1.1.1 On the Future

The above provides no value to developers. What concerns companies is what products will be the successful new products of the future. There are only vague hints in the previous pages, and they are subject to abrupt change without notice. More cannot be done. John Hanson Beadle wrote the following in 1893 (as reported in *New Scientist*, October 15, 1994):

> All history goes to show that the progress of society has invariably been on lines quite different from those laid down in advance, and generally by reasons of inventions and discoveries which few or none had predicted.

Predicting more of the same products with some modifications is doomed to failure because "of inventions and discoveries which few or none had predicted" as Beadle wrote more than a 100 years ago. Predictions depend

- On the next enactment of legislation that changes the rules of the game
- On the next breakthrough in knowledge or advance in technology that dramatically alters our understanding of what we thought we knew, especially in the food microcosm
- On the next ecological or environmental finding establishing damage to the environment by either of the above

These unknowns, combined with a fickle, quixotic public whose collective activities are not clearly understood sociologically, psychologically, or anthropologically, underscore the pitfalls of development and predicting its future.

In 1967, I entered, in a contest, a paper describing what our future food and food production might be in the year 2000. Although the paper won an honorable mention, it was hopelessly inaccurate at describing any of the actual events that did occur by 2000. It failed for all the reasons expounded in this and earlier chapters. The technologies were all there in 1967; all my predictions were possible except they did not happen by 2000. I ignored other events and developments that surround the food microcosm. I ignored the needs of customers, consumers, and the realities of the new marketplaces that were to come and the purveyors in them. I satisfied only my blinkered view as a technologist.

The predictions forming part of this chapter were made by technocrats or those applying their specific disciplines to their predictions; thus, they satisfied personal needs. They figuratively stuck their thumbs into a Christmas pie of esoteric research, pulled out a technical plum and, like little Jack Horner, said "What a good boy am I." This is not likely to satisfy tomorrow's customers and consumers. They will not beat a path to products simply because products are technological marvels: that was something I learned that just wasn't so. The application of technology per se to create new products will not be any guarantee of product success in the marketplace. Food manufacturers must satisfy the needs of an increasingly skeptical and increasingly knowledgeable public in highly competitive marketplaces.

12.4.1.1.2 What I Have Learned So Far

What I have learned so far is as follows:

- Companies cannot continue to expect growth with strategies undertaken 30, 20, 10, or even 5 years ago, no matter how successful they have been. Eventually, management has to develop new strategies based on who they are today and not on who they were and on how they best service their customers. They may have to go in a different direction, with abrupt changes in thinking based on knowledge of their customers' and consumers' needs. Products have finite life cycles, and a company's present product mix cannot continue upward growth forever without maintenance to accommodate the new customers and consumers. Senior strategists within the food microcosm must themselves be innovative and flexible in their development of strategy constantly searching for new ways that ensure survival in the business world.

- New product development is an art, not a science. It probably will never be more than a "soft science." Like all practitioners of the arts, product developers must learn well the individual skills that are required in product development and be able to apply these. Just as the painter must learn the properties of paper, canvas, and coloring materials; learn how to mix paints; know perspective; study human anatomy, even know how to use a brush for effect, so too must product developers know and understand fully their tools before they can even begin to have any success at product development.

- The skills reflected in the soft and hard sciences of product development must be learned well and used dispassionately. Every available avenue must be used to gain knowledge of the customer, consumer, the competition, and the retailer and the retailing environment. New product development is no place for "gut feel" for projects. Bad projects should be cut quickly.

- Trends and the fads they spin off should be observed closely and understood for the substance they provide about the company's customers and consumers. Developers need to think ahead to where the trend is going and not "jump on the bandwagon" hoping to be carried along. What trends *suggest* is happening in the longer term to the customer and consumer is the important objective.

- Greed for short-term profits is dangerous business philosophy. Management must be patient in developing markets and educating customers to the value of their products. Criteria for judging a successful introduction and a stable product must be reasonable. A dominant market share is not necessarily the same as a successful product.

The final words shall be left to Sir Isaac Newton, who wrote in *Mathematical Principles of Natural Philosophy*:

> Errors are not in the art but in the artificers.

References

Abd El-baky, H., El-baz, F., and El-baroty, G., Natural preservative ingredients from marine alga *Ulva lactuca* L. *Int. J. Food Sci. Technol.*, 44(9), 1688–1695, 2009.

Abu-nasr, D., Healthy chips. *The Gazette, Montreal*, W7, October 3, 1998.

Adams, J. P., Peterson, W. R., and Otwell, W. S., Processing of seafood in institutional-sized retort pouches. *Food Technol.*, 37, 123, April 1983.

Ahmad, J. I., Free radicals and health: Is vitamin E the answer? *Food Sci. Technol. Today*, 10(3), 147, 1996.

Ahmad, J. L., Omega-3 fatty acids—The key to longevity. *Food Sci. Technol. Today*, 12(3) 139–146, 1998.

AIC/CIFST joint statement on food irradiation, Ottawa, 1989.

Akre, E., Green politics and industry. *Eur. Food Drink Rev.*, 5, Winter 1991.

Allaith, A., Antioxidant activity of Bahraini date palm (*Phoenix dactylifera* L.) fruit of various cultivars. *Int. J. Food Sci. Technol.*, 43(6), 1033–1040, 2008.

Álvarez, I., Condón, S., and Raso, J., Microbial inactivation by pulsed electric fields. In *Pulsed Electric Fields Technology for the Food Industry: Fundamentals and Applications*, eds. J. Rasa and V. Heinz. Springer Science & Business Media, LLC: New York, 2006, Chap. 4.

American Meat Institute, Just released European Colon Cancer Study no cause for diet changes, June 14, 2005.

American Meat Institute, American Meat Institute calls WCRF (World Cancer Research Fund) panel recommendations on Meat Consumption extreme and unfounded, October 31, 2007.

Ames, B. N., Dietary carcinogens and degenerative diseases. *Science*, 221, 1256, 1983.

Amoriggi, G., The marvellous mango bar. *Ceres*, 24, 25, July/August 1992.

Anderson, J. R., Boyle, C. F., and Reiser, B. J., Intelligent tutoring systems. *Science*, 228, 456, 1985.

Anderssen, E., They know when you are sleeping, they know when you're awake—and whether you like sushi. *The Globe and Mail*, F1, F8, December 18, 2004.

Andrieu, J., Stamatopoulos, A., and Zafiropoulos, M., Equation for fitting desorption isotherms of durum wheat pasta. *J. Food Technol.*, 20, 651, 1985.

Ang, J. F. and Miller, W. B., The case for cellulose powder. *Cereal Foods World*, 36, 562, 1991.

Anon., *The Pocumtuc Housewife*. The Women's Alliance of the First Church of Deerfield: Deerfield, MA, first published in 1805, revised edition 1897, 4th printing 1956.

Anon., R & D lame ducks. *Chem. Ind.*, 913, 1971.

Anon., The future as seen by agricultural engineers. *Food Eng.*, 49, 120, September 1977.

Anon., Surgeon General says get healthy, eat less meat. *Science*, 205, 1112, 1979.

Anon., Analogs, ethnic and geriatric products offer top growth potential; "natural" to wane. *Food Prod. Dev.*, 16, 52, March 1980.

Anon., Irradiation for fruits & vegetables. *Food Eng.*, 53, 152, December 1981.

Anon., Introducing the hamburger. *The Gazette*, Montreal. 16, September, 1986.

Anon., The creative approach. *Roy. Bank Lett.*, 69(2), 1988a.

Anon., Fatal reaction. *Can. Consum.*, 18, 6, July 1988b.

Anon., Showcase: Fiber ingredients. *Prep. Foods*, 157, 151, July 1988c.

Anon., Modified atmosphere packaging: A. An extended shelf life packaging technology, B. Investment decisions, C. The consumer perspective, Report Series, Food Development Division, Agriculture Canada, Ottawa, Canada, 1990a.

Anon., Psyllium stabilizer: Label friendly and functional. *Prep. Foods*, 159, 127, August 1990b.

Anon., Natural oxidation inhibitor breathes shelf life into meats. *Prep. Foods*, 159, 71, June 1990c.

Anon., Food industry source book™. *Prep. Foods*, 160(13), 1991.

Anon., Nutraguide. *Food Process.*, 59(5), 38, 1998d.

Anon., Kentucky grilled: 'I think the Colonel would be happy.' *The Globe and Mail*, B12, April 29, 2009a.

Anon. or NAS, Underexploited tropical plants with promising economic value, Report of an Ad Hoc Panel of the Advisory Committee on Technology Innovation, National Academy of Sciences, Washington, DC, 1975.

Anon., Russia unveils belt-tightening diet for those who can't. *Guardian News Service: Reported in The Globe and Mail*, B10, March 3, 2009b.

Anscombe, A., Regulating botanicals in food. *Food Technol.*, 57, January 18, 2003.

Anthony, S., Clearing the air about MAP. *Prep. Foods*, 158(9), 176, 1989.

A.O.A., Obesity in the U.S. http:// www.obesity.org/subs/fastfacts/obesity_U.S.shtml, November 12, 2002.

Aram, J. D., Innovation via the R & D underground. *Res. Manag.*, 24, November 1973.

Archer, D. L., The true impact of foodborne infections. *Food Technol.*, 42, 53, July 1988.

Archer, D. L., The need for flexibility in HACCP. *Food Technol.*, 44, 174, May 1990.

Arendt, J., Regulating the body's internal clock. *Food Technol. Int. Eur.*, 25, 1989.

Argote, L. and Epple, D., Learning curves in manufacturing. *Science*, 247, 920, 1990.

Assaf, A., The popularity of foodservice systems in Australian hospitals. *J. Foodservice*, 20(1), 47–51, 2009.

Astorg, P., Food carotenoids and cancer prevention: An overview of current research. *Trends Food Sci. Technol.*, 8, 406, December 1997.

Avery, D., The hidden dangers in organic food, November 1, 1998. http://www.hudson. org/index.cfm?fuseaction=publications_details&id=1196 (accessed April 20, 2010).

Avery, A. and Avery, D., Organic food campaign goes sharply negative. *Knight-Ridder Tribune*, October 18, 2002. http://www.hudson.org/index. cfm?fuseaction=publications_details&id=2075 (accessed April 20, 2010).

Babayan, V. K. and Rosenau, J. R., Medium-chain-triglyceride cheese. *Food Technol.*, 45, 111, February 1991.

Babic, I., Hilbert, G., Nguyen-The, C., and Guirard, J., The yeast flora of stored ready-to-use carrots and their role in spoilage. *Int. J. Food Sci. Technol.*, 27, 473, 1992.

Bailey, C., Scaling up in R & D for production. *Food Technol. Int. Eur.*, 115, 1988.

Baird, B., SteiGenics to build new irradiation plant. *Food Technol.*, 53(2), 12, 1999.

Banks, G. and Collison, R., Food waste in catering. *Inst. Food Sci. Technol. Proc.*, 14, 181, 1981.

Barinaga, M., Amino acids: How much excitement is too much? *Science*, 242, 20, 1990.

Barnby-Smith, F. M., Bacteriocins: Applications in food preservation. *Trends Food Sci. Technol.*, 3, 132, 1992.

Barnham, C., The structure and dynamics of a brand. *Int. J. Market Res.* 51(5), 593–610, 2009.

Bascaramurty, D., Meet Dr. Freeze. *The Globe and Mail*, L1, 3, September 2, 2009.

Bastos, D. H. M., Domenech, C. H., and Arêas, J. A. G., Optimization of extrusion cooking of lung proteins by response surface methodology. *Int. J. Food Sci. Technol.*, 26, 403, 1991.

Bauman, H., HACCP: Concept, development, and application. *Food Technol.*, 44, 156, May 1990.

Baxter, J., Blood, R. M., and Gibbs, P. A., Assessment of antimicrobial effects of lactic acid bacteria. *Br. Food Manuf. Ind. Res. Assoc. Res. Rep.*, No. 425, 1983.

BBC News, Ramsay orders seasonal-only menu, 2009. http:// www.newsvote.bbc. co.uk/mpapps/pagetools/print/news.bbc.co.uk/2/hi/uk_news/7390959 (accessed April 5, 2009).

Beauchamp, G. K., Research in chemosensation related to flavor and fragrance perception. *Food Technol.*, 44, 98, January 1990.

Beckers, H. J., Incidence of foodborne diseases in The Netherlands: Annual summary 1982 and an overview from 1979 to 1982. *J. Food Prot.*, 51, 327, 1998.

Behrens, J. H., Barcellos, M., Frewer, L., Nunes, T., and Landgraf, M., Brazilian consumer views on food irradiation. *Innovat. Food Sci. Emerg. Technol.*, 10(3), 383–389, 2009.

Bender, F. E., Douglas, L. W., and Kramer, A., *Statistical Methods for Food and Agriculture*. AVI Publishing Co.: Westport, CT, 1982, Chap. 3.

Bengtsson, N., Contact grilling. In *Proc. IUFoST Int. Symp. Progress in Food Preparation Processes*. SIK, Swedish Food Institute: Göteborg, Sweden, 1986, p. 129.

Berger, K. G., Tropical oils in the U.S.A.: Situation report. *Food Sci. Technol. Today*, 3, 232, 1989.

Berner, L. A., McBean, L. D., and Lofgren, P. A., Calcium and chronic disease prevention: Challenges to the food industry. *Food Technol.*, 44, 50, March 1990.

Bertin, O., Labatt stumbled on to low-cal "fat." *The Globe and Mail*, Toronto, Canada, B18, August 3, 1991.

Best, D., Marketing technology through the looking glass. *Activities Rep. Res. Dev. Assoc.*, 41(1), 86, 1989a.

Best, D., Analogues restructure their market. *Prep. Foods*, 158(11), 72, 1989b.

Best, D., New perspectives on water's role in formulation. *Prep. Foods*, 161(9), 59, 1992.

Best, D. and O'Donnell, C. D., Food products for the next millenium. *Prep. Foods*, 161(7), 48, 1992.

Beuchat, L. and Golden, D., Antimicrobials occurring naturally in foods. *Food Technol.*, 43, 134, January 1989.

Beuchat, L. R., Sensitivity of *Vibrio parahaemolyticus* to spices and organic acid. *J. Food Sci.*, 41, 899, 1976.

Binkerd, E. F., The luxury of new product development. *Food Can.*, 35, 31, November 1975.

Birmingham, C. L., Muller, J. L., Palepu, A., Spinelli, J. L., and Anis, A. H., The cost of obesity in Canada. *Can. Med. Assoc. J.*, 160(4), 483, 1999.

Bishop, D. G., Spratt, W. A., and Paton, D., Computer plotting in 3-dimensions: A program designed for food science applications. *J. Food Sci.*, 46, 1938, 1981.

Biss, C. H., Coombes, S. A., and Skudder, P. J., The development and application of ohmic heating for the continuous heating of particulate foodstuffs. In *Process Engineering in the Food Industry*, eds. R. W. Field and J. A. Howell. Elsevier: London, U.K., 1989, p. 17.

Blades, M., Food allergy and food intolerance. *Food Sci. Technol. Today*, 10(2), 82, 1996.

Blanchfield, J. R., How the new food product is designed. *Food Sci. Technol. Today*, 2, 54, 1988.

Block, E., The chemistry of garlic and onions. *Sci. Am.*, 252, 114, 1985.

Bogaty, H., Development of new consumer products—Ways to improve your chances. *Res. Manag.*, 17, 26–30, July 1974.

Boileau, T. W.-M., Liao, Z., Kim, S., Lemeshow, S., Erdman Jr., J. W., and Clinton, S. K., Prostate carcinogenesis in *N*-methyl-*N*-nitrosourea (NMU)-testosterone-treated rats fed tomato powder, lycopene, or energy-restricted diets. *J. Natl. Cancer Inst.*, 95(21), 1578, November 5, 2003.

Bolaffi, A. and Lulay, D., The Foodservice industry: Continuing into the future with an old friend. *Food Technol.*, 43, 258, September 1989.

Bond, S., New products in the market place. *Food Technol. Int. Eur.*, 109, 1992.

Bone, B., The importance of consumer language in developing product concepts. *Food Technol.*, 41, 58, 1987.

Boudouropoulos, I. D. and Arvanitoyannis, I. S., Current state and advances in the implementation of ISO 14000 by the food industry. Comparison of ISO 14000 to ISO 9000 to other environmental programs. *Trends Food Sci. Technol.*, 9(11–12), 395, 1998.

Brackett, R. E., Microbiological safety of chilled foods: Current issues. *Trends Food Sci. Technol.*, 3, 81, 1992.

Bradbury, F. R., McCarthy, M. C., and Suckling, C. W., Patterns of innovation: Part I. *Chem. Ind.*, 22, 1972.

Bramsnae, F., Maintaining the quality of frozen foods during distribution. *Food Technol.*, 35(4), 38, 1981.

Branswell, H., Kava linked to liver failure: Health Canada bans supplement. *The Gazette, Montreal*, A12, August 22, 2002.

Brassart, D. and Schiffrin, E. J., The use of probiotics to reinforce mucosal defence mechanisms. *Trends Food Sci. Technol.*, 8(10), 321, 1997.

Brat, I., The "neuromarketing" behind Campbells' attempt to sell more soup. *The Globe and Mail*, B15, February 17, 2010.

Brighthouse (sic) Institute, Brighthouse Institute for Thought Sciences launches first "neuromarketing" research company: Company uses neuroimaging to unlock the consumer mind. Available from http://www.thoughtsciences.com/pressrelease060302.htm (accessed December 12, 2002).

Bristol, P., Packaging freshness. *Food Can.*, 50, 30, May 1990.

Brodley, C. E., Lane, T., and Stough, T., Knowledge discovery and data mining. *Am. Sci.*, 87, 54–61, 1999.

Brody, A. L., Chilled foods distribution needs improvement. *Food Technol.*, 51(10), 120, 1997.

Brody, A. L., The return of the retort pouch. *Food Technol.*, 57(2), 76, 2003.

Broihier, K., A thirst for nutraceuticals. *Food Process.*, 60, 42, 1999.

Bronowski, J., The creative process, in *Scientific American*, September, 1958. In *Readings from Scientific American: Scientific Genius and Creativity*. W. H. Freeman and Company: New York, 1987, pp. 2–9, Chap. 1.

Brooker, J. R. and Nordstrom, R. D., Developments in engineered seafoods for commercial and military markets. *Activ. Rep. Res. Dev. Assoc.*, 39(2), 56, 1987.

Brouns, F. and Kovacs, E., Functional drinks for athletes. *Trends Food Sci. Technol.*, 8(12), 414, 1997.

Brown, G., Is it safe to eat out? *Food Sci. Technol.*, 16(3), 50, 2002.

Brücher, H., Amaranth, an old Amerindian crop. *Dragoco Rep.*, 35(2), 1983.

Bruhn, C. and Schutz, H. G., Consumer awareness and outlook for acceptance of food irradiation. *Food Technol.*, 43, 93, July 1989.

Bryan, F. L., Hazard analysis of food service operations. *Food Technol.*, 35, 78–87, February 1981.

Bryan, F. L., Application of HACCP to ready-to-eat chilled foods. *Food Technol.*, 44, 70, 72, 74–77, July 1990.

Buchanan, R. L., Predictive food microbiology. *Trends Food Sci. Technol.*, 4, 6, 1993.

Buchanan, R. L. and Doyle, M. P., Foodborne disease significance of *Escherichia coli* 0157:H7 and other enterohemorrhagic *E. coli*. *Food Technol.*, 51(10), 69, 1997.

Burch, N. L. and Sawyer, C., Hospital foodservice requirements: Special diet convenience foods. *Food Technol.*, 40, 131, July 1986.

Busch, L., Biotechnology: Consumer concerns about risks and values. *Food Technol.*, 45, 96, April 1991.

Bush, P., Expert systems: A coalition of minds. *Prep. Foods*, 158(9), 162, 1989.

Buxton, A., Going online. *Trends Food Sci. Technol.*, 2, 266, 1991.

Campbell, I., Consumer attitudes to food safety. *Visions*, 2(4), 1991.

Campbell, L. B. and Pavlasek, S. J., Dairy products as ingredients in chocolate and confections. *Food Technol.*, 41, 78, October 1987.

Caragay, A. B., Cancer-preventive foods and ingredients. *Food Technol.*, 46, 65, April 1992.

Cardoso, G. and Labuza, T. P., Prediction of moisture gain and loss for packaged pasta subjected to a sine wave temperature/humidity environment. *J. Food Technol.*, 18, 587, 1983.

Carlin, F., Nguyen-The, C., Chambroy, Y., and Reich, M., Effects of controlled atmospheres on microbial spoilage, electrolyte leakage and sugar content of fresh "ready-to-use" grated carrots. *Int. J. Food Sci. Technol.*, 25, 110, 1991.

Carson, R., *Silent Spring*. Houghton Mifflin: Boston, MA, 1962.

Cauvain, S. P., Improving the control of staling in frozen bakery products. *Trends Food Sci. Technol.*, 9(2), 56, 1998.

Cerf, O., Davey, K. R., and Sadoudi, A. K., Thermal inactivation of bacteria—A new predictive model for the combined effect of three environmental factors: Temperature, pH and water activity. *Food Res. Int.*, 29(3–4), 219, 1996.

Chapman, S. and McKernan, B. J., Heat conduction into plastic food containers. *Food Technol.*, 17, 79, September 1963.

de Chernatony, L., Harris, F., and Riley, F., Added value: Its nature, roles and sustainability. *Eur. J. Market.* 34(1/2), 39–56, 2000.

Chinachoti, P., Water mobility and its relation to functionality of sucrose-containing food systems. *Food Technol.*, 47, 134, January 1993.

Chirife, J. and Favetto, G. J., Some physico-chemical basis of food preservation by combined methods. *Food Res. Int.*, 25, 389, 1992.

Chung, K.-T., Wei, C.-I., and Johnson, M. G., Are tannins a double-edged sword in biology and health? *Trends Food Sci. Technol.*, 9(4), 168, 1998.

Church, N., Developments in modified-atmosphere packaging and related technologies. *Trends Food Sci. Technol.*, 5(11), 345, 1994.

Chuzel, G. and Zakhia, N., Adsorption isotherms of gari for estimation of packaged shelf-life. *Int. J. Food Sci. Technol.*, 26, 583, 1991.

Clark, J. P., Processing equipment covered at Food Expo®. *Food Technol.*, 56(8), 102, 2002.

Clarke, D., Chilled foods—The caterer's viewpoint. *Food Sci. Technol. Today*, 4, 227, 1990.

Clausi, A. S., *The Story of Dry Cereal and Freeze Dried Fruit*. Presented at School of Food Science, Cornell University, Ithaca, NY, April 21, 1971.

Clausi, A. S., *The Role of Technical Research in Product Development Programs*. Presented at Institute of Food Technologists Short Course on Ingredient Technology for Product Development, New Orleans, LA, May 12–15, 1974.

Cleveland, W. and McGill, R., The proper display of data, reported in Kolata, G., *Science*, 226, 156, October 12, 1984.

Clifford, M., Are polyphenols good for you? *Food Sci. Technol.*, 15(3), 24, 2003.

Clough, A. R., Bailie, R., and Currie, B., Liver function test abnormalities in users of aqueous kava extracts. *Clin. Toxicol.*, 41(6), 821–829, 2003.

Cocup, R. O. and Sanderson, W. B., Functionality of dairy ingredients in bakery products. *Food Technol.*, 41, 86, October 1987.

Coghlan, A., Out of the frying pan. *New Sci.*, 157, 14, 1998.

Cohen, L. A., Diet and cancer. *Sci. Am.*, 257, 42, 1987.

Cohen, J. C., Applications of qualitative research for sensory analysis and product development. *Food Technol.*, 44, 164, 1990a.

Cohen, J., What I have learned (so far). *Am. Psychol.*, 45(12), 1304, 1990b.

Colby, M. and Savagian, J., Irradiation: Progress or peril? Con: Consumers say no! *Prep. Foods*, 158(9), 62, 1989.

Cole, M. B., Databases in modern food microbiology. *Trends Food Sci. Technol.*, 2, 293, 1991.

Converse, P. E. and Traugott, M. W., Assessing the accuracy of polls and surveys. *Science*, 234, 1094, November 28, 1986.

Cooper, L., A computer system for consumer complaints. *Food Technol. Int. Eur.*, 247, 1990.

Coppock, J. B. M., Has food technology outstripped food science? *Inst. Food Sci. Technol. Proc.*, 11, 193, 1978.

Cremers, H. C., Fishy cure for craggy features. *New Sci.*, 137(1863), 8, 1993.

Crockett, J. G., The new products manager. *Food Technol.*, 23, 25, July 1969.

Crone, G., Welcome to the war room. *Financial Post*, C 15, February 11, 1999.

Culhane, C. T., Resounding roar. *Food Can.*, 59, 16, June 1999.

Cullinane, D., In the know. *Food Sci. Technol.*, 16(3), 48, 2002.

Curiale, M. S., Shelf-life evaluation analysis. *Dairy Food Environ. Sanit.*, 11, 364, 1991.

Currie, A., Customer-centric product development—Supermarket private label brands. In case studies in product development. Earle, M. and Earle, R., eds., Chapter 17, pp. 317–330, CRC Press, 2008.

Daeschel, M. A., Antimicrobial substances from lactic acid bacteria for use as food preservatives. *Food Technol.*, 43, 164, January 1989.

Dahm, L., Kuhn, M. E., and Toops, D., The widening world of nutraceuticals. *Food Processing's Functional Foods*. Supplement to *Food Processing*, F4–F19, July 1999.

Daniel, S. R., How to develop a customer complaint feedback system. *Food Technol.*, 38, 41, September 1984.

Daniels, R. W., Home food safety. *Food Technol.*, 52(2), 54, 1998.

Datta, A. K. and Hu, W., Optimization of quality in microwave heating. *Food Technol.*, 46(12), 53, 1992.

Davey, K. R. and Daughtry, B. J., Validation of a model for predicting the combined effect of three environmental factors on both exponential and lag phases of bacterial growth: Temperature, salt concentration and pH. *Food Res. Int.*, 28(3), 233, 1995.

Davidson, C., Drinking by numbers. *New Sci.*, 158, 36, April 11, 1998.

Davies, R., Birch, G. G., and Parker, K. J., eds., *Intermediate Moisture Foods*. Applied Science Publishers Ltd.: Essex, U.K., 1976

Davies, A., Kochar, S. P., and Weir, G. S., Studies on the efficacy of natural antioxidants. *Leatherhead Food Res. Assoc. Tech. Circ.*, No. 695, 1979.

Day, B., Extension of shelf-life of chilled foods. *Eur. Food Drink Rev.*, 47, Autumn 1989.

Day, B., A perspective of modified atmosphere packaging of fresh produce in Western Europe. *Food Sci. Technol. Today*, 4, 215, 1990.

Dean, R. C., The temporal mismatch—Innovation's pace vs management's time horizon. *Res. Manag.*, 12, May 1974.

Decareau, R. V., Cooking by microwaves. In *Proc. IUFoST Int. Symp. Progress in Food Preparation Processes*. SIK, Swedish Food Institute: Göteborg, Sweden, 1986, p. 173.

Deis, R. C., The facts on functional foods. *Food Prod. Des.*, 13, 41, July 2003.

Demetrakakes, P., Zap! *Food Process.*, 59, 20, February 1998a.

Demetrakakes, P., Information overload. *Food Process.*, 59, 17, April 1998b.

Dempster, J. F., Hawrysh, Z. J., Shand, P., Lahola-Chomiak, L., and Corletto, L., Effect of low-dose irradiation (radurization) on the shelf life of beefburgers stored at 3°C. *J. Food Technol.*, 20, 145, 1985.

Denton, D. K., Four steps to resolving conflicts. *Qual. Prog.*, 29, April 1989.

van Dessel, M., The ZMET technique: A new paradigm for improving marketing and marketing research. In: Broadening the Boundaries: Marketing Research and Research Methodologies: ANZMAC (Australia and New Zealand Marketing Academy) conference Proceedings, December 5–7, 2005. http://www.smib.vuw.ac.nz:8081/WWW/ANZMAC2005/cd-site/pdfs/11-MR-Qualitative/11-van-Dessel.pdf (accessed and copied August 14, 2009).

Dethmers, A. E., Utilizing sensory evaluation to determine product shelf life. *Food Technol.*, 33, 40, September 1979.

De Vries, U., Velthuis, H., and Koster, K., Baking ovens and product quality—A computer model. *Food Sci. Technol. Today*, 9(4), 232, 1995.

Dimou, M., Marnasidis, S., Antoniadou, I., Pliatsika, M., and Besseris, G., The application of taguchi method to determine the optimum blend of unifloral honeys to most closely match thyme honey quality. *Int. J. Food. Sci. Technol.*, 44(10), 1877–1886, 2009.

Dörnenburg, H. and Knorr, D., Monitoring the impact of high-pressure processing on the biosynthesis of plant metabolites using plant cell cultures. *Trends Food Sci. Technol.*, 9(10), 355, 1998.

Downer, L., *Japanese Vegetarian Cooking*. Pantheon Books: New York, 1986.

Drewnowski, A., Henderson, S., Hann, C., Berg, W., and Ruffin, M., Genetic taste markers and preferences for vegetables and fruit of female breast care patients. *J. Am. Diet. Assoc.*, 100(2), 191, 2000.

Duffy, V. and Bartoshuk, L., Food acceptance and genetic variation in taste. *J. Am. Diet Assoc.*, 100(6), 637, 2000.

Duthie, G. G., Antioxidant hypothesis of cardiovascular disease. *Trends Food Sci. Technol.*, 2, 205, 1991.

Duxbury, D., New products, flavors, and uses of under-utilized species. *Activities Rep. Res. Dev. Assoc.*, 39(2), 51, 1987.

Dziezak, J. D., Fats, oils and fat substitutes. *Food Technol.*, 43, 66, July 1989.

Dziezak, J. D., Taking the gamble out of product development. *Food Technol.*, 44, 110, June 1990.

Dziezak, J. D., A focus on gums. *Food Technol.*, 45, 116, March 1991.

Eagle, S., Pushing the envelope. *Food Can.*, 59, 23, June 1999.

Earle, M. D., Changes in the food product development process. *Trends Food Sci. Technol.*, 8(1), 19, 1997a.

Earle, M. D., Innovation in the food industry. *Trends Food Sci. Technol.*, 8(5), 166, 1997b.

Eilperin, J., Sugar industry aims to block WHO report. *The Gazette, Montreal*, A21, April 25, 2003.

Eisner, M., *Introduction into the Technique and Technology of Rotary Sterilization*, 2nd edn., private author's edition, copyright 1988, M. Eisner, Brockdorf, Germany.

Elliott, J. G., Application of antioxidant vitamins in foods and beverages. *Food Technol.*, 53(2), 46, 1999.

Engel, C., Natural colours. Their stability and application in food. *Br. Food Manuf. Ind. Res. Assoc. Sci. Tech. Surveys*, 117, 1979.

Engelund, E., Breum, G., and Friis, A., Optimisation of large-scale food production using Lean Manufacturing principles. *J. Foodservice*, 20(1), 4–14, 2009.

Erhard, D., Nutrition education for the "now" generation. *J. Nutr. Educ.*, 135, Spring, 1971.

Erickson, D., Brain food. *Sci. Am.*, 265, 124, November 1991.

Erkmen, O., Predictive modelling of *Listeria monocytogenes* inactivation under high pressure carbon dioxide. *Lebensm. Wiss. Technol.*, 33(7), 514, 2000.

European Union (EU). Regulation (EC) No 258/97 of the European Parliament and of the Council of 27 January 1997 concerning novel foods and novel food ingredients. *Off. J.*, L043, P0001–P0006, February 14, 1997.

Ewaidah, E. H. and Hassan, B. H., Prickly pear sheets: A new fruit product. *Int. J. Food Sci. Technol.*, 27, 353, 1992.

Falagas, M. and Kompoti, M., Obesity and infection. *Lancet Infect Dis.*, 6(7), 438–446, 2006.

Falk, D., The ads have eyes. University of Toronto Magazine, Winter. URL to article: http://www.magazine.utoronto.ca/leading-edge/cognovisiion-advertising-view-tracking-software/, 2010.

Farber, J. M., Foodborne pathogenic microorganisms: Characteristics of the organisms and their associated diseases I. Bacteria. *Can. Inst. Food Sci. Technol. J.*, 22, AT/311, 1989.

Farr, D., High pressure technology in the food industry. *Trends Food Sci. Technol.*, 1, 14, 1990.

FDD (Food Development Division), Ag. Canada, Modified atmosphere packaging: (A) An extended shelf life technology; (B) Investment decisions; (C) The consumer perspective, Food Development Division, Agricultural Development Branch, Agriculture Canada, Ottawa, Canada, 1990.

Feng, R., Ni, H.-M., Yang, S. et al., Cyanidin-3-rutinoside, a natural polyphenol antioxidant, selectively kills leukemic cells by induction of oxidative stress. *J. Biol. Chem.*, 282(18), 13429–13437, 2007.

Fernandez de Tonella, M. L., Taylor, R. R., and Stull, J. W., Properties of a chocolate-flavored beverage from chick-pea. *Cereal Foods World*, 26, 528, 1981.

Figueiredo, A. A., Mesquite: History, composition and food uses. *Food Technol.*, 44, 118, November 1990.

Flavelle, D., Consumers in the wild. *Tor. Star*, B1, 2, January 16, 2010.

Fletcher, J., Frisvold, D., and Tefft, N., Can soft drink taxes reduce population weight? *Contemp. Econ. Policy*, October 15, 2009 (online). doi: 10.1111/J.1465-7287.2009.00182.X.

Floros, J. D. and Chinnan, M. S., Computer graphics-assisted optimization for product and process development. *Food Technol.*, 42, 72, February 1988.

Foot, D. K. and Stoffman, D., *Boom, Bust & Echo: Profiting from the Demographic Shift in the 21st Century*. Stoddart: Toronto, Canada, 2001.

Francis, F. J., Natural food colorants. *Cereal Foods World*, 26, 565, 1981.

Francis, F. J., A new group of food colorants. *Trends Food Sci. Technol.*, 3, 27, 1992.

Frenzen, P. D., Majchrowicz, A., Buzby, J. C., and Imhoff, B., Consumer acceptance of irradiated meat and poultry products, USDA, Economic Research Service. *Agric. Res. Bull.*, No. 757/August 2000.

Fuchs, C. S., Giovannucci, E. L., Colditz, G. A., Hunter, D. J., Stampfer, M. J., Rosner, B., Speizer, F. E., and Willett, W. C., Dietary fiber and the risk of colorectal cancer and adenoma in women. *N. Engl. J. Med.*, 340, 169, 1999.

Fuller, G. W., Ingredients and "green" labels. *Food Technol.*, 47, 68–71, August 1993.

Fuller, G. W., *New Food Product Development: From Concept to Marketplace*. CRC Press: Boca Raton, FL, 1994.

Fuller, G. W., *Getting the Most Out of Your Consultant: A Guide to Implementation*. CRC Press: Boca Raton, FL, 1999.

Fuller, G. W., *Food, Consumers, and the Food Industry: Catastrophe or Opportunity?* CRC Press: Boca Raton, FL, 2001.

Fuller, G. W., Additives. In *Encyclopedia of Food and Culture*, editor in chief, S. H., Katz, associate editor, Weaver, W. W. Charles Scribner's Sons: Thomson Gale, New York City, 2003, pp. 7–14.

Gabriel, S. L., Separation and identification of the anthocyanins in the sweet potato (*Ipomoea batatas*) using high pressure liquid chromatographic methods. MS thesis, Food Science and Nutrition, University of Massachusetts: Amherst, MA, 1989.

Gachovska, T., Simpson, M. V., Ngadi, M., and Raghavon, G., Pulsed electric field treatment of carrots before drying and rehydration. *J. Sci. Food. Agric.*, 89(14), 2372–2376, 2009.

Gains, N. and Thomson, D., Contextual evaluation of canned lagers using repertory grid method. *Int. J. Food Sci. Technol.*, 25, 699, 1990.

Gaisford, S. E., Information, databases and the food technologist. *Food Technol. Int. Eur.*, 33, 1989.

Garcia, D. J., Omega-3 long-chain PUFA nutraceuticals. *Food Technol.*, 52(6), 44, 1998.

Garetto, J. M., Protecting trademarks and servicemarks. *Food Technol.*, 57, 42, February 2003.

Gaunt, I. F., Food irradiations—Safety aspects. *Inst. Food Sci. Technol. Proc.*, 19, 171, 1985.

Geake, E. and Coghlan, A., Industry "does not need research." *New Sci.*, 133(1808), 15, 1992.

Geeson, J. D., Smith, S. M., Everson, H. P., George, P. M., and Browne, K. M., Modified atmosphere packaging to extend the shelf life of tomatoes. *Int. J. Food Sci. Technol.*, 22, 659, 1987.

Geeson, J. D., Genge, P., Smith, S. M., and Sharples, R. O., The response of unripe Conference pears to modified atmosphere retail packaging. *Int. J. Food Sci. Technol.*, 26, 215, 1991.

Gehm, B. D., McAndrews, J. M., Pei-Yu, C., and Jameson, J. L., Resveratrol, a polyphenolic compound found in grapes and wine, is an agonist for the estrogen receptor. *Proc. Natl. Acad. Sci. U.S.A.*, 94(25), 14138–14143, 1997.

Gélinas, P., Freezing and fresh bread. *Alimentech*, 4, 12, June 1991.

George, K. C., Hebbar, S. A., Kale, S. P., and Kesavan, P. C., Caffeine protects mice against whole-body lethal dose of γ-irradiation. *J. Radiol. Prot.*, 19, 171–176, June 1999.

Gerlach, D., Alleborn, N., Baars, A., Delgado, A., Moritz, J., and Knorr, D., Numerical simulations of pulsed electric fields for food preservation: A review. *Innovat. Food Sci. Emerg. Tech.*, 9(4), 408–417, 2008.

Gershman, M., *Getting it Right the Second Time: How American Ingenuity Transformed Forty-Nine Failures into some of our Most Successful Products*. Addison-Wesley Publishing Company, Inc.: Reading, MA, 1990.

Gibbons, M., Greer, J. R., Jevons, F. R., Langrish, J., and Watkins, D. S., Value of curiosity-oriented research. *Nature*, 225, 1005, 1970.

Gibbs, P. A. and Williams, A. P., Using mathematics for shelf life prediction. *Food Technol. Int. Eur.*, 287, 1990.

Gibson, L. D., The psychology of food: Why we eat what we eat when we eat it. *Food Technol.*, 35, 54, February 1981.

Giddings, G. G., Sterilization of spices: Irradiation vs gaseous sterilization. *Activ. Rep. Res. Dev. Assoc.*, 36(2), 20, 1984.

Giddings, G. G., Irradiation: Progress or peril? Pro: Safety is no longer an issue. *Prep. Foods*, 158(9), 62, 1989.

Giese, J., University centers ease product development. *Food Technol.*, 53, 98, November 1999.

Giese, J., Additional university research centers. *Food Technol.*, 54, 63, February 2000.

Giese, J., Selecting an outside food testing laboratory. *Food Technol.*, 55, 70, November 2001.

Giese, J. H., Alternative sweeteners and bulking agents. *Food Technol.*, 47, 114, January 1993.

Ginsberg, C. and Ostrowski, A., The market for vegetarian foods, 2010. http://www.vrg.org/nutshell/market.htm#market (accessed January 29, 2010.).

Giovannucci, E., Tomatoes, tomato-based products, lycopene and cancer: Review of the epidemiologic literature. *J. Natl. Cancer Inst.*, 91(4), 317, February 17, 1999.

Gitelman, P., Opportunities in the food industry. *Can. Inst. Food Sci. Technol. J.*, 19, xix, 1986.

Glew, G., Introduction to catering production planning. In *Proc. IUFoST Int. Symp. Progress in Food Preparation Processes*. SIK, Swedish Food Institute: Göteborg, Sweden, 1986, p. 203.

Glyer, J., Diet healing: A case study in the sociology of health, *J. Nutr. Educ.*, 4(4), 163, 1972.

Godfrey, W., A retailing perspective. *Food Sci. Technol. Today*, 2, 56, 1988.

Goff, H. D., Low-temperature stability and the glassy state in frozen foods. *Food Res. Int.*, 25, 317, 1992.

Goldenfield, I., The regulations affecting product development—Industry view. *Food Technol.*, 31, 80, July 1977.

Goldman, A., A study of product development management practice among food manufacturing companies located in southern Ontario, Working Paper No. 84-303, ISSN 0826-8878, Department of Consumer Studies, University of Guelph, Guelph, Canada, 1983.

Gómez, L., Úbeda, J., Arévalo-Villena, M., and Briones, A., Novel alcoholic beverages: Production of spirits and liqueurs using maceration of melon fruits in melon distillates. *J. Sci. Food. Agric.*, 6, 1018–1022, 2009.

Gómez-López, V., Devlieghere, F., Bonduelle, V., and Debevere, J., Intense light pulses decontamination of minimally processed vegetables and their shelf-life. *Int. J. Food. Microbiol.*, 103(1), 79–89, 2005.

Goodlad, R., Dietary fibre and the risk of colorectal cancer. *Gut*, 48(5), 587–589, 2001.

Gordon, M. H., Finding a role for natural antioxidants. *Food Technol. Int. Eur.*, 187, 1989.

Gould, G., Predictive mathematical modelling of microbial growth and survival in foods. *Food Sci. Technol. Today*, 3, 89, 1989.

Gowland, H., Allergic customers and food suppliers: Bridging the information gap. *Food Sci. Technol.*, 16(4), 44, 2002.

Graf, E. and Saguy, I. S., In *Food Product Development: From Concept to the Marketplace*, eds. E. Graf and I. S. Saguy. Van Nostrand Reinhold: New York, 1991, Chap. 3.

Graham, D., Quality programs and consumer complaints. *Food Technol. Int. Eur.*, 245, 1990.

Green, C., Getting the best value from outsourcing. *Food Can.*, 56(1), 8, 1996.

Green, S., Basaran, N., and Swanson, B. G., High-intensity light. In *Food Preservation Techniques*, eds. P. Zeuthen and L. Bogh-Sørensen. CRC Press: Boca Raton, FL, 2003, pp. 284–302, Chap. 15.

Grodner, R. and Hinton Jr., A., Low dose gamma irradiation of *Vibrio cholerae* in crab-meat (*Callinectes sapidus*). In *Proc. of 11th Annu. Tropical and Subtropical Fisheries Conf. of the Americas*, eds. R. Grodner and A. Hinton, Jr. Texas A&M Univ.: College Station, TX, 1986, p. 219.

Gutteridge, C. S., New methods for finding the right market niche. *Food Technol. Int. Eur.*, 127, 1990.

Halden, K., De Alwis, A. A. P., and Fryer, P. J., Changes in the electrical conductivity of foods during ohmic heating. *Int. J. Food Sci. Technol.*, 25, 9, 1990.

Hamm, D. J., Preparation and evaluation of trialkoxytricarbalkylate, trioxycitrate, trialkoxyglycerylether, jojoba oil and sucrose polyester as low calorie replacements of edible fats and oils. *J. Food Sci.*, 49, 419, 1984.

Hammer, M., Why projects fail.... *Ceres*, 26(1), 32, 1994.

Han, J. H., Antimicrobial food packaging. *Food Technol.*, 54, 56, March 2000.

Hang, Y. D. and Woodams, E. E., Enzymatic production of soluble sugars from corn husks. *Lebensm. Wiss. Technol.*, 32(4), 208, 1999.

Hang, Y. D. and Woodams, E. E., Corn husks: A potential substrate for production of citric acid by *Aspergillus niger*. *Lebensm. Wiss. Technol.*, 33(7), 520, 2000.

Hang, Y. D. and Woodams, E. E., Enzymatic production of reducing sugars from corn cobs. *Lebensm. Wiss. Technol.*, 34(3), 140, 2001.

Hannigan, K. J., Dried citrus juice sacs add moisture to food products. *Food Eng.*, 54, 88, March 1982.

Hardas, N., Danviriyakul, S., Foley, J. L., Nawar, W. W., and Chinachoti, P., Accelerated stability studies of microencapsulated anhydrous milk fat. *Lebensm. Wiss. Technolo.*, 33(7), 506, 2000.

Hardy, K. G., Fickle tastebuds. *Bus. Q.*, 40, Spring 1991.

Harlander, S., Food technology: Yesterday, today, and tomorrow. *Food Technol.*, 43, 196, September 1989.

Harlander, S., Social, moral and ethical issues in food biotechnology. *Food Sci. Technol. Today*, 6, 66, 1992.

Harlfinger, L., Microwave sterilization. *Food Technol.*, 46(12), 57, 1992.

Harris, J. M., Food product introductions continue to decline in 2000. *Food Rev. Consum. Driven Agric.*, 25(1), 24, May 2002.

Harvey, P., Joint ventures—Problems and opportunities. *Chem. Ind.*, 3949, December 1977.

Hasler, C. M., Functional foods: Their role in disease prevention and health promotion. *Food Technol.*, 52(11), 63, 1998.

Hauck, K., Alone on the range. *Prep. Foods*, 161(11), 32, 1992.

Hawkins, J., Canadians' attitudes toward survey research—Where does the Canadian industry stand? *Appl. Market. Res.*, 31, 27–31, Spring/Summer 1991.

Hayashi, R., Application of high pressure to food processing and preservation: Philosophy and development. In *Engineering and Food*, vol. 2, eds. W. E. L. Spiess and H. Schubert. Elsevier Applied Science: London, U.K., 1989, p. 815.

Head, A. W., The technical manager: Present and future role. *Chem. Ind.*, 716, 1971.

Headford, L., The environmental policies of the food industries in Europe and USA, *Food Sci. Technol. Today*, 10, 99–106, June 1996.

Health Canada, Health Canada issues a stop-sale order for all products containing Kava, August 21, 2002. http://www.hc-sc.gc.ca/english/protection/warnings/2002/2002_56e.htm

Heath, H., *Flavor Technology: Profiles, Products, Applications*. AVI Publishing Company, Inc.: Westport, CT, 1978.

Hegenbart, S., The R&D and marketing melee. *Prep. Foods*, 159, 117, March 1990.

Helander, I. M., von Wright, A., and Mattila-Sandholm, T.-M., Potential of lactic acid bacteria and novel antimicrobials against gram-negative bacteria. *Trends Food Sci. Technol.*, 8(5), 146, 1997.

Heldman, D. R. and Newsome, R. L., Kinetic models for microbial survival during processing. *Food Technol.*, 57, 40, August 2003.

Henika, R. G., Simple and effective system for use with response surface methodology. *Cereal Sci. Today*, 17, 309, 1972.

Herrod, R. A., Industrial applications of expert systems and the role of the knowledge engineer. *Food Technol.*, 43, 130, May 1989.

Hibler, M., Fast foods. *Can. Consum.*, 18, 19, July 1988.

Hill, S., The IFIS food science and technology bibliographic databases. *Trends Food Sci. Technol.*, 2, 269, 1991.

Hill Jr., C. G. and Grieger-Block, R. A., Kinetic data: Generation, interpretation, and use. *Food Technol.*, 34, 56, February 1980.

Hollingsworth, P., The perils of product development. *Food Technol.*, 48, 80, June 1994.

Hollingsworth, P., Slotting fees under fire. *Food Technol.*, 54, 30, November 2000.

Hollingsworth, P., Federal R&D: A boon to food companies. *Food Technol.*, 55, 45, June 2001.

Hollingsworth, P., The "self-care shopper" emerges. *Food Technol.*, 57, 20, July 2003.

Holmes, A. W., The control of research for profit. *Br. Food Manuf. Ind. Res. Assoc. Tech. Circ.*, 412, 1968.

Holmes, A. W., Securing innovation in the food industry. *Br. Food Manuf. Ind. Res. Assoc. Tech. Circ.*, 636, 1977.

Hoover, D. G., Bifidobacteia: Activity and potential benefits. *Food Technol.*, 47(6), 120, 1993.

Hoover, D. G., Metrick, C., Papineau, A. M., Farkas, D., and Knorr, D., Biological effects of high hydrostatic pressure on food microorganisms. *Food Technol.*, 43, 99, March 1989.

Houston, M. and Johnson, S., Buyer-supplier contracts versus joint ventures: Determinants and consequences of transaction structure. *J. Market. Res.*, 37, 1–15, 2000.

Hozawa, A., Kuriyama, S., Nakaya, N. et al., Green tea consumption is associated with lower psychological distress in a general population: The Ohsaki Cohort 2006 study. *Am. J. Clin. Nutr.*, 90(5), 1390–1396, 2009.

Hsieh, Y. P. C., Pearson, A. M., and Magee, W. T., Development of a synthetic meat flavor mixture by using surface response methodology. *J. Food Sci.*, 45, 1125, 1980.

Huggett, A. C. and Conzelmann, C., EU regulation on novel foods: Consequences for the food industry. *Trends Food Sci. Technol.*, 8(5), 133, 1997.

Hughes, D. B. and Hoover, D. G., Bifidobacteria: Their potential for use in American dairy products. *Food Technol.*, 45, 74, April 1991.

Huizenga, T. P., Liepins, K., and Pisano Jr., D. J., Early involvement. *Qual. Prog.*, 20, 81, June 1987.

Hunt, J. T., Moving target. *National Post Business Magazine*, 48, October 2000.

IFST(U.K.), *Guidelines for the Handling of Chilled Foods*, 2nd edn. The Institute of Food Science & Technology: London, U.K., 1990.

IFST(U.K.), Cyclospora: An IFST information statement. *Food Sci. Technol.* 17(3), 8, 2003.

IFT, Kinetics of microbial inactivation for alternative food processing technologies. A report of the Institute of Food Technologists for the Food and Drug Administration of the USDHHS, June 2, 2000. http://www.fda.gov/Food/ScienceResearch/ResearchAreas/SafePracticesforFoodProcesses/ucm100158.htm

Ishani, W., MacDonald, R., Stark, G., Mulrow, C., and Lau, J., B-sitosterols for benign prostatic hyperplasia. *Cochrane Database Syst. Rev.*, CD001043(3), 1999. doi: 10.1002/14651858.CD001043.

Ishibashi, N. and Shimamura, S., Bifidobacteria: Research and development in Japan. *Food Technol.*, 47(6), 126, 1993.

James, W., *The Principles of Psychology.* Cosimo, Inc.: New York, 1890, Chap. 2.

Jantsch, E., *Technological Forecasting in Perspective*, 112pp et seq., O.E.C.D., Paris, 1967. http://www.travail-societe.cnam.fr/lipsor/eng/data/prevtechen.pdf (accessed August 11, 2009).

Jay, J. M., Do background microorganisms play a role in the safety of fresh foods? *Trends Food Sci. Technol.*, 8(12), 421, 1997.

Jenkins, D., Dietary fibre and its relation to nutrition and health. *Inst. Food Sci. Technol. Proc.*, 13, 51, 1980.

Jeremiah, L. E., Penney, N., and Gill, C. O., The effects of prolonged storage under vacuum or CO on the flavor and texture profiles of chilled pork. *Food Res. Int.*, 25, 9, 1992.

Jezek, E. and Smyrl, T. G., Volatile changes accompanying dehydration of apples by the Osmovac process. *Can. Inst. Food Sci. Technol. J.*, 13, 43, 1980.

Johnston, W. A., Surimi—An introduction. *Eur. Food Drink Rev.*, 21, Autumn 1989.

Jolly, D. A., Schutz, H. G., Diaz-Knauf, K., V., and Johal, J., Organic foods: Consumer attitudes and use. *Food Technol.*, 43, 60, November 1989.

Jones, K., The EEC flavoring and food labelling directives. *Dragoco Rep.*, 103(3), 1992.

Josephson, E. S., Military benefits of food irradiation. *Activities Rep. Res. Dev. Assoc.*, 36(2), 30, 1984.

Juven, B. J., Schved, F., and Linder, P., Antagonistic compounds produced by a chicken intestinal strain of *Lactobacillus acidophilus. J. Food Prot.*, 55, 157, 1992.

Kabat, G. C., Cross, A. J., Park, Y. et al., Meat intake and meat preparation in relation to risk of postmenopausal breast cancer in the NIH-AARP diet and health study. *Int. J. Cancer.*, 124(10), 2430–2435, 2009.

Kardinaal, A. F. M., Waalkens-Berendsen, D. H., and Arts, C. J. M., Pseudo-oestrogens in the diet: Health benefits and safety concerns. *Trends Food Sci. Technol.*, 8(10), 327, 1997.

Karlin, R., Governor wants new rules on disclosing calorie counts. *Albany Times*, May 19, 2009.

Katz, R., Skills of an effective administrator. *Harv. Bus. Rev.*, 33(1), 33–42, 1955.

Katz, F., The changing role of water binding. *Food Technol.*, 51(10), 64, 1997.

Katz, F., That's using the old bean. *Food Technol.*, 52(6), 42, 1998.

Katz, F., Top product development trend in Europe. *Food Technol.*, 53(1), 38, 1999.

Katzenstein, A. W., The food update Delphi survey: Forecasting the food industry 10 years from now. *Food Prod. Dev.*, 11, June 1975.

Kennedy, J. P., Structured lipids: Fats of the future. *Food Technol.*, 45, 76, November 1991.

Kernon, J. M., The Foodline scientific and technical, marketing and legislation databases. *Trends Food Sci. Technol.*, 2, 276, 1991.

Kesterton, M., Cooking: A minority art? *The Globe and Mail*, A 18, January 15, 2003.

King, V. A.-E. and Zall, R. R., A response surface methodology approach to the optimization of controlled low-temperature vacuum dehydration. *Food Res. Int.*, 25, 1, 1992.

Kinsella, J. E., Frankel, E., German, B., and Kanner, J., Possible mechanisms for the protective role of antioxidants in wine and plant foods. *Food Technol.*, 47(4), 85, 1993.

Kirca, A. and Arslan, E., Antioxidant capacity and total phenolic content of selected plants from Turkey. *Int. J. Food Sci. Technol.*, 44(11), 2038–2046, 2008.

Kirk, D. and Osner, R. C., Collection of data on food waste from catering outlets in a University and a Polytechnic. *Inst. Food Sci. Technol. Proc.*, 14, 190, 1981.

Kirkpatrick, K. J. and Fenwick, R. M., Manufacture and general properties of dairy ingredients. *Food Technol.*, 41, 58, October 1987.

Klapthor, J., When it comes to healthy menu choices, reality bites. *Inst. Food. Technol.*, 2005. http://www.ift.org/cms/?pid=1001334&printable=1

Kläui, H., Naturally occurring antioxidants. *Inst. Food Sci. Technol. Proc.*, 6, 195, 1973.

Klensin, J. C., Information technology and food composition databases. *Trends Food Sci. Technol.*, 2, 279, 1991.

Knorr, D., Recovery and utilization of chitin and chitosan in food processing waste management. *Food Technol.*, 45, 114, 1991.

Knorr, D., Plant cell and tissue cultures as a model system for monitoring the impact of unit operations on plant food. *Trends Food Sci. Technol.*, 5, 328, October 1994.

Knorr, D., Technology aspects related to microorganisms in functional foods. *Trends Food Sci. Technol.*, 9(8–9), 295, November 8–9, 1998.

Knorr, D., Beaumont, M. D., Caster, C. S., Dörnenburg, H., Gross, B., Pandya, Y., and Romagnoli, L. G., Plant tissue culture for the production of naturally derived food ingredients. *Food Technol.*, 44, 71, June 1990.

Knorr, D., Schlueter, O., and Heinz, V., Impact of high hydrostatic pressure on phase transitions of foods. *Food Technol.*, 52(9), 42, 1998.

Kolata, G., Food affects human behaviour. *Science*, 218, 1209, 1982.

Konuma, H., Shinagawa, K., Tokumaru, M., Onove, Y., Konno, S., Fujino, N., Shigehisa, T., Kurata, H., Kuwabara, Y., and Lopes, C. A. M., Occurrence of *Bacillus cereus* in meat products, raw meat and meat product additives. *J. Food Prot.*, 51, 324, 1988.

Kraushar, P. M., *New Products and Diversification*. Business Books Limited: London, U.K., 1969.

Kritchevsky, D., The effect of dietary garlic on the development of cardiovascular disease. *Trends Food Sci. Technol.*, 2, 141, 1991.

Krizmanic, J., Here's who we are!, *Veg. Times*, 182, 72, 1992.

Kroll, B. J., Evaluating rating scales for sensory testing with children. *Food Technol.*, 44, 78, November 1990.

Kuhn, M. E., Functional foods overdose? *Food Process.*, 59(5), 21, 1998a.

Kuhn, M. E., Partner power. *Food Process.*, 59(8), 67, 1998b.

LaBarge, R. G., The search for a low-caloric oil. *Food Technol.*, 42, 84, January 1988.

Labuza, T. P., The effect of water activity on reaction kinetics of food deterioration. *Food Technol.*, 34, 36, April 1980.

Labuza, T. P. and Hyman, C. R., Moisture migration and control in multi-domain foods, *Trends Food Sci. Technol.*, 9(2), 47, 1998.

Labuza, T. P. and Riboh, D., Theory and application of Arrhenius kinetics to the prediction of nutrient losses in foods. *Food Technol.*, 36, 66, October 1982.

Labuza, T. P. and Schmidl, M. K., Accelerated shelf-life testing of foods. *Food Technol.*, 39, 57, September 1985.

Lachance, P. A., Human obesity. *Food Technol.*, 48(2), 127, 1994.

Lampert, A., Montreal schoolchildren speak in many tongues. *The Gazette, Montreal*, A1, A2, October 9, 2002.

Land, E., The second great product of industry: The rewarding working life, presented at Science and Human Progress: 50th Anniversary of Mellon Institute, Pittsburgh, PA, 1963, p. 107.

Lanier, T. C., Functional properties of surimi. *Food Technol.*, 40, 107, March 1986.

Larsson, S. and Wolk, A., Excess body fatness: An important cause of most cancers. *Lancet* 371(9612), 536–537, 2008.

Lawrie, R. A. and Symons, H., The food industry—Changes foreseen and unseen 1965–2000. *Food Sci. Technol.*, 15(3), 50, 2001.

Leadley, C., Developments in non-thermal processing. *Food Sci. Technol.*, 17(3), 40, 2003.

Lechowich, R. V., Microbiological challenges of refrigerated foods. *Food Technol.*, 42, 84, December 1988.

Lee, C. M., Surimi process technology. *Food Technol.*, 38, 69, November 1984.

Lee, J., The development of new products for the European Market. *Food Sci. Technol. Today*, 5, 155, 1991.

Lee, K., Food neophobia: Major causes and treatment. *Food Technol.*, 43, 62, 1989.

Lee, Y.-K. and Salminen, S., The coming of age of probiotics. *Trends Food Sci. Technol.*, 6(7), 241, 1995.

Leeder, J., What's cooking? A kitchen makeover. *The Globe and Mail*, B3, January 14, 2010.

Leistner, L., Hurdle technology applied to meat products of the shelf stable product and intermediate moisture food types. In *Properties of Water in Foods*, eds. D. Simatos and J. L. Multon. Martinus Nijhoff Publishers: Dordrect, the Netherlands, 1985, p. 309.

Leistner, L., Shelf stable products and intermediate moisture foods based on meat. Presented at *IFT-IUFoST Basic Symp. Water Activity: Theory and Applications*, Dallas, TX, June 13–14, 1986, Chap. 13.

Leistner, L., Food preservation by combined methods. *Food Res. Int.*, 25, 151, 1992.

Leistner, L. and Rödel, W., Inhibition of micro-organisms in food by water activity. In *Inhibition and Inactivation of Vegetative Microbes*, eds. F. A. Skinner and W. B. Hugo. Academic Press: London, U.K. 1976a, p. 219.

Leistner, L. and Rödel, W., The stability of intermediate moisture foods with respect to micro-organisms. In *Intermediate Moisture Foods*, eds. R. Davies, G. G. Birch, and K. J. Parker. Applied Science Publishers: London, U.K., 1976b, Chap. 10.

Leistner, L., Rödel, W., and Krispien, K., Microbiology of meat and meat products in high- and intermediate-moisture ranges. In *Water Activity: Influences on Food Quality*, eds. L. B. Rockland and G. F. Stewart. Academic Press: New York, 1981, p. 855.

Lenz, M. K. and Lund, D. B., Experimental procedures for determining destruction kinetics of food components. *Food Technol.*, 34, 51, February 1980.

Levitt, T., Marketing myopia. *Harv. Bus. Rev.*, 38, 45, July/August 1960.

Levitt, T., Marketing myopia. *Harv. Bus. Rev.*, 53, 26, September/October 1975.

Lewandowski, R., Corporate confidential. *The Financial Post Mag.*, 18, March 1999.

Lieber, H., Preserved radio-active organic matter and food. U.S. Patent 788.480, 1905.

Light, N., Young, H., and Youngs, A., Operating temperatures in chilled food vending machines and risk of growth of food poisoning organisms. *Food Sci. Technol. Today*, 1, 252, 1987.

Lightbody, M. S., New technological approaches to reducing uniformity in processed foods. *Food Sci. Technol. Today Proc.*, 4, 37, 1990.

Lilieveld, H. L. M., Notermans, S., and de Haan, S. W. H., eds., *Food Preservation by Pulsed Electric Fields: From Research to Application*. Woodhead Publishing: Abington, U.K., 2007.

Lingle, R., AmeriQual Foods: At ease with MRE's. *Prep. Foods*, 158(9), 144, 1989.

Lingle, R., Degradable plastics: All sizzle and no steak? *Prep. Foods*, 159(1), 144, 1990.

Lingle, R., Streamlining package design through computers. *Prep. Foods*, 160(8), 86, 1991.

Lingle, R., A sign of changing times. *Prep. Foods*, 161(2), 52, 1992.

Linnemann, A. R., Meerdink, G., Meulenberg, M. T. G., and Jongen, W. M. F., Consumer-oriented technology development. *Trends Food Sci. Technol.*, 9(11–12), 409, 1998.

Livingston, G. E., Foodservice: Older than Methuselah. *Food Technol.*, 44, 54, July 1990.

Loaharanu, P., International trade in irradiated foods: Regional status and outlook. *Food Technol.*, 43, 77, 1989.

Loaharanu, P., Cost/benefit aspects of food irradiation. *Food Technol.*, 48(1), 104, 1994.

Lone, T., Pence, D., Levi, A., Chan, K., and Bianco-Simeral, S., Marketing healthy food to the least interested consumer. *J. Foodservice*, 20(2), 90–99, 2009.

Looney, J. W., Crandall, P. G., and Poole, A. K., The matrix of food safety regulations. *Food Technol.*, 55, April, 60, 2001.

Lord, J. B., Launching the new product. In: *Developing New Food Products for a Changing Marketplace*, eds. A. L. Brody and J. B. Lord. Technomic Publishing Co., Inc.: Lancaster, U.K., 2000, Chap. 17.

Lovel, J., Neuromarketing firm launched by Atlanta ad veteran. In: *Atlanta Business Chronicle*, June 14, 2002. Available from http://www.thoughtsciences.com/abc061402.htm (accessed December 12, 2002).

Lülfs-Baden, F. and Spiller, A., Students' perceptions of school meals: A challenge for schools, school-meal providers, and policy makers. *J. Foodservice*, 20(1), 31–46, 2009.

Lund, B., *Cyclospora cayetanensis* update. *Food Sci. Technol.*, 16(4), 48, 2002.

Lund, D. B., Considerations in modeling food processes. *Food Technol.*, 37, 92, January 1983.

MacAulay, J., Obesity: An issue of national importance. *Food Technol.*, 57, February 20, 2003.

MacDonald, G. A. and Lanier, T., Carbohydrates as cryoprotectants for meats and surimi. *Food Technol.*, 45(3), 150, 1991.

MacFie, H., Factors affecting consumers' choice of food. *Food Technol. Int. Eur.*, 123, 1990.

MacNulty, C. A. R., Food products and the future. *Food Technol. Int. Eur.*, 19, 1989.

Malpas, R., Chemical technology—Scaling greater heights in the next ten years? *Chem. Ind.*, 111, 1977.

Mañas, P., Barsotti, L., and Cheffel, J. C., Microbial inactivation by pulsed electric fields in a batch treatment chamber: Effects of some electrical parameters and food constituents. *Innovat. Food Sci. Emerg. Tech.*, 2(4), 239–249, 2001.

Mans, J., Kyotaru's bridge across the Pacific. *Prep. Foods*, 161(12), 85, 1992.

Mantonakis, A., Rodero, P., Lesschaeve, I., and Hastie, R., Order in choice: Effects of serial position on preferences. *Psychol. Sci.* 20(11), 1309–1312, November 2009.

Marco, M., KFC has a bacon sandwich that uses fried chicken as "bread." *The Consumerist* 12:40 pm, August 21, 2009. http://www.consumerist.com/5342699/kfc-has-a-bacon-sandwich-that-uses-fried-chicken-as-bread

Marcotte, M., Irradiated strawberries enter the U.S. market. *Food Technol.*, 46, 80, May 1992.

Marcus, J. B., Unleashing the power of umami. *Food Technol.*, 63(11), 22–24, 27–28, 30, 32, 34–36, 2009.

Mardon, J., Cripps, W. C., and Matthews, G. T., The organisation and administration of technical departments in large multi-plant companies. *Chem. Ind.*, 450, 1970.

Marechal, P. A., Martínez de Marnañón, I., Poirier, I., and Gervais, P., The importance of the kinetics of application of physical stresses on the viability of microorganisms: Significance for minimal food processing. *Trends Food Sci. Technol.*, 10(1), 15, 1999.

Marlow, P., Qualitative research as a tool for product development. *Food Technol.*, 41, 74, November 1987.

Marquez, V. O., Mittal, G., and Griffiths, M. W., Destruction and inhibition of bacterial spores by high voltage pulsed electrical fields. *J. Food. Sci.*, 62, 399–401, March 1997.

Martin, D., The impact of branding and marketing on perception of sensory qualities. *Food Sci. Technol. Today Proc.*, 4, 44, 1990.

Mason, L. H., Church, I. J., Ledward, D. A., and Parsons, A. L., The sensory quality of foods produced by conventional and enhanced cook-chill method. *Int. J. Food Sci. Technol.*, 25, 247, 1990.

Matthews, M. E., Foodservice in health care facilities. *Food Technol.*, 36, 53, July 1982.

Mattson, P., Eleven steps to low cost product development. *Food Prod. Dev.*, 6, 106, 1970.

Mattson, B. and Soneson, U., eds., *Environmentally-friendly Food Processing*, Woodhead Publishing Limited: Abington, Cambridge, U.K., 2003.

Maugh II, T. H., "Cancer is not inevitable." *Science*, 212, 36, 1982.

Mayo, J. S., AT&T: Management questions for leadership in quality. *Qual. Prog.*, 34, April 1986.

Mazza, G., Development and consumer evaluation of a native fruit product. *Can. Inst. Food Sci. Technol. J.*, 12, 166, 1979.

McCaskill, D. R. and Zhang, F., Use of rice bran oil in foods. *Food Technol.*, 53(2), 50, 1999.

McDermott, B. J., Identifying consumers and consumer test subjects. *Food Technol.*, 44, 154, November 1990.

McDermott, R. L., Functionality of dairy ingredients in infant formula and nutritional specialty products. *Food Technol.*, 41, 91, 1987.

McGinn, C. J. P., Evaluation of shelf life. *Inst. Food Sci. Technol. Proc.*, 15, 153, 1982.

McKenna, J., Functional ingredients close-up. *Food Processing's Functional Foods*: Supplement to *Food Processing*, F20–F21, July 1999.

McLellan, M. R., An introduction to artificial intelligence and expert systems. *Food Technol.*, 43, 120, May 1989.

McLellan, M. R., Hoo, A. F., and Peck, V., A low-cost computerized card system for the collection of sensory data. *Food Technol.*, 41, 66, November 1987.

McPhee, M., Organically grown-up? *Prep. Foods*, 161(3), 17, 1992.

McProud, L. M. and Lund, D. B., Thermal properties of beef loaf produced in food-service systems. *J. Food Sci.*, 48, 677, 1983.

McWatters, K. H., Resurreccion, A. V. A., and Fletcher, S. M., Response of American consumers to akara, a traditional West African food made from cowpea paste. *Int. J. Food Sci. Technol.*, 25, 551, 1990.

McWatters, K. H., Enwere, N. J., and Fletcher, S. M., Consumer response to akara (fried cowpea paste) served plain or with various sauces. *Food Technol.*, 46, 111, February 1992.

McWeeny, D. J., Long term storage of some dry foods: A discussion of some of the principles involved. *J. Food Technol.*, 15, 195, 1980.

Megremis, C. J., Medium-chain triglycerides: A nonconventional fat. *Food Technol.*, 45, 108, February 1991.

Meheriuk, M., Girard, B., Moyls, L., Beveridge, H. J. T., MCKenzie, D.-L., Harrison, J., Weintraub, S., and Hocking, R., Modified atmosphere packaging of "Lapins" sweet cherry. *Food Res. Int.*, 28(3), 239, 1995.

Meisel, H. and Schlimme, E., Milk proteins: Precursors of bioactive peptides. *Trends Food Sci. Technol.*, 1, 41, August 1990.

Meltzer, R., Value added products: A noteworthy niche. *Visions*, 2(3), 1991.

Mermelstein, N. H., Retort pouch earns 1978 IFT food technology industrial achievement award. *Food Technol.*, 32, 22, June 1978.

Mermelstein, N. H., Software for food processing. *Food Technol.*, 54, 56, January 2000.

Mermelstein, N. H., Military and humanitarian rations. *Food Technol.*, 55, 73, November 2001.

Mertens, B. and Knorr, D., Developments of nonthermal processes for food preservation. *Food Technol.*, 46, 124, May 1992.

Messens, W., Van Camp, J., and Huyghebaert, A., The use of high pressure to modify the functionality of food proteins. *Trends Food Sci. Technol.*, 8, 107, April 1997.

Metrick, C., Hoover, D. G., and Farkas, D. F., Effects of high hydrostatic pressure on heat-resistant and heat-sensitive strains of *Salmonella*. *J. Food Sci.*, 54, 1547, 1989.

Meyer, R. S., Eleven stages of successful new product development. *Food Technol.*, 38, 71, July 1984.

Mills, E. N. C., Alcocer, M. J. C., and Morgan, M. R. A., Biochemical interactions of food-derived peptides. *Trends Food Sci. Technol.*, 3, 64, 1992.

Mintz, S., *Sweetness and Power: The Place of Sugar in Modern History*. Penguin Books: London, England, U.K., 1985.

Monagas, M., Nasiruddin, K., Andres-Lacueva, A. et al., Effect of cocoa powder on the modulation of inflammatory biomarkers in patients at high risk of cardio-vascular disease. *Am. J. Clin. Nutr.*, 90(5), 1144–1150, 2009.

Morris, C. E., Why new products fail. *Food Eng.*, 65(6), 130, 1993.

Moskowitz, H. R., Benzaquen, I., and Ritacco, G., What do consumers really think about your product? *Food Eng.*, 53, 80, September 1981.

Moskowitz, H., Katz, R., and Cappuccio, R., Consumer-driven development: Coffee beverage mindsets for three commercial venues. *J. Foodservice*, 18(1), 7–22, 2007.

Mossel, D. A. A. and Ingram, M., The physiology of the microbial spoilage of food. *J. App. Bacteriol.*, 18, 232, 1955.

Mullen, K. and Ennis, D. M., Rotatable designs in product development. *Food Technol.*, 33, 74, July 1979a.

Mullen, K. and Ennis, D. M., Mathematical system enters development realm as aid in achieving optimum ingredient levels. *Food Prod. Dev.*, 15, 50, November 1979b.

Mullen, K. and Ennis, D., Fractional factorials in product development. *Food Technol.*, 39, 90, May 1985.

Muller, R. A., Innovation and scientific funding. *Science*, 209, 880, 1980.

Mundy, C. C., Accessing the literature of food science. *Trends Food Sci. Technol.*, 2, 272, 1991.

Nathan, I., Hackett, A., and Kirby, S., Vegetarianism and health: Is a vegetarian diet adequate for the growing child? *Food Sci. Technol. Today*, 8(1), 13, 1994.

Nazario, S. L., Big firms get high on organic farming. *Wall St. J.*, B1, March 21, 1989.

NCHS., Prevalence of overweight and obesity among adults: United States, 1999–2000, 1999. http://www.cdc.gov/nchs/products/pubs/pubd/hestats/obese/obse99.htm

NCI (National Cancer Institute/Office of Cancer Communications), Cancer prevention research: Chemoprevention and diet, *Backgrounder*, December 1984.

Neaves, P., Gibbs, P. A., and Patel, M., Inhibition of *Clostridium botulinum*, by preservative interactions. *Br. Food Manuf. Ind. Res. Assoc. Res. Rep.*, No. 378, 1982.

Neff, J., Timid steps. *Food Process.*, 59(5), 33, 1998.

Neugebauer, W., "Electronic noses"—Possibilities and limitations of chemical sensor systems. Dragoco Report, 6/1998, p. 257.

Newiss, H., The patenting of genetically modified foods. *Trends Food Sci. Technol.*, 9, 368, October 1998.

Newsome, R., Organically grown foods. *Food Technol.*, 44, 26, June 1990.

Nguyen, M. and Schwartz, S., Lycopene: Chemical and biological properties. *Food Technol.*, 53(2), 38, 1999.

NHANES, Overweight among U.S. children and adolescents. http://www.cdc.Gov/nchs/nhanes.htm, undated.

Niu, K., Hozawa, A., Kuriyama, S. et al., Green tea consumption is associated with depressive symptoms in the elderly. *Am. J. Clin. Nutr.*, 90(6), 1615–1622, 2009.

Noel, T. R., Ring, S. G., and Whittam, M. A., Glass transitions in low-moisture foods. *Trends Food Sci. Technol.*, 1, 62, 1990.

Norback, J., Techniques for optimization of food processes. *Food Technol.*, 34, 86, February 1980.

Normand, F. L., Ory, R. L., and Mod, R. R., Binding of bile acids and trace minerals by soluble hemicelluloses of rice. *Food Technol.*, 41, 86, February 1987.

Norwig, J. F. and Thompson, D. R., Making accurate energy efficiency measurements in foodservice equipment. *Activities Rep. Res. Dev. Assoc.*, 36(2), 37, 1984.

Nuttal-Smith, C., Where the buffalo roam. *Report on Business*, 26(1), 54, September 2009.

O'Brien, J., An overview of online information resources for food research. *Trends Food Sci. Technol.*, 2, 301, 1991.

O'Donnell, C. D., Computers: What's the use? *Prep. Foods*, 160(8), 62, 1991.

O'Donnell, C. D., Aphrodisiacs. *Prep. Foods*, 162(3), 69, 1992.

O'Donnell, C. D., The formulation challenge: Formulating for the big freeze. *Prep. Foods*, 162, 55, 1993.

Ohlsson, T., Boiling in water. In: *Proc. IUFoST Int. Symp. Progress in Food Preparation Processes*. SIK, Swedish Food Institute: Göteborg, Sweden, 1986, p. 89.

Ohlsson, T., Minimal processing—Preservation methods of the future: An overview. *Trends Food Sci. Technol.*, 5(11), 341, 1994.

Ohlsson, T. and Bengtsson, N. In *Minimal Processing Technologies in the Food Industries*, eds. T. Ohlsson and N. Bengtsson. CRC Press: Boca Raton, FL; 2002.

Ohr, L. M., Fats for healthy living. *Food Technol.*, 57, 91, July 2003.

Ohshima, T., Recovery and use of nutraceutical products from marine resources. *Food Technol.*, 52(6), 50, 1998.

Oickle, J. G., *New Product Development and Value Added*. Food Development Division, Agriculture Canada: Ottawa, Canada, May 1990.

Okamoto, M., Kawamura, Y., and Hayashi, R., Application of high pressure to food processing: Textural comparison of pressure- and heat-induced gels of food proteins. *Agric. Biol. Chem.*, 54(1), 183, 1990.

Okoli, E. C. and Ezenweke, L. O., Formulation and shelf-life of a bottled pawpaw juice beverage. *Int. J. Food Sci. Technol.*, 25, 706, 1990.

Olano, A. and Corzo, N., Lactulose as a food ingredient. *J. Sci. Food. Agric.*, 89(12), 1987–1990, 2009.

Olson, A., Gray, G. M., and Chiu, M., Chemistry and analysis of soluble dietary fiber. *Food Technol.*, 41, 71, February 1987.

Olson Zaltman Associates, Overview of ZMET research process, Olson Zaltman Associates, State College, PA 16803, undated.

Oms-Oliu, G., Martin-Belloso, O., and Soliva-Fortuny, R., Pulsed light treatments for food preservation: A review. *Food and Bioprocess Tech.*, 2008. doi: 10: 1007/211947-008-0147.

O'Neill, M., Geneticists' latest discovery: Public fear of "Frankenfood." The *New York Times*, 1, June 28, 1992.

Orpana, H., Berthelot, J.-M., Kaplan, M., Feeny, D., McFarland, B., and Ross, N., BMI and mortality: Results from a national study of Canadian adults. *Obesity*, 18, 214–218, June 18, 2009. doi: 10.1038/oby.2009.191.

O'Sullivan, M. G., Thornton, G., O'Sullivan, G. C., and Collins, J. K., Probiotic bacteria: Myth or reality? *Trends Food Sci. Technol.*, 3, 309, 1992.

Palmer, G. M., *Wine Production from Cheese Whey, EPA-600/2-79-189*. Industrial Environmental Research Laboratory, Office of Research and Development, U.S. Environmental Protection Agency: Cincinnati, OH, October, 1979.

Panickar, K., Polansky, M., and Anderson, R., Cinnamon polyphenols attenuate cell swelling and mitochondrial dysfunction following oxygen-glucose deprivation of glial cells. *Exp. Neurol.*, 216(2), 420–427, 2009.

Park, C. E., Szabo, R., and Jean, A., A survey of wet pasta packaged under a CO:N (20:80) mixture for Staphylococci and their enteriotoxins. *Can. Inst. Food Sci. Technol. J.*, 21, 109, 1988.

Parrott, D. L., Use of ohmic heating for aseptic processing of food particulates. *Food Technol.*, 46(12), 68, 1992.

Parry, C., KFC tests out bunless burger to rave reviews in America's heartland. *Vancouver Sun*, August 25, 2009. http://www.vancouversun.com/health/tests+artery+burger+rave+reviews+America+heartland/1917736/story.html

Parsons, R., *Statistical Analysis: A Decision-making Approach*, 2nd edn. Harper and Row: New York, 1978, Chaps. 5 and 6.

Paster, N., Juven, B. J., Gagel, S., Saguy, I., and Padova, R., Preservation of a perishable pomegranate product by radiation pasteurization. *J. Food Technol.*, 20, 367, 1985.

Patel, T., German syringes turn up in French quarry. *New Sci.*, 135(1835), 7, 1992.

Patel, T., Energy booster "a con" says French food watchdog. *New Sci.*, 137(1863), 8, 1993.

Patel, M. and Thompson, P., Phytosterols and vascular disease. *Atherosclerosis*, 186(1), 12–19, 2006.

Patterson, J. and Haas, S., A rationale for outsourcing. *Food Technol.*, 53, 52, October 1999.

Pearce, S. J., Quality assurance involvement in new product design. *Food Technol.*, 41, 104, 1987.

Pearce, F., Silence of the experts. *New Sci.*, 150, 50, November 30, 1996.

Peck, M. W., *Clostridium botulinum* and the safety of refrigerated processed foods of extended durability. *Trends Food Sci. Technol.*, 8, 186, June 1997.

Pehanich, M., Top tips for working with chains, in *Foodservice Annual*, in *Food Product Design*, 45, April 2003.

Pennington, J. A. T. and Butrum, R. R., Food descriptions using taxonomy and the "Langual" system. *Trends Food Sci. Technol.*, 2, 285, 1991.

Penny, C., Detailing dietary fibre. *Food Ingred. Process. Int.*, 14, November 1992.

Peryam, D. R., Sensory evaluation—The early days. *Food Technol.*, 44, 86, January 1990.

Peters, J. W., Foodservice R & D at Gino's—Spotting the winners faster. *Food Prod. Dev.*, 16, 28, October 1980.

Peters, D., Power, influence, and group dynamics in sensory evaluation. *Food Technol.*, 41, 62, November 1987.

Petersen, K., Nielsen, P. V., Berteses, G., Lawther, M., Olsen, M. B., Nilsson, N. H., and Mortensen, G., Potential of biobased materials for food packaging. *Trends Food Sci. Technol.*, 10(2), 52, 1999.

Peterson, T. and Van Fleet, D., The ongoing legacy of R. L. Katz: An updated typology of management skills. *Manag. Decis.*, 42(10), 1297–1308, 2004.

Petesch, B. L. and Sumiyoshi, H., Recent advances on the nutritional benefits accompanying the uses of garlic as a supplement. *Trends Food Sci. Technol.*, 9(11/12), 415, 1999.

Picard, A., Fat nation heavier wider weaker. *The Globe and Mail*, L1, 4, January 14, 2010.

Picart, L. and Cheftel, J.-C., Pulsed electric fields. In: *Chapter 18 in Food Preservation Techniques*, eds. P. Zeuthen and L. Bogh-Sørensen. CRC Press: Boca Raton, FL, 2003, pp. 360–427.

Piggott, J. R., Simpson, S. J., and Williams, S. A. R., Sensory analysis. *Int. J. Food Sci. Technol.*, 33, February 7, 1998.

Pine, R. and Ball, S., Productivity and technology in catering operations. *Food Sci. Technol. Today*, 1, 174, 1987.

Pirie, N. W., *Leaf Protein and its By-products in Human and Animal Nutrition*, 2nd edn., Cambridge University Press: London, U.K., 1987.

Pokorný, J., Natural antioxidants for food use. *Trends Food Sci. Technol.*, 2, 223, 1991.

Pollan, M., *In Defense of Food*. The Penguin Press: New York, 2008.

Port, O. and Carey, J., Getting to "Eureka!": Researchers are tracking how breakthroughs are made. *Business Week*, November 10, 1997. http://www.businessweek.com.

Porter, W. L., Storage life prediction under noncontrolled environmental temperatures: Product-sensitive environmental call-out. In: *Proc. Food Processors' Institute Shelf-life: A Key to Sharpening your Competitive Edge*. Food Processors Institute: Washington, DC, 1981, p. 1.

Porter, J. and Cant, R., Exploring hospital patients' satisfaction with cook-chill food service systems: A preliminary study using a validated questionnaire. *J. Foodservice*, 20(2), 81–89, 2009.

Poste, L. M., Mackie, D. B., Butler, G., and Larmond, E., *Laboratory Methods for Sensory Analysis of Food*. Publication 1864/E, Research Branch, Agriculture Canada: Ottawa, Canada, 1991.

Pothakamury, U. R., Barbosa-Cánovas, G. V., and Swanson, B. G., Magnetic-field inactivation of microorganisms and generation of biological changes. *Food Technol.*, 47, 85, 1993.

Powers, J. J., Uses of multivariate methods in screening and training sensory panelists. *Food Technol.*, 42, 123, November 1988.

Prospective Studies Collaboration, Body-mass index and cause-specific mortality in 900 000 adults: Collaborative analyses of 57 prospective studies. *Lancet*, 373(9669), 1083–1096, 2009. doi: 10.1016/50140-6736(09)60318-4.

Pszczola, D. E., Food irradiation: Countering the tactics and claims of opponents. *Food Technol.*, 44, 92, 1990.

Pszczola, D. E., Irradiated poultry makes U.S. debut in Midwest and Florida markets. *Food Technol.*, 47(11), 89, 1993.

Puzo, D. P., First irradiated fruit on market sells quickly. *Los Angeles Times*, Los Angeles, CA, 1986.

Pyke, M., A food scientist's reflections. *Food Sci.*, 9(1), 10, 1971.

Qureshe, I., Fang, Y., Haggerty, N., and Compeau, D., Knowledge sharing through computer mediated social ties. In: *PACI (Pacific Asia Conference on Information Systems) 2009 Proceedings*, 2009. http://www.aisel.aisnet.org/pacis2009/38 (accessed April 26, 2010 on PACIS home site).

Rajaram, S., Haddad, E.H., Mejia, A., and Sabaté, J., Walnuts and fatty fish influence different serum fractions in normal to mildly hyperlipidemic individuals: A randomized controlled study. *Am. J. Clin. Nutr.*, 89(5), 1657s–1663s, May, 2009.

Ramarathnam, N. and Osawa, T., International conference on food factors: Chemistry and cancer prevention. Report of a conference held in Hamamatsu, Japan, December 10–15, 1995. *Trends Food Sci. Technol.*, 7(2), 64, 1996.

Ravishankar, S., Zhu, L., Olsen, C., McHugh, T., and Friedman, M., Edible apple film wraps containing plant microbials inactivate foodborne pathogens on meat and poultry products. *J. Food. Sci.*, 74(8), M440–M445, 2009.

Rawls, P., Extra pounds are going to cost you: Obese Alabama workers forced to pay for insurance that's normally free. Associated Press Release: Reported in *Guelph Mercury*, B9, August 29, 2008.

Ray, R. and Sivakumar, P., Traditional and novel fermented foods and beverages from tropical root and tuber crops: A review. *Int. J. Food. Sci. Technol.*, 44(6), 1073–1087, 2009.

Reiser, S., Metabolic effects of dietary pectins related to human health. *Food Technol.*, 41, 91, February 1987.

Renehan, A., Tyson, M., Egger, M., Heller, R., and Zwahlen, M., Body-mass index and incidence of cancer: A systematic review and meta-analysis of prospective observational studies. *Lancet*, 371(612), 569–578, 2008.

Reuters, Worldwide increase in obesity could become disastrous, doctors say. *The Gazette, Montreal*, B15, May 17, 1996.

Reuters, Reported in *Financial Post*, C12, September 26, 2000.

Rezai-Zadeh, K., Shytle, D., Sun, N. et al., Green tea epigallocatechin-3-gallate (EGCG) modulates amyloid precursor protein cleavage and reduces cerebral amyloidosis in Alzheimer transgenic mice. *J. Neurosci.* 25, 8807–8814, 2005.

Righelato, R. C., Biotechnology and food manufacturing. *Food Technol. Int. Eur.*, 155, 1987.

Ripsin, C. M. and Keenan, J. M., The effects of dietary oat products on blood cholesterol. *Trends Food Sci. Technol.*, 3, 137, 1992.

Rizvi, S. S. H. and Acton, J. C., Nutrient enhancement of thermostabilized foods in retort pouches. *Food Technol.*, 36, 105, April 1982.

Robbins, M. P., *The Cook's Quotation Book*, ed. M. P., Robbins. Robert Hale Limited: London, U.K, 1987.

Roberts, T. A., Combinations of antimicrobials and processing methods. *Food Technol.*, 43, 156, January 1989.

Roberts, T. A., Predictive modelling of microbial growth. *Food Technol. Int. Eur.*, 231, 1990.

Robertson, G. L., Shelf life of packaged foods, its measurement and prediction. In *Developing New Food Products for a Changing Marketplace*, eds. A. L. Brody and J. B. Lord, Technomic Publishing Co.: Lancaster, PA, 2000, Chap. 13, p. 329.

Robinson, D. S., Irradiation of foods. *Inst. Food Sci. Technol. Proc.*, 19, 165, 1985.

Rockland, L. B. and Nishi, S., Influence of water activity on food product quality and stability. *Food Technol.*, 34, 42, April 1980.

Rojas-Graü, M., Soliva-Fortuny, R., and Martin-Belloso, O., Edible coatings to incorporate active ingredients to fresh-cut fruits: A review. *Trends Food. Sci. Technol.*, 20, 438–447, 2009.

Roller, S., Ingredients as preservatives and the role of biotechnology. *Food Sci. Technol. Today*, 9(2), 116, 1995.

Roos, Y. and Karel, M., Applying state diagrams to food processing and development. *Food Technol.*, 45, 66, December 1991a.

Roos, Y. and Karel, M., Amorphous state and delayed ice formation in sucrose solutions. *Int. J. Food Sci. Technol.*, 26, 553, 1991b.

Roser, B., Trehalose, a new approach to premium dried foods. *Trends Food Sci. Technol.*, 2, 166, 1991.

Ross, C., Personal communication, 1980.

Rothfeder, J., *McIlhenny's Gold: How a Louisiana Family Built the Tabasco Empire.* HarperCollins Publishers: New York, 2007.

Rowley, D. B., Significance of sublethal injury of foodborne pathogenic and spoilage bacteria during processing. *Activities Rep. Res. Dev. Assoc.*, 36(1), 41, 1984.

Rusoff, I., Nutrition and dietary fiber. *Cereal Foods World*, 29, 668, 1984.

Rutledge, K. R., Accelerated training of sensory descriptive flavor analysis panelists. *Food Technol.*, 46, 114, November 1992.

Rutledge, K. P. and Hudson, J. M., Sensory evaluation: Method for establishing and training a descriptive flavor analysis panel. *Food Technol.*, 44, 78, December 1990.

Ryan, J. K., Jelen, P., and Sauer, W. C., Alkaline extraction of protein from spent honey bees. *J. Food Science*, 48, 886, 1983.

Ryley, J., The impact of fast foods on U.K. nutrition. *Inst. Food Sci. Technol. Proc.*, 16, 58, 1983.

Ryu, C. H. and West, A., Development of kimchi recipes suitable for British consumers and determination of vitamin C changes in kimchi during storage. *Food Sci. Technol. Today*, 14(2), 76, 2000.

Ryval, M., The shape of food to come. *Financial Post Magazine*, 38, April 15, 1981.

Saca, S. A. and Lozano, J. E., Explosion puffing of bananas. *Int. J. Food Sci. Technol.*, 27, 419, 1992.

Saguy, I. and Karel, M., Models of quality deterioration during food processing and storage. *Food Technol.*, 34, 78, February 1980.

Saito, Y., The retort pouch is a way of life in Japan. *Activities Rep. Res. Dev. Assoc.*, 35(2), 8, 1983.

Saldana, G., Meyer, R., and Lime, B. J., A potential processed carrot product. *J. Food Sci.*, 45, 1444, 1980.

Salminen, S., Ouwehand, A., Benno, Y., and Lee, Y.-K., Probiotics: How should they be defined? *Trends Food Sci. Technol.*, 10(3), 107, 1999.

Saltmarsh, M. and Hall, A., Developments in vending—AVEX 2007. *J. Foodservice*, 18(4), 161–163, 2007.

Sampedro, S., Rodrigo, D., Martinez, A., Barbosa Cáno Vas, G., and Rodrigo, M., Review: Application of pulsed electric fields in egg and egg derivatives. *Food. Sci. Technol. Int.*, 15(5), 397–405, 2006.

Sampson, S., Waves of Chocolate. *Tor. Star.* February 4, E8, 2010.

Sanders, M. E., Probiotics. *Food Technol.*, 53(11), 67, 1999.

Sastry, S. K. and Palaniappan, S., Ohmic heating of liquid-particle mixtures. *Food Technol.*, 46(12), 64, 1992.

Schaller, E., Bosset, J. O., and Escher, F., Electronic noses and their application to food. *Lebensm.-Wiss. u.-Technol.*, 31(4), 305, 1998.

Schiffmann, R. F., Microwave processing in the U.S. food industry. *Food Technol.*, 46(12), 50, 1992.

Schlegel, W., Commercial pasteurization and sterilization of food products using microwave technology. *Food Technol.*, 46(12), 62, 1992.

Schmidl, M. K., Massaro, S. S., and Labuza, T. P., Parenteral and enteral food systems. *Food Technol.*, 42, 77, July 1988.

Schneeman, B. O., Soluble vs. insoluble fiber—Different physiological responses. *Food Technol.*, 41, 81, February 1987.

Schutz, H. G., Multivariate analyses and the measurement of consumer attitudes and perceptions. *Food Technol.*, 42, 141, November 1988.

Schwarcz, J., 'Natural' remedies come under microscope. *The Gazette, Montreal*, A1, January 6, 2002.

Scott, V. N., Control and prevention of microbial problems in new generation refrigerated foods. *Activities Rep. Res. Dev. Assoc.*, 39(2), 22, 1987.

Scriven, F. M., Gains, N., Green, S. R., and Thomson, D. M. H., A contextual evaluation of alcoholic beverages using the repertory grid method. *Int. J. Food Sci. Technol.*, 24, 173, 1989.

Selman, J. D., Trends in food processing and food processes. *Food Technol. Int. Eur.*, 55, 1991.

Sensory Evaluation Division, IFT, *Sensory Evaluation Guide for Testing Food and Beverage Products*. Institute of Food Technologists: Chicago, IL, 1981.

Sethi, R., Smith, D., and Park, C., Cross-functional product development teams, creativity and innovativeness of new consumer products. *J. Market. Res.*, 38(1), 73–86, 2001.

Settlemyer, K., A systematic approach to foodservice new product development. *Food Technol.*, 40, 120, July 1986.

Shahidi, F., Arachchi, J. K. V., and Jeon, Y.-J., Food applications of chitin and chitosans. *Trends Food Sci. Technol.*, 10(2), 37, 1999.

Shahin, J., They're heeeere. *American Way*, p. 42, December 1, 1995.

Shapton, D. A. and Shapton, N. F. In *Principles and Practices for the Safe Processing of Foods*, eds. D. A. Shapton and N. F. Shapton. Butterworth—Heinemann Ltd.: Oxford, U.K., 1991.

Shelef, L. A., Naglik, O. A., and Bogen, D. W., Sensitivity of some common foodborne bacteria to the spices sage, rosemary, and allspice. *J. Food Sci.*, 45, 1042, 1980.

Sherman, H. C., *Food Products*. MacMillan Company: New York, 1916, p. 397.

Shi, Z., Bassa, I. A., Gabriel, S. L., and Francis, F. J., Anthocyanin pigments of sweet potatoes—*Ipomoea batatas*. *J. Food Sci.*, 57, 755, 1992a.

Shi, Z., Francis, F. J., and Daun, H., Quantitative comparison of the stability of anthocyanins from *Brassica oleracea* and *Tradescantia pallida* in non-sugar drink model and protein model systems. *J. Food Sci.*, 57, 768, 1992b.

Silva, S., Gomes, L., Leitão, F., Coelho, A., and Boas, L., Phenolic compounds and antioxidant activity of *Olea europaea* L. fruits and leaves. *Food Sci. Technol. Int.*, 15(5), 385–395, 2006.

Silveira, E. T. F., Rahman, M. S., and Buckle, K. A., Osmotic dehydration of pineapple: Kinetics and product quality. *Food Res. Int.*, 29(3–4), 227, 1996.

Silver, D., The seven deadly sins of product development. *Food Prod. Des.*, 12(12), 93, March 2003a.

Silver, D., Vegetarian dishes go mainstream. *Food Prod. Des.*, 13(2), 99, May 2003b.

Singhal, R. S., Gupta, A. K., and Kulharni, P. R., Low-calorie fat substitutes. *Trends Food Sci. Technol.*, 2, 241, 1991.

Sinki, G., Technical myopia. *Food Technol.*, 40, 86, December 1986.

Skarra, L., Rollout roulette. *Prep. Foods*, 167(8), 40, August 1998.

Skinner, R. H. and Debling, G. B., Food industry applications of linear programming. *Food Manuf.*, 44, 35, October 1969.

Skjöldebrand, C., Cooking by infrared radiation. In: *Proc. IUFoST Int. Symp. Progress in Food Preparation Processes*. SIK, Swedish Food Institute: Göteborg, Sweden, 1986, p. 157.

Slade, P. J., Monitoring *Listeria* in the food production environment. I. Detection of *Listeria* in processing plants and isolation methodology. *Food Res. Int.*, 25, 45, 1992.

Slade, L. and Levine, H., Beyond water activity: Recent advances based on an alternative approach to the assessment of food quality and safety. *Critical Rev. Food Sci. Nutr.*, 30, 115, 1991.

Slight, H., The storage and transport of chilled foods—A temperature survey. Leatherhead Food Res. Assoc. Res. Rep., No. 340, 1980.

Sloan, E., Top 10 trends to watch and work on: 3rd biannual report. *Food Technol.*, 55, 38, April 2001.

Sloan, E., Weighing in on the weight-control market. *Food Technol.*, 57, 18, February 2003a.

Sloan, E., Top 10 trends to watch and work on: 2003. *Food Technol.*, 57, 30, April 2003b.

Sloan, E., Reinventing restaurants. *Food Technol.*, 62(10), 26–28, 31–32, 35, 36, 39, 40, 42, 2008.

Sloan, E. A., Top ten trends to watch and work on for the millennium. *Food Technol.*, 53, 40, August 1999.

Smyth, H. F., Sufficient challenge *Fd. Cosmet. Toxicol.*, 5(1), 51, 1967.

Snow, C. P., *Science and Government*. Harvard University Press: Cambridge, MA, 1961.

Snow, P., The shape of shopping to come. *Oxford Today*, 5(1), 55, 1992.

Snyder, O. P., A model food service quality assurance system. *Food Technol.*, 35, 70, February 1981.

Snyder Jr., O. P., Minimizing foodservice energy consumption through improved recipe engineering and kitchen design. *Activities Rep. Res. Dev. Assoc.*, 36(2), 49, 1984.

Snyder Jr., O. P., Microbiological quality assurance in foodservice operations. *Food Technol.*, 40, 122, July 1986.

Solberg, M., Buckalew, J. J., Chen, C. M., Schaffner, D. W., O'Neill, K., McDowell, J., Post, L. S., and Boderck, M., Microbiological safety assurance system for food-service facilities. *Food Technol.*, 44, 68, December 1990.

Solhjoo, K., Food technology and the Baháí faith. *Food Sci. Technol. Today*, 8(1), 2, 1994.

Speck, M. L., Dobrogosz, W. J., and Casas, I. A., *Lactobacillus reuteri* in food supplementation. *Food Technol.*, 47(7), 90, 1993.

Spitz, P., The public granary. *Ceres*, 12, 16, November/December 1979.

Stanton, J. L., Diffusion of slotting allowances: A marketing practice going global. http://www.johnlstanton.com, undated.

Stavric, S. and Speirs, J. I., *Escherichia coli* associated with hemorrhagic colitis. *Can. Inst. Food Sci. Technol. J.*, 22, AT/205, 1989.

Steinbock, E., Knowledge is power. *Food Sci. Technol.*, 16(3), 49, 2002.

Stent, G. S., Prematurity and uniqueness in scientific discovery, *Scientific American*, December 1972. In *Readings from Scientific American: Scientific Genius and Creativity*. W. H. Freeman and Company: New York, 1987, pp. 95–104, Chap. 12.

Stevenson, K. E., Implementing HACCP in the food industry. *Food Technol.*, 44, 179, May 1990.

Stiles, M. E. and Hastings, J. W., Bacteriocin production by lactic acid bacteria: Potential for use in meat preservation. *Trends Food Sci. Technol.*, 2, 247, 1991.

St.-Onge, M.-P., Lamarche, B., Mauger, J.-F., and Jones, P., Consumption of a functional oil rich in phytosterols and medium-chain triglyceride oil improves plasma lipid profiles in men. *J. Nutr.*, 133, 1815–1820, 2003.

Strauss, S., Save stomachs: Zap food, Canada advised. *The Globe and Mail*, A1, June 17, 1998.

Stuller, J., The nature and process of creativity. *Sky*, 11(1), 37, 1982.

Sudman, S., Sirken, M., and Cowan, C., Sampling rare and elusive populations. *Science*, 240, 991, May 20, 1988.

Tamime, A. Y., Khaskheli, M., Barclay, M. N., and Muir, D. D., Effect of processing conditions and raw materials on the properties of kishk. 1. Compositional and microbiological qualities. *Lebensm. Wiss. Technol.*, 33(6), 444, 2000.

Taylor, A. W., Scaling-up of process operations in the food industry. *Inst. Food Sci. Technol. Proc.*, 2, 86, 1969.

Taylor, B., Sandwiches in Britain—Some observations. *Food Sci. Technol. Today*, 10, 78, June 1996.

Taylor, G., A côte of many colours. *The Gazette, Montreal*, A1, A17, August 10, 2002.

Taylor, J., Weber, W., Standish, L. et al., Efficacy and safety of Echinacea in treating upper respiratory tract infections in children: A randomized controlled trial. *JAMA*, 290(21), 2824–2830, 2003.

Taylor, L. and Leith, S., TQM. *Food Can.*, 51, 18, June 1991.

Taylor, T. G., Diet and coronary heart disease. *Inst. Food Sci. Technol. Proc.*, 13, 45, 1980.

Teutonico, R. A. and Knorr, D., Amaranth: Composition, properties, and applications of a rediscovered food crop. *Food Technol.*, 39, 49, April 1985.

Thakur, S. and Saxena, D., Formation of extruded snack food (gum based cereal-pulse blend): Optimization of ingredients levels using response surface methodology. *Lebensm. Wiss. Technol.*, 33(5), 354, 2000.

Thibault, J.-F., Asther, M., Ceccaldi, B. et al., Fungal bioconversion of agricultural by-products to vanillin. *Lebensm. Wiss. Technol.*, 31(6), 530, 1998.

Thomson, D., Recent advances in sensory and affective methods. *Food Sci. Technol. Today*, 3, 83, 1989.

Thorne, S., What temperature? *Inst. Food Sci. Technol. Proc.*, 11, 207, 1978.

Tiwari, N. P. and Aldenrath, S. G., Occurrence of *Listeria* species in food and environmental samples in Alberta. *Can. Inst. Food Sci. Technol. J.*, 23, 109, 1990.

Tiwari, B., O'Donnell, C. P., and Cullen, P. J., Effect of non thermal processing technologies on the anthocyanin content of fruit juices. *Trends Food Sci. Technol.*, 20(3/4), 137–145, 2009.

Toma, R. B., Curtis, D. J., and Sobotar, C., Sucrose polyester: Its metabolic role and possible future applications. *Food Technol.*, 42, 93, January 1988.

Tonon, R., Brabet, C., Pallet, D., Brat, P., and Hubinger, M., Physicochemical and morphological characteristics of açai (*Euterpe oleraceae Mart*) powder produced with different carrier agents. *Int. J. Food Sci. Technol.*, 44(10), 1950–1958, 2009.

Traill, C. P., *The Canadian Settler's Guide*. McClelland and Stewart Limited, 1969, 1st published 1855, Toronto, Canada.

Tregunno, N. B. and Goff, H. D., Osmodehydrofreezing of apples: Structural and textural effects. *Food Res. Int.*, 29(5–6), 471, 1996.

Tsen, J.-H., Lin, Y.-P., and King, V., Response surface methodology optimisation of immobilised *Lactobacillus acidophilus* banana purée fermentation. *Int. J. Food. Sci. Technol.*, 44(1), 120–127, 2009.

Tsuji, K., Low-dose Cobalt 60 irradiation for reduction of microbial contamination in raw materials for animal health products. *Food Technol.*, 37, 48, February 1983.

Tung, M. A., Garland, M. R., and Campbell, W. E., Quality comparison of cream style corn processed in rigid and flexible containers. *Can. Inst. Food Sci. Technol. J.*, 8, 211, 1975.

Tuomy, J. M. and Young, R., Retort-pouch packaging of muscle foods for the Armed Forces. *Food Technol.*, 36, 68, February 1982.

Turner, M. and Glew, G., Home-delivered meals for the elderly. *Food Technol.*, 36, 46, July 1982.

Turtoi, M. and Nicolau, A., Intense light pulse treatment as alternative method for mould spores destruction on paper-polyethylene packaging material. *J. Food Eng.*, 83(1), 47–53, 2007.

Tye, R. J., Konjac flour: Properties and applications. *Food Technol.*, 45, 81, March 1991.

Tyler, V. E., *The Honest Herbal: A Sensible Guide to the Use of Herbs and Related Remedies*, 3rd edn. Pharmaceutical Products Press, Haworth Press, Inc.: New York, 1993.

USDA, Food and Nutrient Intakes of Individuals in 1 Day in the United States, Spring 1977, Nationwide Food Consumption Survey 1977–78, Preliminary Report No. 2, U.S. Dept. of Agriculture, September 1980.

USDHHS (U.S. Dept. of Health and Human Services), Nutrition and cancer prevention: A guide to food choices, draft publication, 1984.

USDHHS, (U.S. Dept. of Health and Human services), Cancer prevention research summary: Nutrition, National Institute of Health Publication No. 85–2616, reprinted March 1985.

Vega-Mercado, H., Pothakamury, U. R., Fu-Jung Chang, Barbosa-Cánovas, G. V., and Swanson, B. G., Inactivation of *Escherichia coli* by combining pH, ionic strength and pulsed electric fields hurdles. *Food Res. Int.*, 29(2), 117, 1996.

Vega-Mercado, H., Martín-Bellos, O., Bai-Lin Qin, Fu Jung Chang, Góngora-Nieto, M. M., Barbosa-Cánovas, G. V., and Swanson, B. G., Non-thermal food preservation: Pulsed electric fields. *Trends Food Sci. Technol.*, 8, 151, May 1997.

Vegetarian Resource Group. http://www.vrg.org/nutshell/faq.htm#poll, 2003.

Vegetarian Society. http://www.vegsoc.org/info/statveg.html, 2003.

Vicente, P., Reis, E., and Santos, M., Using mobile phones for survey research: A comparison with fixed phones. *Int. J. Market Res.*, 51(5), 613–633, 2009.

Vidacek, S., De Las Heras, C., Solas, M., Mahillo, A., and Tejada, M., Effect of high hydrostatic pressure on mortality and allergenicity of *Anisakis simplex L3* and on muscle properties of infested hake. *J. Sci. Food Agric.*, 89(13), 2228–2235, 2009.

Voosen, P., Courts force U.S. reckoning with dominance of GM crops. *The New York Times*. October 8, 2009 (this is the first in a series of articles on the economics, ethics and problems with GM crops).

Vorell, M. and Shulman, G., Mapping racial cognitive constructs using ZMET: An analysis of organizational subculture in a university, paper delivered at annual meeting of International Communication Association: New Orleans, LA, May 27, 2004. http://www.allacademic.com/meta/p112636_index.html (accessed August 14, 2009).

Wachsmuth, I. K., *Escherichia coli* 0157:H7—Harbinger of change in food safety and tradition in the industrialized world. *Food Technol.*, 51(10), 26, 1997.

Waite-Wright, M., Chilled foods—The manufacturer's responsibility. *Food Sci. Technol. Today*, 4, 223, 1990.

Walker, S. J. and Jones, J. E., Predictive microbiology: Data and model bases. *Food Technol. Int. Eur.*, 209, 1992.

Walston, J., C.O.D.E.X. spells controversy. *Ceres*, 24, 28, July/August 1992.

Wan, J., Coventry, J., Swiergon, P., Sanguansri, P., and Versteeg, C., Advances in innovative processing technologies for microbial inactivation and enhancement of food safety—Pulsed electric field and low-temperature plasma. *Trends Food. Sci. Technol.*, 20, 414–424, 2009.

Wang, C., Food business: Seven ways brand management systems kill innovation. *Food Process.*, 60, 35, September 1999.

Warburton, J., The electronic "NOSE"—The technology and applications. *Food Sci. Technol. Today*, 10, 91, June 1996.

Warriner, K. and Namvar, A., What is the hysteria with *Listeria*? *Trends Food Sci. Technol.*, 20(6/7), 245–254, 2009.

Waters, K., Designing screening questionnaires to minimize dishonest answers. *Appl. Market. Res.*, 31, 51–53, Spring/Summer, 1991.

Watson, P., Morgan, M., and Hemmington, N., Online communities and the sharing of extraordinary restaurant experiences. *J. Foodservice*, 19(6), 289–302, 2008.

Webb, M., When you're confined...things spread easier: Analysis. *Tor. Star*, A30, 31, October 10, 2009.

Webster, S. N., Fowler, D. R., and Cooke, R. D., Control of a range of food related microorganisms by a multi-parameter preservation system. *J. Food Technol.*, 20, 311, 1985.

Weingärtner, O., Böhm, M., and Laufs, U., Controversial role of plant sterol esters in the management of hypercholesterolaemia. *European Heart J.*, 30(4), 404–409, 2009.

West, A., Good catering practice: Use of manufactured meals in catering. *Food Sci. Technol. Today*, 8(3) 172, 1994.

Whitehead, R., What'll we eat in 1999? *Ind. Week*, 30, May 17, 1976.

Whitney, L. F., What expert systems can do for the food industry. *Food Technol.*, 43, 135, May 1989.

Wiley, H. W., *Foods and their Adulteration*, 3rd edn. P. Blakiston's Sons & Co.: Philadelphia, PA, 1917, p. 455.

Wilhelmi, F., Product safety and how to ensure it. *Dragoco Rep.*, 14(1), 1988.

Williams, A., Another challenge for the food industry? *Food Sci. Technol.*, 17(2), 41, 2003.

Williams, D., Partnering for successful product development. *Food Technol.*, 56, 28, 2002.

Williams, E. F., Weak links in the cool and cold chain. *Inst. Food Sci. Technol. Proc.*, 11, 211, 1978.

Williams, M., Electronic databases. *Science*, 228, 445, 1985.

Williams, R., Vending in the Canadian foodservice industry. *Visions*, 2(5), 1991.

Williams, R., A profile of the Canadian fast-food sector. *Visions*, 3(2), 1992.

Williams, A. P., Blackburn, C., and Gibbs, P., Advances in the use of predictive techniques to improve the safety and extend the shelf-life of foods. *Food Sci. Technol. Today*, 6, 148, 1992.

Williamson, M., Tasting tests carried out at the Leatherhead Food Research Association. *Leatherhead Food Res. Assoc. Tech. Circ.*, 749, 1981.

Wilson, C., Ethical shopping. *Food Can.*, 52, 7, November/December 1992.

Wilson, D., Marketing mycoprotein: The Quorn Foods story. *Food Technol.*, 55(7), 48, 2001.

Wolfe, K. A., Use of reference standards for sensory evaluation of product quality. *Food Technol.*, 33, 43, September 1979.

Wood, P. J., Oat ß-glucan—Physicochemical properties and physiological effects. *Trends Food Sci. Technol.*, 2, 311, 1991.

Wood, S., Food law enforcement—Where next? An industry viewpoint. *Inst. Food Sci. Technol Proc.*, 18, 89, 1985.

Worsfold, D. and Griffith, C., Food safety behaviour in the home. *Brit. Food J.*, 99(3), 97, 1997.

Wrick, K. L., Development opportunities for functional foods. *Food Prod. Des.*, 13, July, 73, 2003.

Wurtman, R. J. and Wurtman, J. J., Carbohydrates and depression. *Sci. Am.*, 260, 68, January 1989.

Yankelovich Partners, Today's market? *The Globe and Mail*, A14, August 3, 2000.

Young, J., A perspective on functional foods. *Food Sci. Technol. Today*, 10(1), 18, 1996.

Zaika, L. L. and Kissinger, J. C., Inhibitory and stimulatory effects of oregano on *Lactobacillus plantarum* and *Pediococcus cerevisiae*. *J. Food Sci.*, 46, 1205, 1981.

Zaika, L., Kissinger, J. C., and Wasserman, A. E., Inhibition of lactic acid bacteria by herbs. *J. Food Sci.*, 48, 1455, 1983.

Zaltman, G., *The Dimensions of Brand Equity for Nestlé Crunch Bar: A Research Case.* Harvard Business School: Cambridge, MA, N9-500-083, January 27, 2000a.

Zaltman, G., Consumer researchers: Take a hike! *J. Consum. Res.*, 26, 423, March 2000b.

Zhou, Y. and Lee, A., Mechanism for the suppression of the mammalian stress response by genistein, and anticancer phytoestrogen from soy. *J. Natl. Cancer Inst.*, 90(5), 381, 1998.

Zind, T., Phytochemicals: The new vitamins? *Food Process.*, 59(11), 29, 1998.

Zind, T., The functional foods frontier. *Food Process.*, 60(4), 45, 1999.

Index